D0887081

Introduction to Algebraic K-Theory

CHAPMAN AND HALL
MATHEMATICS SERIES

Edited by Professor R. Brown
Head of the Department of Pure Mathematics,
University College of North Wales, Bangor
and Dr M. A. H. Dempster
University Lecturer in Industrial Mathematics
and Fellow of Balliol College, Oxford

A Preliminary Course in Analysis
R. M. F. Moss and G. T. Roberts

Elementary Differential Equations
R. L. E. Schwarzenberger

A First Course on Complex Functions
G. J. O. Jameson

Rings, Modules and Linear Algebra
B. Hartley and T. O. Hawkes

Regular Algebra and Finite Machines
J. H. Conway

Complex Numbers
W. H. Cockcroft

Galois Theory
Ian Stewart

Topology and Normed Spaces
G. J. O. Jameson

Introduction to Optimization Methods
P. R. Adby and M. A. H. Dempster

Graphs, Surfaces and Homology
P. J. Giblin

Linear Estimation and Stochastic Control
M. H. A. Davis

Mathematical Programming and Control Theory
B. D. Craven

Algebraic Number Theory
Ian Stewart and David Tall

Independence Theory in Combinatorics
V. Bryant and H. Perfect

Introduction to Algebraic *K*-Theory

JOHN R. SILVESTER
King's College
University of London

LONDON AND NEW YORK
CHAPMAN AND HALL

First published 1981 by
Chapman and Hall Ltd
11 New Fetter Lane, London EC4P 4EE
Published in the USA by
Chapman and Hall
in association with Methuen, Inc.
733 Third Avenue, New York NY 10017

Printed in Great Britain by
Whitstable Litho Ltd., Whitstable, Kent

ISBN 0 412 22700 2 (cased)
ISBN 0 412 23740 7 (paperback)

British Library Cataloguing in Publication Data

Silvester, John R.
Introduction to algebraic *K*-theory.
— (Chapman and Hall mathematics series)
1. Algebra, Linear
I. Title
512′.5 QA184

ISBN 0-412-22700-2
ISBN 0-412-23740-7 Pbk

CONTENTS

PREFACE

Algebraic K-theory is the name given to a body of theory which may be regarded in the first instance as an attempt to generalize parts of linear algebra, notably the theory of dimension of vector spaces, and determinants, to modules over arbitrary rings. The subject gets its name from its notation - given a ring R, one constructs groups called K_0R, K_1R, K_2R, ... - and is called algebraic K-theory, rather than just K-theory, because it derives from (topological) K-theory, which is to do with the topology of vector bundles.

The intention of this text is to make algebraic K-theory accessible at a more elementary level than heretofore. The only absolute prerequisites are standard undergraduate first courses in linear algebra and in groups and rings, although an acquaintance with the beginnings of the theory of group presentations, modules, and categories, would be helpful. From time to time, some algebraic topology is used, but these sections can easily be omitted at a first reading.

I have tried to make the text as self-contained as possible, so I have included proofs of many standard results on such topics as, for example, tensor products and modules of fractions.

I have also reduced to a minimum the number of proofs left to the reader; at the risk of tedium, most proofs are given in full detail. The more sophisticated reader is encouraged to skip the proofs (or, better, provide his own) whenever the propositions seem obvious.

The text is an expanded version of a London M.Sc. lecture course given at King's College in 1976. The approach inevitably owes much to the standard texts by Bass, Milnor, and Swan (listed in the Bibliography). The choice of material is somewhat arbitrary, being limited by space and by the requirement that it be elementary, and so I have been content to establish the most basic properties of the functors K_0, K_1, K_2, and to do a few explicit computations. Many important topics, such as the K-theory of polynomial extensions, and localization exact sequences, are omitted; Quillen's higher K-theory is also beyond our scope.

My thanks are due to many people for their interest and encouragement, and I wish to thank especially Keith Dennis and Michael Stein for conversations and correspondence which have strongly influenced later sections of the text. I wish also to thank the students and colleagues who attended the original course of lectures, especially Philip Higgins and Tony Barnard, whose penetrating questions and insistence on clarity were a constant stimulus.

London J.R.S.
March, 1981

INDEX OF NOTATION

CHAPTER ONE
Modules

The word *ring* will mean an associative, but not necessarily commutative, ring with a multiplicative identity, written 1. Let R, S be rings. A map $f : R \to S$ is a (ring) *homomorphism* if, for all $r, s \in R$, we have $f(r + s) = f(r) + f(s)$ and $f(rs) = f(r)f(s)$, and also $f(1) = 1$. Note that in this last equation the symbol 1 has two meanings: on the left, $1 \in R$, and on the right, $1 \in S$.

Let R be a ring. A (*left*) *R-module* is an abelian group M, written additively, with a map $R \times M \to M$, called *scalar multiplication* and written $(r, m) \mapsto rm$ ($r \in R$, $m \in M$), such that, for all $r, s \in R$ and $m, n \in M$, we have

$$r(m + n) = rm + rn \tag{i}$$

$$(r + s)m = rm + sm \tag{ii}$$

$$(rs)m = r(sm) \tag{iii}$$

$$1m = m. \tag{iv}$$

Elements of R are then referred to as *scalars*. A *right* R-module is defined similarly, except that (iii) is replaced by

$$(rs)m = s(rm). \tag{iii'}$$

Alternatively, and more naturally, the scalars may be written on the right, so that (iii') reads $m(sr) = (ms)r$. The word *module* will usually mean a *left* module. Of course, if the ring R is commutative, the notions of left and right R-module coincide; if R is a field, an R-module is just a vector space over R. If $R = \mathbb{Z}$, the ring of rational integers, an R-module is just an (additive) abelian group.

Let M be an R-module. Remember M is a group: a subgroup N of M is called a *submodule* if it is closed under scalar multiplication, that is, if $rn \in N$ for all $r \in R$ and $n \in N$. More generally, if N is any non-empty subset of M, write RN for the set of all finite sums $\Sigma_i r_i n_i$ ($r_i \in R$, $n_i \in N$). RN is clearly a submodule of M, and is the smallest submodule of M containing N. Thus N is a submodule if and only if $RN = N$. Obviously M itself is a submodule, and so is $\{0\}$, where 0 denotes the additive identity of M. This follows from the fact that $r0 = 0$, all $r \in R$, which is easily deduced from the module axioms.

Let N be a submodule of M. The quotient group M/N becomes an R-module if we define the scalar multiplication by

$$r(m + N) = rm + N \ (r \in R,\ m \in M).$$

It is easy to see that this is well-defined and satisfies the axioms. M/N is then called the *quotient module* of M by N.

Let M, N be R-modules. A map $f : M \to N$ is called a (module) *homomorphism* if, for all m, $n \in M$ and $r \in R$, we have

$$f(m + n) = f(m) + f(n)$$

and

$$f(rm) = r[f(m)].$$

The *kernel*

$$\ker f = f^{-1}\{0\} = \{m \in M : f(m) = 0\}$$

is a submodule of M, and the *image*

$$f(M) = \{f(m) : m \in M\}$$

is a submodule of N. The homomorphism f is a *monomorphism* if it is an injection, or equivalently if $\ker f = 0$; here we are writing 0 for the zero (sub)module: $0 = \{0\}$. The homomorphism f is an *epimorphism* if it is a surjection, that is, if $f(M) = N$. We sometimes write $f : M \hookrightarrow N$ for a monomorphism and $f : M \twoheadrightarrow N$

for an epimorphism. If f is both a monomorphism and an epimorphism, it is an *isomorphism*; if such an f exists we say M, N are *isomorphic*, and write $M \simeq N$. In this case, $f^{-1} : N \to M$ is also an isomorphism. An *endomorphism* is a homomorphism $M \to M$, and an *automorphism* is an isomorphism $M \to M$. The identity map $M \to M$ is an automorphism. The *first isomorphism theorem* states that if $f : M \to N$ is a homomorphism then $M/\ker f \simeq f(M)$. The proof is left to the reader.

1.1 *Direct sums*

Let M be an R-module, with submodules M_1, M_2. If every element of M can be written uniquely in the form $m_1 + m_2$ ($m_1 \in M_1$, $m_2 \in M_2$) we say M is the *direct sum* of M_1 and M_2, and write $M = M_1 \oplus M_2$. If, for arbitrary submodules M_1, M_2 of M we write

$$M_1 + M_2 = \{m_1 + m_2 : m_1 \in M_1, m_2 \in M_2\}$$

then it is clear that $M_1 + M_2$ and $M_1 \cap M_2$ are submodules of M, and that $M = M_1 \oplus M_2$ if and only if $M = M_1 + M_2$ and $M_1 \cap M_2 = 0$.

More generally, if M is an R-module with submodules M_λ ($\lambda \in \Lambda$), where the index set Λ may be infinite, and if every element $m \in M$ can be written uniquely, except for zeros and the order of the terms, as a finite sum $m = \Sigma_\lambda m_\lambda$ ($m_\lambda \in M_\lambda$), then we say M is the *direct sum* of the M_λ, and write $M = \bigoplus_{\lambda\in\Lambda} M_\lambda$.

Now suppose we are given R-modules M_λ ($\lambda \in \Lambda$). We shall construct their direct sum. Let M be the subset of the cartesian product $\times_{\lambda\in\Lambda} M_\lambda$ consisting of all Λ-tuples (m_λ) with $m_\lambda = 0$ for almost all λ (that is, for all but a finite number of values of λ). M becomes an R-module if we define addition and scalar multiplication componentwise; explicitly,

$$(m_\lambda) + (n_\lambda) = (m_\lambda + n_\lambda) \text{ and } r(m_\lambda) = (rm_\lambda)$$

where m_λ, $n_\lambda \in M_\lambda$ and $r \in R$. For each $\lambda \in \Lambda$, M contains a submodule

$$M'_\lambda = \{(m_\mu) : m_\mu = 0 \text{ for all } \mu \neq \lambda\}$$

which is isomorphic to M_λ. If we identify M_λ with M'_λ in the obvious way, then $M = \bigoplus_{\lambda \in \Lambda} M_\lambda$.

Let $M = \bigoplus_{\lambda \in \Lambda} M_\lambda$ and let $i_\lambda : M_\lambda \hookrightarrow M$ be the natural monomorphism which embeds M_λ as a submodule of M. Let $\pi_\lambda : M \twoheadrightarrow M_\lambda$ be the epimorphism given by $\pi_\lambda(\Sigma_\mu m_\mu) = m_\lambda$, where $m_\mu \in M_\mu$, all $\mu \in \Lambda$. We have

$$\pi_\lambda i_\lambda = 1 : M_\lambda \to M_\lambda$$

and $$\pi_\mu i_\lambda = 0 : M_\lambda \to M_\mu \quad (\lambda \neq \mu)$$

where we are writing 1 for the identity isomorphism and 0 for the zero homomorphism (that is, the homomorphism whose image is the zero submodule). It is hoped that it will always be clear from the context when the symbols 0 and 1 are being used to denote maps and when they are being used to denote elements of a ring or module. Note that, since we are writing maps on the left of the elements on which they act, the map $\pi_\lambda i_\lambda$ means the composite map obtained by applying first i_λ and then π_λ. Now let

$$N = \bigcup_{\lambda \in \Lambda} M_\lambda = \bigcup_{\lambda \in \Lambda} i_\lambda(M_\lambda).$$

Clearly N generates M.

Conversely, given R-modules M, M_λ $(\lambda \in \Lambda)$ and homomorphisms $i_\lambda : M_\lambda \to M$, $\pi_\lambda : M \to M_\lambda$ such that $\pi_\lambda i_\lambda = 1$ for each λ and $\pi_\mu i_\lambda = 0$ for $\lambda \neq \mu$, and such that $\bigcup_{\lambda \in \Lambda} i_\lambda(M_\lambda)$ generates M, then $M \simeq \bigoplus_{\lambda \in \Lambda} M_\lambda$. For we can construct a map $\bigoplus_{\lambda \in \Lambda} M_\lambda \to M$ by $(m_\lambda) \mapsto \Sigma_\lambda i_\lambda(m_\lambda)$, and this is clearly an isomorphism.

Given an R-module N and homomorphisms $f_\lambda : M_\lambda \to N$, there is a unique homomorphism $f : \bigoplus_{\lambda \in \Lambda} M_\lambda \to N$ such that $fi_\lambda = f_\lambda$ for all λ: it is given by $f[(m_\lambda)] = \Sigma_\lambda f_\lambda(m_\lambda)$. The last expression makes sense since $m_\lambda = 0$ for almost all λ.

Note that R itself is a left R-module in a natural way; the submodules of R are precisely the left ideals of R. Given an R-module M and $m \in M$ there is a unique homomorphism $f : R \to M$ with $f(1) = m$: it is given by $f(r) = rm$, all $r \in R$.

1.2 *Free modules*

The R-module M is *free* if there is a subfamily $\{m_\lambda\}_{\lambda\in\Lambda}$ of M such that every element $m \in M$ can be written uniquely, except for zeros and the order of the terms, as a finite sum $m = \Sigma_\lambda r_\lambda m_\lambda$, where $r_\lambda \in R$, all λ. The set $\{m_\lambda\}_{\lambda\in\Lambda}$ is called a *basis*, or *R-basis*, of M. Given such M, it is clear that if $M_\lambda = R\{m_\lambda\}$ then M_λ is a submodule of M, $M_\lambda \simeq R$, all λ, and $M = \oplus_{\lambda\in\Lambda} M_\lambda$. Conversely, if $M = \oplus_{\lambda\in\Lambda} M_\lambda$, where $M_\lambda \simeq R$ for all λ, then M is free. The proof consists of choosing a suitable basis, and the details are left to the reader.

If M is free with basis $\{m_\lambda\}_{\lambda\in\Lambda}$ and N is an R-module, then any map $f : \{m_\lambda\}_{\lambda\in\Lambda} \to N$ extends to a unique homomorphism $f : M \to N$. For f extends to $M_\lambda = R\{m_\lambda\}$ by $f(rm_\lambda) = rf(m_\lambda)$, all $r \in R$, and the result follows since $M = \oplus_{\lambda\in\Lambda} M_\lambda$.

Suppose M, N_1, N_2 are R-modules, where M is free, and that $g : M \to N_2$ is a homomorphism and $h : N_1 \twoheadrightarrow N_2$ is an epimorphism. We show how to construct a homomorphism $f : M \to N_1$ with $hf = g$, that is, so that the diagram

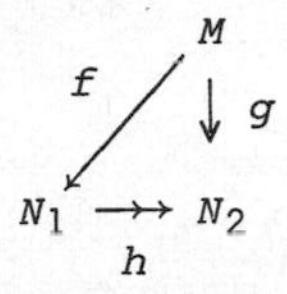

commutes. (Such f will not in general be unique.) For, if $\{m_\lambda\}_{\lambda\in\Lambda}$ is a basis of M, then $g(m_\lambda) \in N_2$, each $\lambda \in \Lambda$, and since h is surjective we can choose $n_\lambda \in N_1$ with $h(n_\lambda) = g(m_\lambda)$, each $\lambda \in \Lambda$. Since M is free, we can define f by $f(m_\lambda) = n_\lambda$, all λ, and then extend to the whole of M as above; the fact that $hf = g$ follows from the fact that $hf(m_\lambda) = g(n_\lambda)$, all $\lambda \in \Lambda$.

1.3 *Projective modules*

We see now that the above property of free modules is shared by a larger class of modules, called projective modules. The R-module P is *projective* if it satisfies the following equivalent conditions:

(i) There exists a module Q such that $P \oplus Q$ is free

(ii) Given modules N_1, N_2 and homomorphisms $g : P \to N_2$ and $h : N_1 \twoheadrightarrow N_2$, there is a homomorphism $f : P \to N_1$ with $hf = g$

(iii) Given a module N and an epimorphism $\pi : N \twoheadrightarrow P$, then π *splits*, that is, there is a monomorphism $i : P \hookrightarrow N$ with $\pi i = 1 : P \to P$.

To show the equivalence of these conditions, we prove (i) => (ii) => (iii) => (i).

(i) => (ii). By the previous argument, there is a homomorphism $f_1 : P \oplus Q \to N$ with $hf_1 = g\pi_P$, where $\pi_P : P \oplus Q \to P$ is the natural epimorphism. Put $f = f_1 i_P$, where $i_P : P \to P \oplus Q$ is the natural monomorphism. Then $hf = h(f_1 i_P) = (hf_1)i_P = (g\pi_P)i_P = g(\pi_P i_P) = g1_P = g$, as required. The appropriate diagram is:

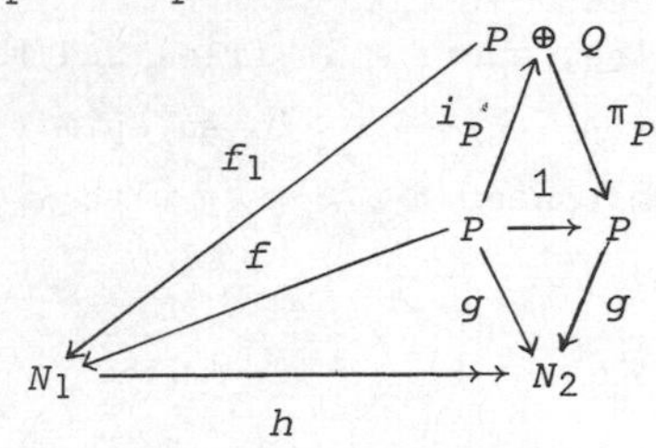

(ii) => (iii). The proof is immediate:

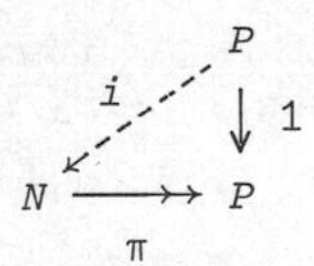

(The dotted arrow indicates the map whose existence is asserted.)

(iii) => (i). From (iii), if

$$N \xrightarrow{\pi} P, \text{ we obtain } P \xhookrightarrow{i} N$$

with $\pi i = 1_P$, and it follows that

$$N = \ker \pi \oplus i(P) \simeq \ker \pi \oplus P.$$

For we have the inclusion $\ker \pi \hookrightarrow N$, and the map $N \twoheadrightarrow \ker \pi$ is given by $m \mapsto m - i\pi(m)$, all $m \in N$. Now, given P, choose a generating set $\{m_\lambda\}_{\lambda\in\Lambda}$, and let N be free with basis $\{n_\lambda\}_{\lambda\in\Lambda}$. Then π is defined by extending the map $n_\lambda \overset{\pi}{\longmapsto} m_\lambda$ $(\lambda \in \Lambda)$, and the result follows.

Note: if P, P_1 are projective, so is $P \oplus P_1$.

A module M is *finitely generated* if there is a finite subset N of M with $RN = M$. If we write $R^n = R \oplus R \oplus \ldots \oplus R$ (n terms), then R^n is finitely generated (it is free with a finite basis), and M is finitely generated if and only if there is a natural number n and an epimorphism $R^n \twoheadrightarrow M$.

Note: if M, M_1 are finitely generated, so is $M \oplus M_1$. Further, the module P is finitely generated and projective if and only if there is a module Q and a natural number n such that $P \oplus Q \simeq R^n$. Of course Q is then finitely generated and projective also. In particular, R^n is finitely generated and projective.

CHAPTER TWO

The Grothendieck Group, K_0R

Algebraic K-theory is concerned with properties of certain groups K_0R, K_1R, K_2R,..., constructed from a ring R. We are now in a position to describe the first of these groups.

Let R be a ring. We define the *Grothendieck group* K_0R as the abelian group given by the following generators and relations: we take one generator $[P]$ for each isomorphism class of finitely generated projective R-modules P (so $[P] = [P_1]$ if $P \simeq P_1$) and one relation $[P] + [Q] = [P \oplus Q]$ for each pair P, Q of finitely generated projective R-modules.

More formally, write $\mathcal{F} = \mathcal{F}(R)$ for the free abelian group whose free generators $\langle P \rangle$ are the isomorphism classes of finitely generated projective R-modules P (so $\langle P \rangle = \langle P_1 \rangle$ if *and only if* $P \simeq P_1$) and let $\mathcal{R}$ be the subgroup of $\mathcal{F}$ generated by all expressions $\langle P \rangle + \langle Q \rangle - \langle P \oplus Q \rangle$. Then $K_0R = \mathcal{F}/\mathcal{R}$.

Lemma 1. Each element of K_0R can be written in the form $[P] - [Q]$.

Proof. Just use the relations to collect together the positive terms and the negative terms.

Example (i). Let F be a field. Then F-modules are vector spaces, necessarily free and hence projective. Thus the finitely generated projective F-modules are the finite-dimensional vector spaces over F. We have a map $f : \mathcal{F} \to \mathbb{Z}$ induced by $\langle P\rangle \mapsto \dim P$; further, since $\dim (P \oplus Q) = \dim P + \dim Q$, we have $\mathcal{R} \subset \ker f$, so f induces $\bar{f} : K_0F \to \mathbb{Z}$. Also we have $\mathbb{Z} \to K_0F$ induced by $n \mapsto [F^n]$ $(n > 0)$ and this map and $\bar{f}$ are inverse group homomorphisms, so $K_0F \simeq \mathbb{Z}$. Note that here we are only concerned with the additive structure of $\mathbb{Z}$; later we shall see that a multiplication can sometimes be defined to make K_0R into a ring.

Remark: Example (i) works equally well if F is a skew field.

We have seen that the calculation of K_0 of a field amounts to the observation that the modules (vector spaces) in question have well-defined dimensions which behave sensibly under the formation of direct sums. One might regard that part of K-theory which is concerned with K_0 as an attempt to generalize these elementary facts from linear algebra to modules over an arbitrary ring; calculation of K_0R measures in some sense the extent to which the finitely generated projective R-modules have a dimension theory like that for vector spaces.

Definition. The R-modules P, Q are *stably isomorphic* if, for some natural number n, $P \oplus R^n \simeq Q \oplus R^n$.

Proposition 2. Let P, Q be finitely generated projective R-modules. Then $[P] = [Q]$ in $K_0R \iff P, Q$ are stably isomorphic.

Proof. $\Leftarrow$: If $P \oplus R^n \simeq Q \oplus R^n$, then $[P \oplus R^n] = [Q \oplus R^n]$, so $[P] + [R^n] = [Q] + [R^n]$, whence $[P] = [Q]$.

$\Rightarrow$: Working in $\mathcal{F}(R)$, we have $\langle P\rangle - \langle Q\rangle \in \mathcal{R}$, so there is an expression of the form

$$\langle P\rangle - \langle Q\rangle = \Sigma_i\ (\langle A_i\rangle + \langle B_i\rangle - \langle A_i \oplus B_i\rangle) + \Sigma_j\ (\langle C_j \oplus D_j\rangle - \langle C_j\rangle - \langle D_j\rangle).$$

Collecting terms of like sign,

$$\langle P\rangle + \Sigma_i \langle A_i \oplus B_i\rangle + \Sigma_j \langle C_j\rangle + \Sigma_j \langle D_j\rangle$$

$$= \langle Q\rangle + \Sigma_i \langle A_i\rangle + \Sigma_i \langle B_i\rangle + \Sigma_j \langle C_j \oplus D_j\rangle.$$

Since $\mathcal{F}$ is free abelian, the terms on each side of this equation are the same, except possibly for the order in which they appear. So if we put $A = \oplus_i A_i$, $B = \oplus_i B_i$, $C = \oplus_j C_j$, and $D = \oplus_j D_j$, we have

$$P \oplus A \oplus B \oplus C \oplus D \simeq Q \oplus A \oplus B \oplus C \oplus D$$

or

$$P \oplus M \simeq Q \oplus M$$

where

$$M = A \oplus B \oplus C \oplus D.$$

Now M is a finitely generated projective R-module, since it is a finite direct sum of such modules, and so there is an R-module N such that $M \oplus N \simeq R^n$, say, and thus $P \oplus M \oplus N \simeq Q \oplus M \oplus N$, or $P \oplus R^n \simeq Q \oplus R^n$, as required.

Example (ii). $K_0\mathbb{Z} \simeq \mathbb{Z}$. For a $\mathbb{Z}$-module is just an abelian group, and a finitely generated projective $\mathbb{Z}$-module is thus a free abelian group of finite rank. Two such are stably isomorphic if and only if they are isomorphic, and so $[P] \mapsto \text{rank } P$ gives the isomorphism. (Here we have used the structure theorem for finitely generated abelian groups.)

Exercise: Show that, if R is a principal ideal domain, then $K_0R \simeq \mathbb{Z}$.

Example (iii). Let V be a vector space of infinite dimension over the field F. So $V \oplus V \simeq V$. Put $R = \text{End}_F(V) = \text{Hom}_F(V, V) =$ the set of all F-linear maps $V \to V$. R is a ring if, for $\alpha, \beta \in R$ we define $\alpha + \beta$, $\alpha\beta$ by $(\alpha + \beta)v = \alpha(v) + \beta(v)$, $(\alpha\beta)v = \alpha(\beta v)$. As R-modules,

$$R \oplus R \simeq \text{Hom}_F(V, V) \oplus \text{Hom}_F(V, V)$$

$$\simeq \text{Hom}_F(V \oplus V, V) \simeq \text{Hom}_F(V, V) = R.$$

Explicitly, if $v \mapsto (v_1, v_2)$ is an isomorphism $V \to V \oplus V$, then the isomorphism $R \oplus R \to R$ is given by $(\alpha, \beta) \mapsto \overline{(\alpha, \beta)}$, where $\overline{(\alpha, \beta)}(v) = \alpha v_1 + \beta v_2$. So $R^2 \simeq R$, and more generally $R^n \simeq R$, all $n > 0$. In K_0R, we then have $[R] = [R \oplus R] = [R] + [R]$, so $[R] = 0$, and hence $[R^n] = 0$, all $n > 0$. We shall show in fact that $K_0R = 0$.

Let P be a finitely generated projective R-module. So $P \oplus Q \simeq R^n$, for some Q; but $R^n \simeq R$, so $P \oplus Q \simeq R$, and thus we may regard P, Q as left ideals of R. Write $1 = p + q$ ($p \in P$, $q \in Q$). If $r \in P$, then $r = rp + rq$, so $r - rp = rq \in P \cap Q = 0$. Thus $rq = 0$ and $rp = r$: in particular, $p^2 = p$ and $pq = 0$. Similarly, if $r \in Q$, $rp = 0$ and $rq = r$, so $q^2 = q$ and $qp = 0$.

We show that, as left R-modules, $P \simeq \mathrm{Hom}_F(pV, V)$. For we can map $P \to \mathrm{Hom}_F(pV, V)$ by restriction: we have $P \subset R = \mathrm{Hom}_F(V, V)$ and $pV \subset V$. If $r \in P$ and $r(pV) = 0$, then $(rp)V = 0$, so $rp = 0$. But $rp = r$, so $r = 0$, and therefore $P \to \mathrm{Hom}_F(pV, V)$ is a monomorphism. To see that it is an epimorphism, choose any $\alpha \in \mathrm{Hom}_F(pV, V)$ and define $r \in R$ by $rv = \alpha(pv)$, all $v \in V$. Then $rqv = \alpha pqv = 0$, all v, so $rq = 0$ and thus $r = rp + rq = rp$, so $r \in P$. Therefore we have an isomorphism; in a similar way we also have $Q \simeq \mathrm{Hom}_F(qV, V)$.

Now, for any $v \in V$, $v = pv + qv$, so $V = pV + qV$. Then if $pv = qu$ (some v, $u \in V$), $pv = p^2v = pqu = 0$, so $pV \cap qV = 0$. Therefore $V = pV \oplus qV$, and since V is infinite-dimensional, we must have $pV \simeq V$ or $qV \simeq V$ (possibly both). If $pV \simeq V$, then $P \simeq \mathrm{Hom}_F(pV, V) \simeq \mathrm{Hom}_F(V, V) = R$, so in K_0R, $[P] = [R] = 0$. If $qV \simeq V$, then by a similar argument $Q \simeq R$, and so in K_0R, $[P] + [R] = [P] + [Q] = [P \oplus Q] = [R]$, so $[P] = 0$ once again. Thus we have shown that $K_0R = 0$.

The above example illustrates the fact that, over certain rings, a free module may have bases of different cardinality. If this is not the case, that is, if the ring R is such that, given $n, m > 0$, $R^n \simeq R^m$ only if $n = m$, we say R has *invariant basis number* (IBN). If $R^n \simeq R^m$, then we can find an $n \times m$ matrix A and an $m \times n$ matrix B, with coefficients in R, such that $AB = I_n$ (the $n \times n$ identity matrix) and $BA = I_m$. So the statement that R has IBN is equivalent to the statement that a matrix over R which has a two-sided inverse is necessarily a square matrix.

Exercise. Show that every non-zero commutative ring has IBN.

Exercise. Show that R has IBN if and only if $[R]$ has infinite order in K_0R.

2.1 *Bimodules and tensor products*

We introduce the tensor product here for two purposes: (i) when R is commutative, the tensor product makes K_0R into a ring; and (ii) if $f : R \to S$ is a homomorphism of rings, we can use the tensor product to construct a group homomorphism $K_0f : K_0R \to K_0S$.

Let R, S be rings. An *R-S-bimodule* is an abelian group M which is:

(i) a left R-module

(ii) a right S-module, and

(iii) satisfies an associative law $(rm)s = r(ms)$ $(r \in R, m \in M, s \in S)$.

Note that a left R-module is an R-$\mathbb{Z}$-bimodule, and a right R-module is a $\mathbb{Z}$-R-bimodule. If R is commutative, an R-module is an R-R-bimodule in a natural way. For bimodules we can define submodules, quotient modules, homomorphisms and direct sums exactly as for modules: for instance, an R-S-homomorphism $f : M \to N$ must satisfy $f(m + n) = f(m) + f(n)$ and $f(rms) = rf(m)s$ $(m, n \in M, r \in R, s \in S)$.

Let R, S, T be rings, let M be an R-S-bimodule, and let N be an S-T-bimodule. We define the *tensor product* $M \otimes_S N$ of M and N over S to be the following R-T-bimodule: as an abelian group it is generated by the symbols $m \otimes n$ (all $m \in M$, $n \in N$), with defining relations:

(i) $(m_1 + m_2) \otimes n = m_1 \otimes n + m_2 \otimes n$ (all $m_1, m_2 \in M$, $n \in N$)

(ii) $m \otimes (n_1 + n_2) = m \otimes n_1 + m \otimes n_2$ (all $m \in M$, $n_1, n_2 \in N$)

(iii) $ms \otimes n = m \otimes sn$ (all $m \in M$, $s \in S$, $n \in N$).

The left R-action is then defined by $r(m \otimes n) = rm \otimes n$, and the right T-action by $(m \otimes n)t = m \otimes nt$ ($r \in R$, $m \in M$, $n \in N$, $t \in T$): this defines the actions on generators, and hence on arbitrary expressions in $M \otimes_S N$, by linearity. (An arbitrary element of $M \otimes_S N$ can be written, though not uniquely, in the form $\Sigma_i (m_i \otimes n_i)$ ($m_i \in M$, $n_i \in N$), i.e. without minus signs. Why?) The relations (i) and (ii), together with the R-, T-actions, may be described by saying that $\otimes : M \times N \to M \otimes_S N$ is a *bilinear* map (that is, it is linear in each of the two variables), and (iii) may be described by saying that $\otimes$ is *balanced*, or *S-balanced*.

The problem of handling a group (or module) which is given by means of generators and relations is a tricky one, and the less experienced reader is warned that there are many pitfalls for the unwary. We shall try as far as possible in this book to work within the group itself rather than in the overlying free group, as this is more natural and avoids tedious formalities. However, we invite the reader, at least until he gains confidence, to supply the formal details himself, for instance in the above definition (rather in the way we gave a second, more formal, definition of K_0R at the beginning of this chapter).

Given bimodules M, N as above, an R-T-bimodule P and a bilinear balanced map $f : M \times N \to P$, that is,

(i) $f(r_1m_1 + r_2m_2, n) = r_1f(m_1, n) + r_2f(m_2, n)$

(ii) $f(m, n_1t_1 + n_2t_2) = f(m, n_1)t_1 + f(m, n_2)t_2$

(iii) $f(ms, n) = f(m, sn)$

for all $r_1, r_2 \in R$, $m, m_1, m_2 \in M$, $s \in S$, $n, n_1, n_2 \in N$, and $t_1, t_2 \in T$, then there is a unique R-T-homomorphism g : $M \otimes_S N \to P$ such that $g\otimes = f$:

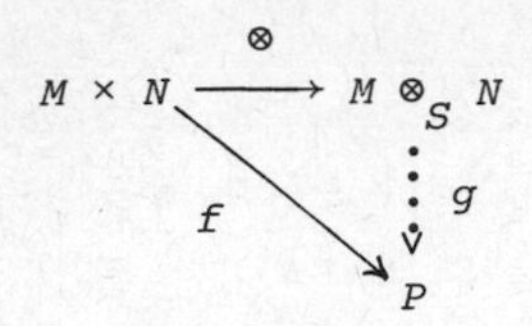

Indeed, g must be defined on the generators of $M \otimes_S N$ by $g(m \otimes n) = f(m, n)$ $(m \in M, n \in N)$. This is the *universal* property of the tensor product: the bilinear balanced map $\otimes$ is universal in that any other bilinear balanced map from $M \times N$ (for instance, f above) can be factored uniquely through $\otimes$.

Exercise. This property characterizes $M \otimes_S N$, up to isomorphism. Explicitly, if $h : M \times N \to Q$ is a bilinear balanced map, where Q is an R-T-bimodule, such that, given any R-T-bimodule P and a bilinear balanced map $f : M \times N \to P$, there is a unique R-T-homomorphism $g : Q \to P$ with $gh = f$, then $Q \simeq M \otimes_S N$.

Let $\alpha : M \to M_1$ be a homomorphism of R-S-bimodules, and let $\beta : N \to N_1$ be a homomorphism of S-T-bimodules. The map $f : M \times N \to M_1 \otimes_S N_1$ given by $f(m, n) = \alpha m \otimes \beta n$ is bilinear and balanced, and so induces a homomorphism $M \otimes_S N \to M_1 \otimes_S N_1$ which we denote by $\alpha \otimes \beta$: explicitly, $(\alpha \otimes \beta)(m \otimes n) = \alpha m \otimes \beta n$. If $\alpha_1 : M_1 \to M_2$ and $\beta_1 : N_1 \to N_2$ are bimodule homomorphisms, then clearly $(\alpha_1 \otimes \beta_1)(\alpha \otimes \beta) = \alpha_1\alpha \otimes \beta_1\beta$.

Exercise. Let R_i be a ring, $i = 0,1,2,3$, and let M_i be an R_{i-1}-R_i-bimodule, $i = 1,2,3$. Show that, as R_0-R_3-bimodules,

$$M_1 \otimes_{R_1} (M_2 \otimes_{R_2} M_3) \simeq (M_1 \otimes_{R_1} M_2) \otimes_{R_2} M_3.$$

Now suppose $M = \bigoplus_{\lambda \in \Lambda} M_\lambda$ is a direct sum of R-S-bimodules and $N = \bigoplus_{\lambda' \in \Lambda'} N_{\lambda'}$ is a direct sum of S-T-bimodules.

Proposition 3

$$M \otimes_S N = \bigoplus_{\lambda \in \Lambda, \lambda' \in \Lambda'} (M_\lambda \otimes_S N_{\lambda'})$$

Proof. Let $\pi_\lambda : M \twoheadrightarrow M_\lambda$, $i_\lambda : M_\lambda \hookrightarrow M$, $\tau_{\lambda'} : N \twoheadrightarrow N_{\lambda'}$, $j_{\lambda'} : N_{\lambda'} \hookrightarrow N$ be the natural maps. Then we have $\pi_\lambda \otimes \tau_{\lambda'} : M \otimes_S N$

$\to M_\lambda \otimes_S N_{\lambda'}$ and $i_\lambda \otimes j_{\lambda'} : M_\lambda \otimes_S N_{\lambda'} \to M \otimes_S N$. Then:

(i) $(\pi_\lambda \otimes \tau_{\lambda'})(i_\lambda \otimes j_{\lambda'}) = \pi_\lambda i_\lambda \otimes \tau_{\lambda'} j_{\lambda'} = 1 \otimes 1 = 1$;

(ii) $(\pi_\lambda \otimes \tau_{\lambda'})(i_{\lambda_1} \otimes j_{\lambda_1'}) = \pi_\lambda i_{\lambda_1} \otimes \tau_{\lambda'} j_{\lambda_1'} = 0$ if $\lambda \neq \lambda_1$ or if $\lambda' \neq \lambda_1'$, since then $\pi_\lambda i_{\lambda_1} = 0$ or $\tau_{\lambda'} j_{\lambda_1'} = 0$;

(iii) $M \otimes_S N$ is generated by the images of $i_\lambda \otimes j_{\lambda'}$ ($\lambda \in \Lambda$, $\lambda' \in \Lambda'$). For $M \otimes_S N$ is generated by all $m \otimes n$ ($m \in M$, $n \in N$), and for each such m, n we can write $m = \Sigma_\lambda\, i_\lambda m_\lambda$, $n = \Sigma_{\lambda'}\, j_{\lambda'} n_{\lambda'}$, where the sums are finite. Then

$$\begin{aligned} m \otimes n &= (\Sigma_\lambda\, i_\lambda m_\lambda) \otimes (\Sigma_{\lambda'}\, j_{\lambda'} n_{\lambda'}) \\ &= \Sigma_\lambda \Sigma_{\lambda'}\, (i_\lambda \otimes j_{\lambda'})(m_\lambda \otimes n_{\lambda'}). \end{aligned}$$

The result follows.

Exercise. Let M be an R-S-bimodule. Then $R \otimes_R M \simeq M \simeq M \otimes_S S$.

Corollary 4

$$R^n \otimes_R R^m \simeq R^{nm}$$

Proof.

$$\begin{aligned} R^n \otimes_R R^m &\simeq (R \otimes_R R)^{nm} \text{ by Proposition 3} \\ &\simeq R^{nm} \text{ by the above exercise (with } M = R = S). \end{aligned}$$

2.2 K_0 *of a commutative ring*

If the ring R is commutative, and P, Q are finitely generated projective R-modules, they are R-R-bimodules, and hence so is $P \otimes_R Q$. Let $P \oplus P_1 \simeq R^n$ and $Q \oplus Q_1 \simeq R^m$. Then by Proposition 3, we have

$$\begin{aligned} &(P \otimes_R Q) \oplus (P \otimes_R Q_1) \oplus (P_1 \otimes_R Q) \oplus (P_1 \otimes_R Q_1) \\ &\simeq (P \oplus P_1) \otimes_R (Q \oplus Q_1) \\ &\simeq R^n \otimes_R R^m \\ &\simeq R^{nm}, \text{ by Corollary 4} \end{aligned}$$

and so $P \otimes_R Q$ is a finitely generated projective R-module. Thus we can make K_0R into a ring by putting $[P][Q] = [P \otimes_R Q]$. More formally, the definition $\langle P\rangle\langle Q\rangle = \langle P \otimes_R Q\rangle$ makes $\mathcal{F}(R)$ into a ring, and it is easy to see that the subgroup $\mathcal{R}$ is an ideal of $\mathcal{F}$, so $\mathcal{F}/\mathcal{R} = K_0R$ is a ring. Clearly $[R] = 1$ (that is, $[R][P] = [P] = [P][R]$, all generators $[P]$).

Note that in Examples (i) and (ii) above, the isomorphisms $K_0F \simeq \mathbb{Z}$ and $K_0\mathbb{Z} \simeq \mathbb{Z}$ are now *ring* isomorphisms, by Corollary 4.

2.3 *The map* $K_0f : K_0R \to K_0S$

Let $f : R \to S$ be a homomorphism of rings (not necessarily commutative). We can use f to make S into a right R-module by the map $S \times R \to S$ given by $(s, r) \mapsto sf(r)$. S is also a left S-module, and it is clear that S thus becomes an S-R-bimodule. If P is an R-module, then regarding it as an R-$\mathbb{Z}$-bimodule, we can form $S \otimes_R P$, which will be an S-$\mathbb{Z}$-bimodule, that is, an S-module. We shall write $S \otimes_f P$ for $S \otimes_R P$.

Now $S \otimes_f R \simeq S$, for the map $S \to S \otimes_f R$, $s \mapsto s \otimes 1$, has inverse induced by $S \times R \to S$, $(s, r) \mapsto sf(r)$ (remember $f(1) = 1$). If P is a finitely generated projective R-module, say $P \oplus Q \simeq R^n$, then

$$\begin{aligned}(S \otimes_f P) \oplus (S \otimes_f Q) &\simeq S \otimes_f (P \oplus Q)\\ &\simeq S \otimes_f (R^n)\\ &\simeq (S \otimes_f R)^n\\ &\simeq S^n\end{aligned}$$

and so $S \otimes_f P$ is a finitely generated projective S-module. We may thus define a map $K_0f : K_0R \to K_0S$ by $[P] \mapsto [S \otimes_f P]$. (More formally, $\langle P\rangle \mapsto [S \otimes_f P]$ induces a homomorphism $\mathcal{F} \to K_0S$, and $\mathcal{R}$ lies in the kernel.) K_0f is a homomorphism of abelian groups (Proposition 3), and further, if R, S are commutative, K_0f is a ring homomorphism, since, for any P, Q,

$$(S \otimes_R P) \otimes_S (S \otimes_R Q) \simeq (P \otimes_R S) \otimes_S (S \otimes_R Q)$$

$$\simeq P \otimes_R (S \otimes_S S) \otimes_R Q$$

$$\simeq P \otimes_R S \otimes_R Q$$

$$\simeq S \otimes_R (P \otimes_R Q).$$

There is some sleight-of-hand in the above argument. Firstly, since the tensor product is associative up to isomorphism, we have not used as many brackets as are strictly necessary. Then, since R, S are commutative, an R-S-bimodule is an S-R-bimodule in a natural way, and in this sense we may write

$S \otimes_R P \simeq P \otimes_R S$ (the proof of which is easy)

and then form the tensor product of either with another module. Note also that $K_0f : 1 \mapsto 1$, since $S \otimes_f R \simeq S$.

If $f : R \to S$ and $g : S \to T$ are homomorphisms of rings, and P is an R-module, then

$$T \otimes_g (S \otimes_f P) \simeq (T \otimes_g S) \otimes_f P \simeq T \otimes_{gf} P$$

and so $(K_0g)(K_0f) = K_0(gf)$. Also the identity map $1 : R \to R$ gives $K_01 = 1 : K_0R \to K_0R$. When we come, later, to discuss categories, we shall describe this situation by saying that K_0 is a *functor* from the category of rings and ring homomorphisms to the category of abelian groups and group homomorphisms.

Proposition 5. Let R, S be rings, and let $f : R \to S$ and $g : S \to R$ be ring homomorphisms with $gf = 1$. Then K_0f is injective, K_0g is surjective, and $K_0S \simeq K_0R \oplus \ker (K_0g)$.

Proof. Immediate, since $(K_0g)(K_0f) = K_0gf = K_01 = 1$.

Example (i). R = any ring, $S = R[x]$, the ring of polynomials in the indeterminate x with coefficients in R. Let $f : R \to S$ be the natural injection, and define $g : S \to R$ by $p(x) \mapsto p(0)$.

Example (ii). R = any ring, G = any (multiplicative) group, $S = RG$, the group-ring of G with coefficients in R, that is, the ring of all finite formal sums $\Sigma_\lambda r_\lambda g_\lambda$ ($r_\lambda \in R$, $g_\lambda \in G$) with the obvious addition and multiplication. Let $f : R \to S$ be the natural injection $r \mapsto r1$ and let $g : S \to R$ be the augmentation map $\Sigma_\lambda r_\lambda g_\lambda \mapsto \Sigma_\lambda r_\lambda$.

Sometimes the map $K_0 f$, in the above examples, turns out to be an isomorphism. For instance, in (i), if R is a field, then $S = R[x]$ is a principal ideal domain, and over such a ring every submodule of a free module is free, and so $K_0 S \simeq \mathbb{Z}$. But $K_0 R \simeq \mathbb{Z}$, and $K_0 f$ is a homomorphism of rings (preserving 1), so it is an isomorphism.

Let R be a ring. We write $J \lhd_\ell R$ to mean that J is a left ideal of R (that is, a submodule of R, as left R-module) and $J \lhd R$ to mean that J is a two-sided ideal of R.

Lemma 6. Let $f : R \twoheadrightarrow S$ be a surjective ring homomorphism. Then $K_0 f : K_0 R \to K_0 S$ is given by $[P] \mapsto [\bar{P}]$, where $\bar{P} = P/JP$, $J = \ker f$.

Proof. We show that $\bar{P} \simeq S \otimes_f P$. Now $J \lhd_\ell R$ (in fact, $J \lhd R$) so JP is a submodule of P, and so P/JP is an R-module. Consider the map $F : S \times P/JP \to P/JP$ given by $(s, \bar{p}) \mapsto \overline{rp}$, where $f(r) = s$ and $\bar{p} = p + JP$. F is well defined since if $f(r_1) = f(r)$, then $r - r_1 \in J$, so $rp - r_1 p \in JP$, and thus $\overline{rp} = \overline{r_1 p}$. Firstly, F makes P/JP into an S-module. Secondly, F together with the natural map $P \to P/JP$ gives a bilinear balanced map $S \times P \to P/JP$, and hence induces $g : S \otimes_f P \to P/JP$ by the universal property. But now g^{-1} is given by $\bar{p} \mapsto 1 \otimes p$; this is well defined since, if $\bar{p} = \bar{p}_1$, then $p - p_1 \in JP$, so

$$\begin{aligned} 1 \otimes p - 1 \otimes p_1 &= 1 \otimes (p - p_1) \\ &= 1 \otimes (\Sigma_\lambda r_\lambda p_\lambda) \quad (r_\lambda \in J, \text{ all } \lambda) \\ &= \Sigma_\lambda (1 \otimes r_\lambda p_\lambda) \\ &= \Sigma_\lambda (f(r_\lambda) \otimes p_\lambda) \\ &= \Sigma_\lambda (0 \otimes p_\lambda), \text{ since } r_\lambda \in J \\ &= 0. \end{aligned}$$

2.4 *Products of rings*

Let S, T be rings. We make the cartesian product $R = S \times T$ into a ring by defining addition and multiplication componentwise:

$$(s, t) + (s', t') = (s + s', t + t')$$
$$(s, t)(s', t') = (ss', tt').$$

With this structure R is called the *product* of S and T. There are natural ring homomorphisms $\pi_1 : R \twoheadrightarrow S$, $(s, t) \mapsto s$, and $\pi_2 : R \twoheadrightarrow T$, $(s, t) \mapsto t$. Note, however, that the obvious map $S \to R$, $s \mapsto (s, 0)$ is not a ring homomorphism, since $1_S \mapsto (1_S, 0)$, which is not the same as $1_R = (1_S, 1_T)$, except in the trivial case $T = 0$. (Indeed, there may be no ring homomorphism $S \to R$ at all; for instance, $\mathbb{Z}/6\mathbb{Z} \simeq \mathbb{Z}/2\mathbb{Z} \times \mathbb{Z}/3\mathbb{Z}$, but there is no ring homomorphism $\mathbb{Z}/2\mathbb{Z} \to \mathbb{Z}/6\mathbb{Z}$, since 1 has additive order 2 on the left and 6 on the right.) If we identify S with $S \times \{0\}$ by $s \mapsto (s, 0)$, and similarly T with $\{0\} \times T$, then S, T are not subrings of R, since $1_R \notin S$, $1_R \notin T$, but $S \lhd R$, $T \lhd R$. In particular, S, T are R-submodules of R, and as such $S \oplus T = R$.

Conversely, let R be a ring, $S \lhd R$, $T \lhd R$, and $R = S \oplus T$ as left R-modules. Write $1 = s + t$, $s \in S$, $t \in T$. Then $Rs \subset S$, $Rt \subset T$, and $R = Rs + Rt$, so $Rs = S$, $Rt = T$. Next, $ST \subset S \cap T = 0$, so $ST = 0$, and similarly $TS = 0$. It follows that S, T are rings, with $1_S = s$ and $1_T = t$, and that $R = S \times T$. Note that we do need $S \lhd R$, $T \lhd R$, not just $S \lhd_\ell R$, $T \lhd_\ell R$. For instance, let R be the ring of all 2×2 matrices over a field, and put

$$s = \begin{pmatrix} 1 & 0 \\ 0 & 0 \end{pmatrix}, \quad t = \begin{pmatrix} 0 & 0 \\ 0 & 1 \end{pmatrix}, \quad S = Rs, \text{ and } T = Rt.$$

Then $S \lhd_\ell R$, $T \lhd_\ell R$ and $R = S \oplus T$, but $R \neq S \times T$ since $ST \neq 0$.

Proposition 7. $K_0(S \times T) \simeq K_0S \oplus K_0T$.

Proof. We have $\pi_1 : K_0(S \times T) \to K_0S$ by taking K_0 of the natural map $S \times T \to S$, that is, $\pi_1 : [P] \mapsto [P/TP]$, by Lemma 6. The map $i_1 : K_0S \to K_0(S \times T)$ is defined as follows: if P is a finitely generated projective S-module, we make it an $(S \times T)$-module by making T act trivially, i.e., $(s, t)p = sp$ ($s \in S$, $t \in T$, $p \in P$). Writing $\tilde{P}$ for this new module, we put $i_1([P]) = [\tilde{P}]$. It is easy to see i_1 is well defined, and induces $i_1 : K_0S \to K_0(S \times T)$. If P is an S-module, $T\tilde{P} = 0$, so $\tilde{P}/T\tilde{P} \simeq P$ (as S-modules), whence $\pi_1 i_1 = 1$. If $\pi_2 : K_0(S \times T) \to K_0T$ and $i_2 : K_0T \to K_0(S \times T)$ are defined similarly, then $\pi_2 i_2 = 1$. For any S-module P, $SP = P$, so $S\tilde{P} = \tilde{P}$, and $\tilde{P}/S\tilde{P} = 0$. Therefore $\pi_2 i_1 = 0$, and similarly $\pi_1 i_2 = 0$. Finally, for any $(S \times T)$-module M, $M = SM \oplus TM$, and $[SM] = i_1\pi_1[SM]$, $[TM] = i_2\pi_2[TM]$ in $K_0(S \times T)$ (if M is finitely generated and projective), whence $K_0(S \times T)$ is generated by the union of the images of i_1 and i_2.

Note: (i) The lemma extends easily to $K_0(S_1 \times S_2 \times \ldots \times S_n)$.

(ii) If S, T are commutative rings, then so are K_0S and K_0T, and indeed $K_0(S \times T)$. It is straightforward to show $K_0(S \times T) \simeq K_0S \times K_0T$ (the ring product). For instance, this gives $K_0(\mathbb{Z}/6\mathbb{Z}) \simeq \mathbb{Z} \times \mathbb{Z}$ (as rings).

Let R be a ring. Write $R^{\cdot} = \{\alpha \in R : \alpha\beta = 1 = \beta\alpha$, some $\beta \in R\}$. $R^{\cdot}$ is the *group of units* of R; clearly $R^{\cdot}$ is a multiplicative group with identity 1. When $\alpha\beta = 1 = \beta\alpha$, then as usual we write $\beta = \alpha^{-1}$.

Proposition 8. *(Nakayama's lemma)* Let R be a ring, $J \lhd R$ and $1 + J \subset R^{\cdot}$ (that is, $1 + j \in R^{\cdot}$, all $j \in J$). If M is a finitely generated R-module, and $M \neq 0$, then $JM \neq M$.

Proof. Let $a_1, a_2, \ldots, a_n$ generate M, where $a_1, a_2, \ldots, a_{n-1}$ generate a proper submodule N of M: so $a_n \notin N$. We may assume $n \geq 2$, by taking $a_1 = 0$. If $JM = M$, then $a_n \in JM$, so $a_n = j_1a_1 + j_2a_2 + \ldots + j_na_n$ for some $j_1, j_2, \ldots, j_n \in J$. Thus $(1 - j_n)a_n = j_1a_1 + j_2a_2 + \ldots + j_{n-1}a_{n-1}$, and since $j_n \in J$, we have $1 - j_n \in R^{\cdot}$, and so

$$a_n = (1 - j_n)^{-1}(j_1a_1 + j_2a_2 + \ldots + j_{n-1}a_{n-1}).$$

It follows that $a_n \in N$, which is a contradiction. So $JM \neq M$.

Corollary. Let R, M, J be as above. If N is a submodule of M with $N + JM = M$, then $N = M$.

Proof. Apply Nakayama's lemma to M/N. The conditions imply that $J(M/N) = M/N$, so $M/N = 0$, or $N = M$.

Proposition 9. Let R, S be rings, and let $f : R \twoheadrightarrow S$ be a surjective ring homomorphism with $1 + \ker f \subset R^{\bullet}$. Then $K_0f : K_0R \hookrightarrow K_0S$ is an injection.

Proof. Suppose P, Q are finitely generated projective R-modules, and that $\bar{P} = P/JP$ and $\bar{Q} = Q/JQ$, where $J = \ker f$, are isomorphic as S-modules. We show that P, Q are isomorphic as R-modules. Let $\theta : \bar{P} \to \bar{Q}$ be an S-module isomorphism. Then θ is bijective, so it is an R-module isomorphism (recall that $\bar{P}$, $\bar{Q}$ are defined as S-modules by writing $s\bar{p} = r\bar{p}$, where $f(r) = s$, so $\theta(r\bar{p}) = \theta(s\bar{p}) = s\theta(\bar{p}) = r\theta(\bar{p})$).

Since P is projective, and $\pi_2 : Q \twoheadrightarrow \bar{Q}$ is an epimorphism, $\theta\pi_1 : P \to \bar{Q}$ induces $\tau : P \to Q$ so that the diagram

$$\begin{array}{ccc} P & \xrightarrow{\pi_1} & \bar{P} \\ {\scriptstyle\tau}\vdots & & \downarrow{\scriptstyle\theta} \\ Q & \xrightarrow{\pi_2} & \bar{Q} \end{array}$$

commutes. We claim that τ is an isomorphism. For π_1 and θ are epimorphisms, hence so is $\theta\pi_1 = \pi_2\tau$. Thus $Q = \tau P + \ker \pi_2 = \tau P + JQ$. By the corollary to Nakayama's lemma, $\tau P = Q$, that is, τ is an epimorphism. But Q is projective, so there exists $i : Q \hookrightarrow P$ with $\tau i = 1_Q$, and then $P = iQ \oplus \ker \tau$. Then

$$\ker \tau \subset \ker \pi_2\tau = \ker \theta\pi_1 = \ker \pi_1 = JP$$

so $P = iQ + JP$: the corollary gives $P = iQ$, and so $\ker \tau = 0$, and we have an isomorphism.

We now show that $\ker (K_0f) = 0$. By Lemma 1, any element of K_0R can be written as $[P] - [Q]$ (for some finitely generated projective R-modules P, Q), and if $[P] - [Q] \in \ker (K_0f)$, then $[\bar{P}] - [\bar{Q}] = 0$ in K_0S, that is, $\bar{P}$, $\bar{Q}$ are stably isomorphic over S, by Proposition 2. So for some n, $P \oplus S^n \simeq Q \oplus S^n$, i.e.,

$\overline{P \oplus R^n} \simeq \overline{Q \oplus R^n}$, and by the above argument $P \oplus R^n \simeq Q \oplus R^n$. By Proposition 2, in K_0R, $[P] - [Q] = 0$, and the result follows.

The ring R is a *local* ring if the non-units form an ideal of R: $R - R^{\bullet} \triangleleft R$. Putting $J = R - R^{\bullet}$, it is clear that every proper ideal of R is contained in J, and so J is a maximal ideal of R. Indeed, J is the only maximal ideal of R, and in the commutative case R is local if and only if it has a unique maximal ideal. In any case R/J is a field or skew field, and $1 + J \subset R^{\bullet}$, for if $j \in J$ and $1 + j \notin R^{\bullet}$ then $1 + j \in J$, so $1 \in J$, which is absurd. *Exercise:* The ring R is local if and only if $R - R^{\bullet}$ is an additive subgroup of R.

Corollary. If R is a local ring, $K_0R \simeq \mathbb{Z}$.
Proof. There is a monomorphism of groups $K_0R \hookrightarrow K_0(R/J)$, where $J = R - R^{\bullet}$, and $K_0(R/J) \simeq \mathbb{Z}$, by Example (i). Since the monomorphism maps $[R]$ to $[R/J] \neq 0$, the result follows.

Exercise. Prove the corollary directly, by showing that over a local ring, finitely generated projective modules are free of unique rank (that is, all bases have the same cardinality).

Corollary. $K_0(\mathbb{Z}/m\mathbb{Z}) \simeq \mathbb{Z}^n$ (the ring product of n copies of $\mathbb{Z}$), where n is the number of distinct prime numbers dividing the positive integer m.
Proof. If

$$m = p_1^{\alpha_1} p_2^{\alpha_2} \cdots p_n^{\alpha_n}$$

where $p_1, p_2, \ldots, p_n$ are distinct prime numbers and $\alpha_1, \alpha_2, \ldots, \alpha_n$ are positive integers, then by the Chinese remainder theorem,

$$\mathbb{Z}/m\mathbb{Z} \simeq \mathbb{Z}/p_1^{\alpha_1}\mathbb{Z} \times \mathbb{Z}/p_2^{\alpha_2}\mathbb{Z} \times \ldots \times \mathbb{Z}/p_n^{\alpha_n}\mathbb{Z}.$$

But $\mathbb{Z}/p_i^{\alpha_i}\mathbb{Z}$ is local, all i, and the result follows by Proposition 7.

2.5 *Matrix rings*

Let R be a ring, and let S be the ring of all $n \times n$ matrices with coefficients in R. Put $M = R^n$ (written as column vectors); M is a right R-module, and as such $\mathrm{End}_R(M) = S$. Writing the endomorphisms on the left, M becomes an S-R-bimodule. Similarly, if $N = R^n$ (written as row vectors), N is an R-S-bimodule. Thus $M \otimes_R N$ is an S-S bimodule and $N \otimes_S M$ is an R-R-bimodule. We claim that $M \otimes_R N \simeq S$ and $N \otimes_S M \simeq R$; in fact, we prove something slightly more general:

Proposition 10. Let R be a ring, let S be the ring of all $n \times n$ matrices over R, and let T be the ring of all $m \times m$ matrices over R. Let M be the S-T-bimodule of all $n \times m$ matrices over R, and let N be the T-S-bimodule of all $m \times n$ matrices over R, where the action of the scalars S or T is just matrix multiplication. Then $N \otimes_S M \simeq T$, as T-T-bimodules, and $M \otimes_T N \simeq S$, as S-S-bimodules.

Proof. We show $N \otimes_S M \simeq T$; the other result follows similarly. Define $\mu : N \times M \to T$ by $\mu(A, B) = AB$, the matrix product. This map is clearly bilinear and balanced (the latter since matrix multiplication is associative). Let U be a T-T-bimodule and suppose $f : N \times M \to U$ is a bilinear balanced map. The result follows if we show that there is a unique homomorphism $\nu : T \to U$ with $\nu\mu = f$, by the universal characterization of the tensor product. Since such a map ν is determined uniquely by $\nu(I) \in U$ (as usual, we write I for 1 in a matrix ring), it is enough to show that there is an element $x \in U$ such that $f(A, B) = ABx = xAB$, all $A \in N$, $B \in M$.

Write $E^S_{ij}(a)$ for the matrix in S with a in the i, j position and 0 elsewhere, and similarly in T, M, and N. Let $A \in N$, $B \in M$, so $A = \Sigma_{i,j}\, E^N_{ij}(a_{ij})$ and $B = \Sigma_{k,\ell}\, E^M_{k\ell}(b_{k\ell})$, say. Then

$$\begin{aligned} f(A, B) &= \Sigma_{i,j,k,\ell}\, f(E^N_{ij}(a_{ij}), E^M_{k\ell}(b_{k\ell})) \\ &= \Sigma_{i,j,k,\ell}\, f(E^N_{ij}(a_{ij}), E^S_{k1}(b_{k\ell})E^M_{1\ell}(1)) \\ &= \Sigma_{i,j,k,\ell}\, f(E^N_{ij}(a_{ij})E^S_{k1}(b_{k\ell}), E^M_{1\ell}(1)) \\ &= \Sigma_{i,j,\ell}\, f(E^N_{i1}(a_{ij}b_{j\ell}), E^M_{1\ell}(1)) \\ &= \Sigma_{i,j,k,\ell}\, f(E^T_{ik}(a_{ij}b_{jk})E^N_{\ell 1}(1), E^M_{1\ell}(1)) \\ &= \Sigma_{i,j,k}\, E^T_{ik}(a_{ij}b_{jk})\, \Sigma_\ell\, f(E^N_{\ell 1}(1), E^M_{1\ell}(1)) \\ &= ABx, \text{ where } x = \Sigma_\ell\, f(E^N_{\ell 1}(1), E^M_{1\ell}(1)). \end{aligned}$$

Similarly, $f(A, B) = xAB$, and the proof is complete.

Corollary. $K_0S \simeq K_0T$; in particular, taking $m = 1$, $K_0S \simeq K_0R$.

Proof. Let P be a finitely generated projective left S-module, say $P \oplus Q \simeq S^r$. Then $(N \otimes_S P) \oplus (N \otimes_S Q) \simeq N \otimes_S (P \oplus Q) \simeq N \otimes_S (S^r) \simeq (N \otimes_S S)^r \simeq N^r$. Further, it is clear that, as left T-modules, $N^m \simeq T^n$ (each being the set of all $m \times nm$ matrices over R), so $N \otimes_S P$ is a direct summand of $N^{rm} \simeq T^{rn}$, and so it is a finitely generated projective T-module. By a similar argument, if P, Q are stably isomorphic S-modules, then $N \otimes_S P$, $N \otimes_S Q$ are stably isomorphic T-modules. Thus we can construct a homomorphism $K_0S \to K_0T$ by mapping the generators thus: $[P] \mapsto [N \otimes_S P]$. The remaining formal details are left to the reader. Similarly, there is a homomorphism $K_0T \to K_0S$ with $[P'] \mapsto [M \otimes_T P']$. The two homomorphisms are inverse to each other, since

$$M \otimes_T (N \otimes_S P) \simeq (M \otimes_T N) \otimes_S P \simeq S \otimes_S P \simeq P$$

and

$$N \otimes_S (M \otimes_T P') \simeq (N \otimes_S M) \otimes_T P' \simeq T \otimes_T P' \simeq P'.$$

Note: If F is a field, we have $K_0F \simeq \mathbb{Z}$, generated additively by $[F]$. If S is the ring of $n \times n$ matrices over F, then by the above result, $K_0S \simeq \mathbb{Z}$ also, but this time $[S]$ does not generate K_0S, but a subgroup of index n. K_0S is generated additively by $[F^n]$, where F^n is considered as an S-module, and in the isomorphism $K_0S \to K_0F$, we have $[S] \mapsto n[F]$.

2.6 *The Jacobson radical, Wedderburn-Artin theory, and K_0 of semi-local rings*

Let R be a ring. The *Jacobson radical* of R, $\mathrm{rad}(R)$, is the intersection of all the maximal left ideals of R.

Proposition 11

(i) $\mathrm{rad}(R) \lhd R$

(ii) $\mathrm{rad}(R) = \cup \{J : J \lhd R \text{ and } 1 + J \subset R^{\bullet}\}$

(iii) $\mathrm{rad}(R/\mathrm{rad}(R)) = 0$

Proof. (i) Clearly $\mathrm{rad}(R) \lhd_{\ell} R$. Next we show that, for $m \in \mathrm{rad}(R)$, $R(1 + m) = R$. For $R(1 + m) \lhd_{\ell} R$, and if $R(1 + m) \neq R$, there is a maximal left ideal M with $R(1 + m) \subset M$. But $1 + m \in M$, and $m \in M$ also, since $m \in \mathrm{rad}(R)$, and thus $1 \in M$, a contradiction. Now let $m \in \mathrm{rad}(R)$ and $a \in R$, and suppose $ma \notin \mathrm{rad}(R)$. Then there is a maximal left ideal M with $ma \notin M$. So $M + Rma = R$, and therefore $m_1 + rma = a$ for some $m_1 \in M$, $r \in R$. Thus $m_1 = (1 - rm)a$: but $-rm \in \mathrm{rad}(R)$, so $R(1 - rm) = R$ and therefore $\alpha(1 - rm) = 1$ for some $\alpha \in R$. So $a = \alpha m_1 \in M$: but $ma \in M$ so we have a contradiction. Therefore $ma \in \mathrm{rad}(R)$, and the result follows.

(ii) Let $J \lhd R$ with $1 + J \subset R^{\bullet}$. Let M be a maximal left ideal of R, and suppose $J \not\subset M$. Then $M + J = R$, so $m + j = 1$ for some $m \in M$, $j \in J$. But then $m = 1 - j \in R^{\bullet}$, which is a contradiction. So $J \subset M$, and therefore

$$\cup \{J : J \lhd R \text{ and } 1 + J \subset R^{\bullet}\} \subset \mathrm{rad}(R).$$

It remains to show $1 + \mathrm{rad}(R) \subset R^{\bullet}$. If $m \in \mathrm{rad}(R)$, $R(1 + m) = R$, so $\alpha(1 + m) = 1$ for some $\alpha \in R$. Put $m' = \alpha - 1$: we have $(1 + m')(1 + m) = 1$, or $m + m' + m'm = 0$, or $m' = -m - m'm$, so $m' \in \mathrm{rad}(R)$. So, similarly, we can find m'' with $(1 + m'')(1 + m') = 1$. It is immediate that $m'' = m$, and so $1 + m \in R^{\bullet}$.

(iii) Immediate: any maximal left ideal of $R/\mathrm{rad}(R)$ is of the form $M/\mathrm{rad}(R)$, where M is a maximal left ideal of R, and conversely.

Note that, because of the left-right symmetry of (ii), $\mathrm{rad}(R)$ is the intersection of the maximal *right* ideals of R. Also $r \in \mathrm{rad}(R)$ if and only if $1 + arb \in R^{\bullet}$ for all $a, b \in R$.

Corollary. The natural map $R \twoheadrightarrow R/\mathrm{rad}(R)$ induces a monomorphism $K_0R \hookrightarrow K_0(R/\mathrm{rad}(R))$.

Proof. Immediate, from Proposition 11 (ii) and Proposition 9.

Exercise: Let R be the subring of the field of rational numbers consisting of all $n/(1 + 6m)$ $(n, m \in \mathbb{Z})$. Show that R is a principal ideal domain, and that $\mathrm{rad}(R) = 6R$, and deduce that in this case the monomorphism of the last corollary is not an isomorphism.

Lemma 12. Let R be a ring with $\mathrm{rad}(R) = 0$, and suppose R is (left) Artinian (that is, R satisfies the minimum condition on left ideals). Let P be a minimal left ideal of R. Then there exists $e \in R$ with $e^2 = e$, $P = Re$, and $R = P \oplus R(1 - e)$.

Proof. $PP \subset P$: if $PP = 0$ then $(PR)(PR) \subset PPR = 0$, so we have $PR \lhd R$ with $1 + PR \subset R^{\bullet}$. (In detail, if $x \in PR$ then $x^2 = 0$, so $1 + x$ has inverse $1 - x$.) But $\mathrm{rad}(R) = 0$, so $PR = 0$ by Proposition 11 (ii), and hence $P = 0$, a contradiction. So since $PP \lhd_{\ell} R$ and $PP \subset P$, we must have $PP = P$, by the minimality of P. Thus we can find $x \in P$ with $Px \neq 0$. Now consider the map $P \to P$ given by $p \mapsto px$ (all $p \in P$); this is a bijection, by Schur's lemma. (In detail, a module is *simple* if the only submodules are 0 and the module itself. *Schur's lemma* says that a homomorphism between simple R-modules is either 0 or is an isomorphism. The proof consists of observing that the kernel and image are submodules. In this case P is clearly a simple R-module, and the given map $P \to P$ is not 0.) So we can find $e \in P$ with

$x = ex$, and then $e^2 \in P$ and $e^2x = e(ex) = ex$, so $e^2 = e$, since $p \mapsto px$ is a bijection. Then $Re \subset P$, and $Re \lhd_{\ell} R$. But Re contains $e^2 = e$, and $e \neq 0$ since $x \neq 0$, so $Re \neq 0$ and hence $Re = P$. Then $r = re + r(1 - e)$, all $r \in R$, so $R = P + R(1 - e)$. Finally, if $r, s \in R$ and $re = s(1 - e)$, then $re = re^2 = s(1 - e)e = s(e - e^2) = 0$, so $P \cap R(1 - e) = 0$ and the result follows.

Proposition 13. Let R be as in Lemma 12. Then R is a finite direct sum of minimal left ideals.

Proof. Let M be a left ideal of R, and let $P \subset M$ be a minimal left ideal. Then for some $e \in P$, $P = Re$, by Lemma 12, and then $M = P \oplus M(1 - e)$, by a similar argument to Lemma 12. So choose a minimal left ideal P_1 and write $R = P_1 \oplus M_1$; next choose a minimal left ideal $P_2 \subset M_1$ and write $M_1 = P_2 \oplus M_2$; so $R = P_1 \oplus P_2 \oplus M_2$. Choose a minimal left ideal $P_3 \subset M_2$, and so on: we obtain $R \supset M_1 \supset M_2 \supset \ldots$, which must terminate, since R is Artinian, and the result follows.

Corollary (The Wedderburn-Artin theorem). R, as in the proposition, is isomorphic to a finite direct product of full matrix rings over skew fields.

Proof. We have

$$R \simeq P_1^{n_1} \oplus P_2^{n_2} \oplus \ldots \oplus P_r^{n_r}$$

where the P_i are minimal left ideals of R, pairwise non-isomorphic as R-modules. Now $\mathrm{End}_R(R) \simeq R$ as rings (explicitly, if $f : R \to R$ is an endomorphism of R as an R-module, the isomorphism is given by $f \mapsto f(1)$). Put $D_i = \mathrm{End}_R(P_i)$. D_i is a skew field (possibly a field, of course), by Schur's lemma, since every non-zero endomorphism of P_i is an automorphism, and so has an inverse. Then $\mathrm{End}_R(P_i^{n_i})$ is isomorphic to the ring of all $n_i \times n_i$ matrices over D_i, and $\mathrm{Hom}_R(P_i, P_j) = 0$ for $i \neq j$, by Schur's lemma. The result follows.

Exercise: Establish the converse: if R is a finite product of full matrix rings over skew fields, then it is left Artinian and has zero radical. Deduce that, for any ring R, R is left Artinian with $\mathrm{rad}(R) = 0$ if and only if R is right Artinian with $\mathrm{rad}(R) = 0$.

Corollary. If R is as in the corollary above, then $K_0R \simeq \mathbb{Z}^r$.
Proof. Immediate, from the corollary above, Proposition 7, and the corollary to Proposition 10.

Let R be a ring. If $R/\mathrm{rad}(R)$ is left (equivalently, right) Artinian, we say R is *semi-local*. Examples: any field or skew field; more generally, any local ring (for if R is local, $\mathrm{rad}(R) = R - R^{\bullet}$ and $R/\mathrm{rad}(R)$ is thus a field or skew field). Any left or right Artinian ring is semi-local, and in particular any finite ring is semi-local. But $\mathbb{Z}$, for instance, is not semi-local.

Proposition 14. If R is a semi-local ring, K_0R is a free abelian group of finite rank.
Proof. By the corollary to Proposition 11, K_0R is isomorphic to a subgroup of $K_0(R/\mathrm{rad}(R))$, which is a free abelian group of finite rank by the second corollary to Proposition 13, applied to $R/\mathrm{rad}(R)$. The result follows.

2.7 *The reduced group* $\widetilde{K}_0R$

Given a ring R, there is a unique ring homomorphism $f : \mathbb{Z} \to R$ (remember we must have $f(1) = 1$), and this induces $K_0f : K_0\mathbb{Z} \to K_0R$. We define the *reduced* (Grothendieck) group $\widetilde{K}_0R$ by $\widetilde{K}_0R = K_0R/\mathrm{image}(K_0f)$. Note that the image of K_0f is the cyclic subgroup of K_0R generated by $[R]$.

Suppose there is a ring S with K_0S generated by $[S]$, and $K_0S \simeq \mathbb{Z}$, such that there is a ring homomorphism $g : R \to S$.

Identifying K_0S and $K_0\mathbb{Z}$ with $\mathbb{Z}$, we see that $(K_0g)(K_0f) = K_0(gf) : \mathbb{Z} \to \mathbb{Z}$ is a group homomorphism with $1 \mapsto 1$ (i.e. $[\mathbb{Z}] \mapsto [S]$), and so it is an isomorphism. In this case we deduce $K_0R = \text{image}(K_0f) \oplus \ker(K_0g) \simeq \mathbb{Z} \oplus \tilde{K}_0R$. Of course, this particular decomposition of K_0R as a direct sum depends on the particular choice of S and g. For instance, if $R = \mathbb{Z} \times \mathbb{Z}$, we may take $S = \mathbb{Z}$ and $g : \mathbb{Z} \times \mathbb{Z} \to \mathbb{Z}$ to be projection to the first or second component, but these two choices of g give different decompositions of K_0R as a direct sum $\mathbb{Z} \oplus \tilde{K}_0R$. Detailed verification of this is left to the reader.

Let R be a commutative ring. If $I \lhd R$ is maximal, then put $S = R/I$, and S is a field. So suitable S and g, as above, always exist in this case.

Proposition 15. Let R be a commutative ring and let F be a field. A homomorphism $R \to F$ embeds $\tilde{K}_0R$ as an ideal of the ring K_0R, and $\mathbb{Z} \to R$ embeds $\mathbb{Z}$ as a subring of K_0R, and then $K_0R = \mathbb{Z} \oplus \tilde{K}_0R$, as additive groups.

Proof. Done already, except to remark that K_0f and K_0g, as above, are ring homomorphisms in this case, so image(K_0f) is a subring of K_0R and $\ker(K_0g)$ is an ideal of K_0R.

Note that the kernel of $g : R \to F$, above, is a prime ideal of R. Conversely, given a prime ideal P of R, R/P is an integral domain. If we denote the field of fractions of R/P by F, then there is an obvious homomorphism $g : R \to F$. Then if M is a finitely generated projective R-module, $K_0g[M] \in K_0F = \mathbb{Z}$ (identifying), and the integer obtained is called the *rank* of M at P, written $\text{rank}_P(M)$. It is the dimension of the F-vector space $F \otimes_g M$.

CHAPTER THREE

K_0 of Commutative Rings, and the Picard Group

All rings in this chapter are commutative.

3.1 *Rings of fractions*

Let R be a (commutative) ring, and let S be a multiplicative subset of R, that is, $1 \in S$ and, for all $s_1, s_2 \in S$, we have $s_1 s_2 \in S$. On the set $S \times R$, write $(s, r) \sim (s_1, r_1)$ if $tsr_1 = ts_1 r$ for some $t \in S$; this is an equivalence relation on $S \times R$. Write $\frac{r}{s}$, or r/s, for the equivalence class of (s, r), and $S^{-1}R$ for the quotient set $(S \times R)/\sim$ (the set of equivalence classes). $S^{-1}R$ is a ring if we define

$$\frac{r}{s} + \frac{r'}{s'} = \frac{s'r + sr'}{ss'} \text{ and } \frac{r}{s}\frac{r'}{s'} = \frac{rr'}{ss'}.$$

The reader should check that these operations are well defined and do indeed give a ring. The additive and multiplicative identities are 0/1 and 1/1 respectively; with the usual abuse of notation we denote them by 0 and 1. Next,

$$r/s = 0 \iff (s, r) \sim (1, 0) \iff tr = 0, \text{ some } t \in S.$$

In particular, $1 = 0$ in $S^{-1}R$ if and only if $0 \in S$, that is, $S^{-1}R = 0$ if and only if $0 \in S$. There is a natural ring homomorphism $R \to S^{-1}R$ given by $r \mapsto r/1$. It need not be a monomorphism, for instance if $R \neq 0$ but $0 \in S$.

A special and familiar case of the above construction occurs when R is an integral domain and $S = R - \{0\}$; note that $1 \neq 0$ (we are taking this to be part of the definition of integral domain), so $1 \in S$. Then $S^{-1}R$ is a field, the *field of fractions* of R, and in this case the natural map $R \to S^{-1}R$ *is* a monomorphism.

In general, $R \to S^{-1}R$ is universal for S-inverting homomorphisms. Explicitly, if $f : R \to \bar{R}$ is a homomorphism of rings such that $f(s) \in \bar{R}^{\cdot}$, all $s \in S$, then there is a unique homomorphism $\bar{f} : S^{-1}R \to \bar{R}$ with $\bar{f}(r/1) = f(r)$, all $r \in R$. It is given by $\bar{f}(r/s) = f(r)(f(s))^{-1}$, all $r \in R$, $s \in S$.

3.2 *Modules of fractions*

Let R be a ring and let S be a multiplicative subset of R. Let M be an R-module. On the set $S \times M$, write $(s, m) \sim (s_1, m_1)$ if $tsm_1 = ts_1m$ for some $t \in S$; this is an equivalence relation on $S \times M$. Write $\frac{m}{s}$, or m/s, for the equivalence class of (s, m), and $S^{-1}M$ for the quotient set $(S \times M)/\sim$. $S^{-1}M$ is an $S^{-1}R$-module if we define

$$\frac{m}{s} + \frac{m'}{s'} = \frac{s'm + sm'}{ss'} \quad \text{and} \quad \frac{r}{s}\frac{m}{s'} = \frac{rm}{ss'}.$$

Again, details are left to the reader. Writing $f : R \to S^{-1}R$ for the natural map, we have:

Proposition 16. $S^{-1}M \simeq S^{-1}R \otimes_f M$.

Proof. Define a map $S^{-1}R \times M \to S^{-1}M$ by $(r/s, m) \mapsto rm/s$. This is well defined, since if $r/s = r'/s'$, then $tsr' = ts'r$, some $t \in S$, and then $tsr'm = ts'rm$, so $rm/s = r'm/s'$. It is clearly bilinear and balanced, so it induces a homomorphism $S^{-1}R \otimes_f M \to S^{-1}M$, which is obviously surjective. Then any element of $S^{-1}R \otimes_f M$ can be written as $(r/s) \otimes m$, some $r \in R$, $s \in S$, and $m \in M$, since

$$\frac{r}{s} \otimes m + \frac{r'}{s'} \otimes m' = \frac{s'r}{s's} \otimes m + \frac{sr'}{ss'} \otimes m'$$

$$= \frac{1}{s's} \otimes s'rm + \frac{1}{ss'} \otimes sr'm'$$

$$= \frac{1}{ss'} \otimes (s'rm + sr'm').$$

If $rm/s = 0$, then $(s, rm) \sim (1, 0)$, or $trm = 0$, some $t \in S$ Therefore

$$\frac{r}{s} \otimes m = \frac{tr}{ts} \otimes m = \frac{1}{ts} \otimes trm = \frac{1}{ts} \otimes 0 = 0$$

and so we have an isomorphism.

Let $g : N \to M$ be a homomorphism of R-modules, where R, S are as above. Write $S^{-1}g : S^{-1}N \to S^{-1}M$ for the homomorphism given by $(S^{-1}g)(n/s) = g(n)/s$. (If we identify $S^{-1}N$ with $S^{-1}R \otimes_f N$ and $S^{-1}M$ with $S^{-1}R \otimes_f M$, then $S^{-1}g$ is just $1 \otimes g$.) Now if g is an injection, so is $S^{-1}g$. For if $(S^{-1}g)(n/s) = 0$, then $g(n)/s = 0$, so $tg(n) = 0$, some $t \in S$. Therefore $g(tn) = 0$, and so $tn = 0$, since g is injective. So $n/s = tn/ts = 0/ts = 0$, and we have an injection. Thus, if N is a submodule of M, we can regard $S^{-1}N$ as a submodule of $S^{-1}M$ in a natural way.

Exercise: Let R, S be as above, and let $g : N \to M$ be a homomorphism of R-modules. Then $S^{-1}(\ker(g)) = \ker(S^{-1}g)$ and $S^{-1}(g(N)) = (S^{-1}g)(S^{-1}N)$, on identifying, as above.

Let M be an R-module, and let N be a submodule of M. Then $S^{-1}(M/N) \cong (S^{-1}M)/(S^{-1}N)$; the isomorphism is given by $(m + N)/s \mapsto m/s + S^{-1}N$. This is well defined since

$$\frac{m+N}{s} = \frac{m'+N}{s'} \Rightarrow ts'(m + N) = ts(m' + N), \text{ some } t \in S$$

$$\Rightarrow ts'm + N = tsm' + N$$

$$\Rightarrow ts'm - tsm' \in N$$

$$\Rightarrow \frac{m}{s} - \frac{m'}{s'} = \frac{s'm - sm'}{ss'} = \frac{ts'm - tsm'}{tss'} \in S^{-1}N$$

$$\Rightarrow m/s + S^{-1}N = m'/s' + S^{-1}N$$

and it is now easy to check that it is an isomorphism.

If $g : N \to M$ is a homomorphism, the *cokernel* of g is $\mathrm{coker}(g) = M/g(N)$.

Exercise: $S^{-1}(\mathrm{coker}(g)) = \mathrm{coker}(S^{-1}g)$ (on identifying in the obvious way).

3.3 *Localization*

Localization is a process whereby, from a given (commutative) ring, certain local rings are constructed. The terminology comes from algebraic geometry, where studying local properties of a geometrical object (a variety) corresponds to localizing a ring of functions on the object (the coordinate ring). For a description of the geometrical process, see [8] (numbers in square brackets are references to the Bibliography at the end of the book); we shall describe the process of localization purely algebraically.

First, we need to know some more about the finitely generated projective modules over a local ring: we shall see that they are in fact free. Let R be a local ring, with unique maximal ideal J, and let $\bar{R} = R/J$: so $\bar{R}$ is a field. Let M, N be non-zero R-modules with $M \oplus N = R^n$. On passing to the quotient we have $\bar{M} \oplus \bar{N} = \bar{R}^n$; but $\bar{R}$ is a field, so we can choose a basis $\bar{m}_1, \bar{m}_2, \ldots, \bar{m}_n$ of $\bar{R}^n$ so that $\bar{m}_1, \bar{m}_2, \ldots, \bar{m}_r$, say, is a basis of $\bar{M}$ and $\bar{m}_{r+1}, \bar{m}_{r+2}, \ldots, \bar{m}_n$ is a basis of $\bar{N}$. Now lift to R^n: we can choose $m_i \in R^n$, each i, so that $m_i \mapsto \bar{m}_i$ under the natural map $R^n \to \bar{R}^n$, and so that $m_i \in M$ for $i \leq r$ and $m_i \in N$ for $i > r$. We show $m_1, m_2, \ldots, m_n$ is a basis for the free module R^n. Since $m_i \in R^n$, we have $m_i = (a_{i1}, a_{i2}, \ldots, a_{in})$, some $a_{ij} \in R$ $(1 \leq i \leq n,\ 1 \leq j \leq n)$, and clearly what we must show is that the $n \times n$ matrix $A = (a_{ij})$ has a two-sided inverse. Now $\bar{A} = (\bar{a}_{ij})$ is an invertible matrix over $\bar{R}$, since the $\bar{m}_i = (\bar{a}_{i1}, \bar{a}_{i2}, \ldots, \bar{a}_{in})$, $1 \leq i \leq n$, form a basis of $\bar{R}^n$, and so we can find an $n \times n$ matrix B over R such that $\bar{B}\bar{A} = I$, the $n \times n$ identity matrix, over $\bar{R}$; that is, such that $BA \equiv I$ modulo J, over R. It is now easy to reduce BA to diagonal form by elementary row operations. For the 1, 1 entry of BA is congruent to 1 modulo J, and so is a unit, since R is local, and thus we can use elementary row operations to obtain zeros below this entry in the first column. Then the 2, 2 entry in the new matrix is still congruent to 1 modulo J, and thus we proceed. So we obtain a matrix C, a product of elementary matrices, so that CBA is a diagonal matrix, with diagonal entries congruent to 1 modulo J, hence units. So CBA is an invertible matrix, and so A has a left inverse. By a similar argument A has a right inverse, which must coincide with the left inverse.

Thus $m_1, m_2, \ldots, m_n$ is a basis of R^n; but $m_i \in M$ for $i \leq r$, $m_i \in N$ for $i > r$, and $R^n = M \oplus N$. It is immediate that M is free, with free basis $m_1, m_2, \ldots, m_r$. We have shown that, over a local ring, every finitely generated projective module is free. (The argument works even if R is not commutative; of course, $\bar{R}$ may then be a skew field, not a field.)

We now return to rings of fractions. Let R be a (commutative) ring, and let P be a prime ideal of R. Put $S = R - P$; then S is a multiplicative subset of R. We write $R_P = S^{-1}R$, and, if M is an R-module, we write $M_P = S^{-1}M$. The natural map $R \to R_P$ makes R_P into an R-module, and Proposition 16 now reads $M_P \simeq R_P \otimes_R M$. If $g : N \to M$ is a homomorphism of R-modules, we write $g_P = S^{-1}g$. Considering R as an R-module, P is a submodule, and so

$$P_P = \{p/s : p \in P,\ s \in S\}$$

is a submodule of R_P, as an R_P-module, that is, P_P is an ideal of the ring R_P. If $r \in R$, $s \in S$ and $r/s \notin P_P$, then $r \notin P$, so $r \in S$ and therefore $r/s \in R_P^{\bullet}$ (explicitly, $(r/s)^{-1} = s/r$). Thus $R_P - P_P = R_P^{\bullet}$, so the non-units of R_P form an ideal, P_P, and therefore R_P is a local ring. It is called the *localization* of R at P.

Since P is prime, R/P is an integral domain, and has a field of fractions F. Then the composite homomorphism $R \to R/P \hookrightarrow F$ is S-inverting, and so by the universal property of the ring of fractions R_P, there is a homomorphism $R_P \to F$ so that the diagram

$$\begin{array}{ccc} R & \longrightarrow & R_P \\ \downarrow & & \downarrow \\ R/P & \longrightarrow & F \end{array}$$

commutes.

Exercise: $R_P \to F$ is surjective, and its kernel is P_P.
So we can identify F with the residue field R_P/P_P of R_P. Now $K_0R_P \cong \mathbb{Z}$, generated by $[R_P]$, since R_P is local, by the corollary to Proposition 9. But $K_0F \cong \mathbb{Z}$ also, and so $R_P \to F$ induces a ring homomorphism $K_0R_P \to K_0F$, i.e., $\mathbb{Z} \to \mathbb{Z}$, which must be an isomorphism, since $1 \mapsto 1$.

If M is a finitely generated projective R-module, then by definition, $\text{rank}_P(M)$ is the image of $[M]$ in $\mathbb{Z} = K_0F$. This is the same as the image of $[M]$ in $\mathbb{Z} = K_0R_P$. That is, $\text{rank}_P(M) = n$ means $[R_P \otimes_R M] = [R_P{}^n]$ in K_0R_P. But $R_P \otimes_R M = M_P$, which is free, since R_P is local, and so $\text{rank}_P(M) = n$ means $M_P \cong R_P{}^n$.

Proposition 17. Let P, P_1 be prime ideals of R with $P \subset P_1$. Then for any finitely generated projective R-module M, we have $\text{rank}_P(M) = \text{rank}_{P_1}(M)$.
Proof. If $S = R - P$ and $S_1 = R - P_1$, then $S_1 \subset S$. So $R \to R_P$ is S_1-inverting, since it is S-inverting, and induces $R_{P_1} \to R_P$ so that the diagram

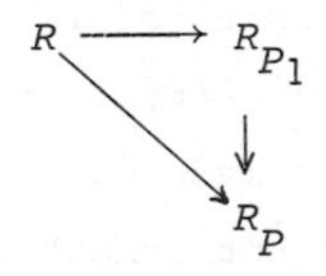

commutes. Then $R_{P_1} \to R_P$ induces a ring homomorphism $K_0R_{P_1} \to K_0R_P$, which is an isomorphism since $K_0R_{P_1}$ and K_0R_P are each isomorphic to $\mathbb{Z}$. (Remember ring homomorphisms preserve 1.) The result follows.

Corollary. If R is an integral domain, $\text{rank}_P(M)$ is the same for all prime ideals P.

Proof. 0 is a prime ideal, and $0 \subset P$, all P, so $\text{rank}_P(M) = \text{rank}_0(M)$, all P.

If R is a ring and M is a finitely generated projective R-module such that $\text{rank}_P(M) = n$ is the same for all prime ideals P, we say M is of *constant rank*, and write $\text{rank}(M) = n$. For instance, R^n has constant rank n (remember that since R is commutative it has IBN, so the rank is uniquely defined).

Corollary. M is of constant rank if $\text{rank}_P(M)$ is the same for all *maximal* ideals P. (Of course, all maximal ideals are prime.)

Proof. Every proper ideal, hence every prime ideal, is contained in a maximal ideal, and the result follows.

Note: this shows that over a local ring every finitely generated projective module is of constant rank, since there is only one maximal ideal. This result is also clear when we recall that such modules are necessarily free.

3.4 *Dual modules*

Let M be an R-module. The *dual* of M is defined to be $M^* = \text{Hom}_R(M, R)$. This is an R-module if, for $f, g \in M^*$, $r \in R$, we define $f + g$, rf, by $(f + g)(m) = f(m) + g(m)$ and $(rf)(m) = r(f(m))$, all $m \in M$. Note that to show that rf is a module homomorphism we need the fact that R is commutative. (We leave it to the reader to show how $M^* = \text{Hom}_R(M, R)$ can be made into a *right* R-module when M is a left module and R is *not* commutative.) Then $R^* = \text{Hom}_R(R, R) = \text{End}_R(R) \simeq R$, and $(M \oplus N)^* \simeq M^* \oplus N^*$, so $(R^n)^* \simeq (R^*)^n \simeq R^n$, and it follows that if M is finitely generated and projective, so is M^*.

Proposition 18. Let M be a finitely generated projective R-module, and let S be a multiplicative set in R. Then $S^{-1}(M^*) \simeq (S^{-1}M)^*$.

Proof. We must show $S^{-1}\mathrm{Hom}_R(M, R) \simeq \mathrm{Hom}_{S^{-1}R}(S^{-1}M, S^{-1}R)$. Given $\alpha \in \mathrm{Hom}_R(M, R)$ and $s \in S$, define

$$[s, \alpha] \in \mathrm{Hom}_{S^{-1}R}(S^{-1}M, S^{-1}R)$$

by

$$[s, \alpha](\tfrac{m}{t}) = \frac{\alpha(m)}{st} \quad (m \in M,\ t \in S).$$

Note that $\alpha(m)/st \in S^{-1}R$, and $[s, \alpha]$ is well defined, since if $m/t = m'/t'$ then $s't'm = s'tm'$, some $s' \in S$, whence $\alpha(s't'm) = \alpha(s'tm')$, or $s't'\alpha(m) = s't\alpha(m')$, and so $\alpha(m)/st = \alpha(m')/st'$. So $[s, \alpha]$ is a map from $S^{-1}M$ to $S^{-1}R$, and it is an $S^{-1}R$-module homomorphism, since for $m, m' \in M$, $r \in R$, and $t, t' \in S$, we have

$$[s, \alpha](\tfrac{m}{t} + \tfrac{m'}{t'}) = [s, \alpha](\frac{t'm + tm'}{tt'}) = \frac{\alpha(t'm + tm')}{stt'}$$

$$= \frac{t'\alpha(m) + t\alpha(m')}{stt'} = \frac{\alpha(m)}{st} + \frac{\alpha(m')}{st'}$$

$$= [s, \alpha](\tfrac{m}{t}) + [s, \alpha](\tfrac{m'}{t'})$$

and

$$[s, \alpha](\tfrac{r}{t}\,\tfrac{m}{t'}) = [s, \alpha](\tfrac{rm}{tt'}) = \frac{\alpha(rm)}{stt'} = \frac{r\alpha(m)}{tst'} = \frac{r}{t}\,\frac{\alpha(m)}{st'}$$

$$= \tfrac{r}{t}[s, \alpha](\tfrac{m}{t'}).$$

Now if, in $S^{-1}\mathrm{Hom}_R(M, R)$, we have $\alpha/s = \alpha'/s'$ ($\alpha, \alpha' \in \mathrm{Hom}_R(M, R)$, $s, s' \in S$), then $t's'\alpha = t's\alpha'$, some $t' \in S$, so $t's'\alpha(m) = t's\alpha'(m)$, all $m \in M$, and therefore $\alpha(m)/st = \alpha'(m)/s't$, all $m \in M$, $t \in S$: that is, $[s, \alpha] = [s', \alpha']$. So we have a map $S^{-1}(M^*) \to (S^{-1}M)^*$ given by $\alpha/s \mapsto [s, \alpha]$. This is an $S^{-1}R$-module homomorphism, for

$$\frac{\alpha}{s} + \frac{\alpha'}{s'} = \frac{s'\alpha + s\alpha'}{ss'}$$

and then

$$[ss', s'\alpha + s\alpha'](\tfrac{m}{t}) = \frac{(s'\alpha + s\alpha')(m)}{ss't} = \frac{s'\alpha(m) + s\alpha'(m)}{ss't}$$

$$= \frac{\alpha(m)}{st} + \frac{\alpha'(m)}{s't}$$

$$= [s, \alpha](\tfrac{m}{t}) + [s', \alpha'](\tfrac{m}{t})$$

$$= ([s, \alpha] + [s', \alpha'])(\tfrac{m}{t})$$

also $(r/s)(\alpha/s') = (r\alpha)/(ss')$, and then

$$[ss', r\alpha](\tfrac{m}{t}) = \frac{(r\alpha)(m)}{ss't} = \frac{r\alpha(m)}{ss't} = \frac{r}{s}\,\frac{\alpha(m)}{s't}$$

$$= \frac{r}{s}([s', \alpha](\tfrac{m}{t})) = (\tfrac{r}{s}[s', \alpha])(\tfrac{m}{t}).$$

Next we show that $\alpha/s \mapsto [s, \alpha]$ is injective. Let $[s, \alpha] = 0$, and let $m_1, m_2, \ldots, m_n$ be a set of generators for M. Then

$$[s, \alpha](m_i/1) = 0,\ 1 \le i \le n, \text{ or } \alpha(m_i)/s = 0,\ 1 \le i \le n.$$

So we can find $s_i \in S$ with $s_i\alpha(m_i) = 0$, $1 \le i \le n$. Put $s_0 = s_1 s_2 \ldots s_n \in S$. Then $s_0\alpha(m_i) = 0$, $1 \le i \le n$, and therefore $s_0\alpha(m) = 0$, all $m \in M$, since $m_1, m_2, \ldots, m_n$ generate M. So

$$s_0\alpha = 0, \text{ and } \alpha/s = s_0\alpha/s_0 s = 0.$$

Thus $\alpha/s \mapsto [s, \alpha]$ is injective; finally, we show that it is surjective. Let $\beta \in (S^{-1}M)^*$: we must find s, α with $\beta = [s, \alpha]$. Let

$$\beta(m_i/1) = r_i/t_i,\ 1 \le i \le n\ (r_i \in R,\ t_i \in S,\ 1 \le i \le n)$$

and put $s = t_1 t_2 \ldots t_n \in S$. So for all $m \in M$, $\beta(sm/1) = r/1$ for some $r \in R$. If $\bar{R}$ is the image of R in $S^{-1}R$, we can define $\bar{\beta} : M \to \bar{R}$ by $\bar{\beta}(m) = \beta(sm/1)$, all $m \in M$. Since M is projective, we can find $\alpha \in M^*$ with $\alpha(m)/1 = \bar{\beta}(m)$, all $m \in M$:

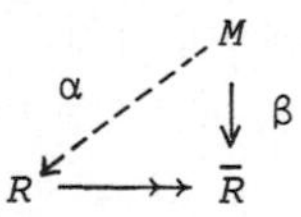

that is, $\alpha(m)/1 = \beta(sm/1)$, or $\alpha(m)/st = \beta(m/t)$, all $m \in M$, $t \in S$. So $\beta = [s, \alpha]$, and the proof is complete.

Corollary. Let R be a ring, P a prime ideal of R, and let M be a finitely generated projective R-module. Then $(M_P)^* \simeq (M^*)_P$, and $\text{rank}_P(M) = \text{rank}_P(M^*)$.

Proof. The isomorphism is immediate, and then if $\text{rank}_P(M) = n$, $M_P \simeq R_P{}^n$, so $M_P{}^* \simeq R_P{}^n$ also, whence the result.

Let M be an R-module, and let N be a subset of M. The *annihilator* of N is $\text{ann}(N) = \{r \in R : rn = 0, \text{ all } n \in N\}$. Clearly $\text{ann}(N) \lhd R$. We write $\text{ann}(m)$ for $\text{ann}(\{m\})$, $m \in M$. Since $1m = m$, all $m \in M$, $\text{ann}(m) = R$ if and only if $m = 0$.

Proposition 19. Let M be an R-module such that $M_P = 0$ for every maximal ideal P of R. Then $M = 0$.

Proof. If $M \neq 0$, choose $m \in M$, $m \neq 0$; so $\text{ann}(m) \neq R$. Thus there is a maximal ideal P with $\text{ann}(m) \subset P$. By hypothesis, $M_P = 0$, so $m/1 = 0$ in M_P, that is $sm = 0$ for some $s \in R - P$. But $\text{ann}(m) \subset P$, so $s \notin \text{ann}(m)$, which is absurd.

Corollary. Let $f : M \to N$ be a homomorphism of R-modules. Then:

(i) f is a monomorphism if and only if f_P is a monomorphism for every prime ideal P of R

(ii) f is an epimorphism if and only if f_P is an epimorphism for every prime ideal P of R

(iii) f is an isomorphism if and only if f_P is an isomorphism for every prime ideal P of R.

Proof. Immediate, since $\ker(f_P) = (\ker(f))_P$ and $\text{coker}(f_P) = (\text{coker}(f))_P$.

Corollary. (i) If M is a finitely generated projective R-module of constant rank 0, then $M = 0$.

(ii) If M, N are finitely generated projective R-modules, both of constant rank n, and if $f : M \twoheadrightarrow N$ is an epimorphism, then f is an isomorphism.

Proof. (i) Immediate.

(ii) Since N is projective, $M \simeq N \oplus \ker(f)$, and since M is finitely generated and projective, so is $\ker(f)$. Localizing,

$\text{rank}_P(M) = \text{rank}_P(N) + \text{rank}_P(\ker(f))$, all P, whence $\ker(f)$ is of constant rank 0, so $\ker(f) = 0$ and the result follows. (Note that we do not actually need M, N to be of constant rank; $\text{rank}_P(M) = \text{rank}_P(N)$, all P, is enough.)

Now let M, N be finitely generated projective R-modules, and let P be a prime ideal of R. Then $M \otimes_R N$ is a finitely generated projective R-module, and $(M \otimes_R N)_P \simeq M_P \otimes_{R_P} N_P$. For

$$(M \otimes_R N)_P \simeq R_P \otimes_R M \otimes_R N \simeq R_P \otimes_{R_P} R_P \otimes_R M \otimes_R N$$
$$\simeq R_P \otimes_{R_P} M_P \otimes_R N \simeq M_P \otimes_{R_P} R_P \otimes_R N \simeq M_P \otimes_{R_P} N_P$$

(using Proposition 16 and the fact that the tensor product is associative and commutative, over a commutative ring, up to isomorphism). Let $\text{rank}_P(M) = m$ and $\text{rank}_P(N) = n$: so $M_P \simeq R_P{}^m$ and $N_P \simeq R_P{}^n$. By Corollary 4, $M_P \otimes_{R_P} N_P \simeq R_P{}^{mn}$, and putting all this together, we have $\text{rank}_P(M \otimes_R N) = \text{rank}_P(M)\text{rank}_P(N)$. It is also true that $\text{rank}_P(M \oplus N) = \text{rank}_P(M) + \text{rank}_P(N)$, since clearly $(M \oplus N)_P \simeq M_P \oplus N_P$.

Let M be an R-module, and consider the *evaluation map* $M^* \times M \to R$, $(\alpha, m) \mapsto \alpha(m)$. This is clearly bilinear and balanced, and so it induces a natural homomorphism $M^* \otimes_R M \to R$. The image of this homomorphism is an ideal of R, and is called the *trace ideal* of M, $\tau(M)$. Note that, if M is free with basis $\{m_\lambda\}_{\lambda\in\Lambda}$, we can define $\alpha \in M^*$ by fixing $\lambda \in \Lambda$ and defining $\alpha(m)$ to be the coefficient of m_λ in the expression for m in terms of the given basis: then $\alpha(m_\lambda) = 1$ so $1 \in \tau(M)$ and it follows that in this case $\tau(M) = R$.

Proposition 20. Let M be a finitely generated projective R-module such that for every prime ideal P of R, $\text{rank}_P(M) > 0$. Then $\tau(M) = R$.

Proof. Let $f : M^* \otimes_R M \to R$ be the natural map: so $f(\alpha \otimes m) = \alpha(m)$ $(m \in M, \alpha \in M^*)$. Then $f_P : (M^* \otimes_R M)_P \to R_P$ is given by

$$f_P\left(\frac{\alpha \otimes m}{s}\right) = \frac{\alpha(m)}{s}$$

where P is a prime ideal of R, $m \in M$, $\alpha \in M^*$, and $s \in S = R - P$. But under the natural isomorphism $(M^* \otimes_R M)_P \simeq M_P^* \otimes_{R_P} M_P$, the element $(\alpha \otimes m)/s$ corresponds to $(\alpha/1) \otimes (m/s) = (\alpha/s) \otimes (m/1)$, and further, $(\alpha/1)(m/s) = \alpha(m)/s$. So if we identify $(M^* \otimes_R M)_P$ with $M_P^* \otimes_{R_P} M_P$, then $f_P : M_P^* \otimes_{R_P} M_P \to R_P$ is precisely the natural map induced by the evaluation map $M_P^* \times M_P \to R_P$. But R_P is local, so M_P is free, and by the remarks preceding this proposition it follows that f_P is an epimorphism. Since this is true for every prime ideal P of R, f itself is an epimorphism, by the first corollary to Proposition 19, and the proof is complete.

Exercise: Give an alternative proof of Proposition 20, as follows. If $\tau(M) \neq R$, there is a maximal ideal P of R with $\tau(M) \subset P$. Then $M_P \simeq R_P{}^n$, where $n > 0$, and we can find a free basis $m_1/1, \ldots, m_n/1$ of M_P. Show that, if $\alpha : M \to M/PM$ is the natural map and $\beta : R^n \to M/PM$ is given by

$$(r_1, \ldots, r_n) \mapsto (r_1m_1 + \ldots + r_nm_n) + PM$$

then $\alpha = \beta\gamma$ for some $\gamma : M \to R^n$. Deduce that $\alpha = 0$ and obtain a contradiction.

Proposition 21. Let M be an R-module. The following conditions are equivalent:

(i) M is finitely generated projective of constant rank 1

(ii) $M^* \otimes_R M \simeq R$

(iii) There exists an R-module N with $N \otimes_R M \simeq R$.

Further, in (iii) it follows that $N \simeq M^*$.

Definition. If M satisfies the above conditions, it is called an *invertible* module.

Proof of Proposition 21. (i) => (ii): M is finitely generated projective of constant rank 1, hence so are M^* and $M^* \otimes_R M$. (Recall $\mathrm{rank}(M^*) = \mathrm{rank}(M)$, so $\mathrm{rank}(M^* \otimes_R M) = (\mathrm{rank}(M))^2 = 1$.) By Proposition 20, the natural map $M^* \otimes_R M \to R$ is an epimorphism, and since R itself is finitely generated projective of constant rank 1, we have an isomorphism, by the second corollary to Proposition 19.

(ii) => (iii): just put $N = M^*$.

(iii) => (i): we have $N \otimes_R M \simeq R \simeq M \otimes_R N$. So let $f : R \to M \otimes_R N$ and $g : N \otimes_R M \to R$ be isomorphisms, and let $f(1) = \Sigma_i m_i \otimes n_i$ ($m_i \in M$, $n_i \in N$, $1 \le i \le n$). There is an isomorphism $\alpha : M \to M$ given by the composite

$$M \to R \otimes_R M \xrightarrow{f\otimes 1} M \otimes_R N \otimes_R M \xrightarrow{1\otimes g} M \otimes_R R \longrightarrow M$$

$$m \mapsto 1 \otimes m \mapsto \Sigma_i m_i \otimes n_i \otimes m \mapsto \Sigma_i m_i \otimes g(n_i \otimes m)$$

$$\mapsto \Sigma_i g(n_i \otimes m)m_i.$$

Define $\beta_i : M \to R$ by $\beta_i(m) = g(n_i \otimes m)$, all $m \in M$, and we have $\alpha(m) = \Sigma_i \beta_i(m)m_i$, all $m \in M$. But α is an isomorphism, so if $m_i' = \alpha^{-1}(m_i)$, all i, we have $m = \Sigma_i \beta_i(m)m_i'$, all $m \in M$. If we define maps

$$M \to R^n \text{ by } m \mapsto (\beta_1(m), \beta_2(m), \ldots, \beta_n(m))$$

and

$$R^n \to M \text{ by } (r_1, r_2, \ldots, r_n) \mapsto r_1m_1' + r_2m_2' + \ldots + r_nm_n'$$

the composite is $1 : M \to M$, so M is isomorphic to a direct summand of R^n, and so is finitely generated and projective; by a similar argument, so is N. If P is a prime ideal of R, $\text{rank}_P(N \otimes_R M) = \text{rank}_P(R) = 1$, so $\text{rank}_P(N)\text{rank}_P(M) = 1$, and therefore $\text{rank}_P(M) = 1$.

Finally, given (iii), and hence (i) and (ii), we have

$$N \simeq R \otimes_R N \simeq M^* \otimes_R M \otimes_R N \simeq M^* \otimes_R R \simeq M^*.$$

If we write $\langle M \rangle$ for the isomorphism class of the invertible module M, it is clear by Proposition 21 that the set of all such $\langle M \rangle$ forms an abelian group under the operation

$$\langle M \rangle \langle N \rangle = \langle M \otimes_R N \rangle$$

with

$$\langle R \rangle = 1 \text{ and } \langle M \rangle^{-1} = \langle M^* \rangle.$$

This group is the *Picard group* of R, Pic R. There is a natural

homomorphism of multiplicative groups, Pic $R \to (K_0R)^{\bullet}$, given by $\langle M \rangle \mapsto [M]$. We shall show in the next section that this is injective; the proof uses exterior powers, which we now introduce.

3.5 *Exterior powers*

Let M be an R-module. Write $\times^n M$ for $M \times M \times \ldots \times M$ (n terms) (we write this rather than the more usual M^n, since M^n means the module $M \oplus M \oplus \ldots \oplus M$), and write $\otimes_R^n M$ for $M \otimes_R M \otimes_R \ldots \otimes_R M$ (n terms), the nth *tensor power* of M. Note that $\otimes_R^1 M = M$; conventionally we write $\otimes_R^0 M = R$, any M. So for all n, $m \geq 0$,

$$(\otimes_R^n M) \otimes_R (\otimes_R^m M) = \otimes_R^{n+m} M.$$

(We are identifying all possible ways of bracketing the tensor powers.) The nth *exterior power* of M, $\wedge_R^n M$, is the quotient of $\otimes_R^n M$ by the submodule generated by all $m_1 \otimes m_2 \otimes \ldots \otimes m_n$ with $m_i \in M$, all i, and $m_i = m_j$ for some $i \neq j$. The image of $m_1 \otimes m_2 \otimes \ldots \otimes m_n$ in $\wedge_R^n M$ is written $m_1 \wedge m_2 \wedge \ldots \wedge m_n$. Note that $\wedge_R^1 M = M$ and $\wedge_R^0 M = R$, all M.

Exercise: Show that the above submodule of $\otimes_R^n M$ is generated by all $m_1 \otimes m_2 \otimes \ldots \otimes m_n$ with $m_i = m_{i+1}$ for some $i < n$.

We have $m_1 \wedge m_2 \wedge \ldots \wedge m_n = 0$ if $m_i = m_j$, some $i \neq j$. In $\wedge_R^2 M$, $0 = (m + n) \wedge (m + n) = m \wedge m + m \wedge n + n \wedge m + n \wedge n = m \wedge n + n \wedge m$, all $m, n \in M$, so $m \wedge n = -(n \wedge m)$, all $m, n \in M$. More generally, let n be a positive integer, and let S_n be the nth symmetric group, the group of all permutations of the symbols $1,2,\ldots,n$. Then in $\wedge_R^n M$,

$$m_{\sigma_1} \wedge m_{\sigma_2} \wedge \ldots \wedge m_{\sigma_n} = \varepsilon(\sigma)(m_1 \wedge m_2 \wedge \ldots \wedge m_n)$$

for any $\sigma \in S_n$ (writing $\sigma : i \mapsto \sigma_i$, $i = 1,2,\ldots,n$), where $\varepsilon(\sigma)$ is the signature of σ, that is, $\varepsilon(\sigma) = 1$ if σ is an even permutation and $\varepsilon(\sigma) = -1$ if σ is an odd permutation.

Let M, N be R-modules. A map $f : \times^n M \to N$ is called n-*linear*, or *multilinear*, if it is linear in each entry, that is,

if

$$f(m_1, \ldots, r_i m_i + r_i' m_i', \ldots, m_n)$$
$$= r_i f(m_1, \ldots, m_i, \ldots, m_n) + r_i' f(m_1, \ldots, m_i', \ldots, m_n)$$

each i. Note that it is a consequence that f is balanced, in the obvious generalized sense, that is,

$$f(rm_1, m_2, \ldots, m_n) = f(m_1, rm_2, \ldots, m_n)$$
$$= \ldots = f(m_1, m_2, \ldots, rm_n).$$

The map f is *alternating* if $f(m_1, m_2, \ldots, m_n) = 0$ whenever $m_i = m_j$ for some $i \neq j$. It is easy to see that $\wedge_R^n M$, together with the natural map $\times^n M \to \wedge_R^n M$, which is clearly multilinear and alternating, is universal for multilinear alternating maps on $\times^n M$; that is, given an R-module N and a multilinear alternating map $f : \times^n M \to N$, there is a unique R-module homomorphism $g : \wedge_R^n M \to N$ such that $g(m_1 \wedge m_2 \wedge \ldots \wedge m_n) = f(m_1, m_2, \ldots, m_n)$, all $m_i \in M$. Further, this property characterizes the exterior power, up to isomorphism.

We have $\wedge_R^0 R = \wedge_R^1 R = R$; in $\wedge_R^2 R$, $r \wedge s = rs(1 \wedge 1) = 0$, all $r, s \in R$, so $\wedge_R^2 R = 0$, and similarly $\wedge_R^n R = 0$ for $n > 2$.

Proposition 22. Let R be a ring, S a multiplicative set in R, and M an R-module. Then $S^{-1}(\wedge_R^n M) \cong \wedge_{S^{-1}R}^n (S^{-1}M)$.

Proof. We have an isomorphism $\alpha : S^{-1}(\otimes_R^n M) \to \otimes_{S^{-1}R}^n (S^{-1}M)$ given by

$$\alpha\left(\frac{m_1 \otimes m_2 \otimes \ldots \otimes m_n}{s}\right) = \frac{1}{s}\left(\frac{m_1}{1} \otimes \frac{m_2}{1} \otimes \ldots \otimes \frac{m_n}{1}\right)$$

(Use Proposition 16, and the associativity and commutativity up to isomorphism of the tensor product operation.) It is now easy to see that α maps the kernel of the natural map $S^{-1}(\otimes_R^n M) \twoheadrightarrow S^{-1}(\wedge_R^n M)$ onto the kernel of the natural map $\otimes_{S^{-1}R}^n (S^{-1}M) \twoheadrightarrow \wedge_{S^{-1}R}^n (S^{-1}M)$, and the result follows.

Corollary. If P is a prime ideal of R and M is an R-module, then $(\wedge_R^n M)_P \simeq \wedge_{R_P}^n M_P$.

Recall that, from Proposition 3, if M, N are R-modules, then

$$\otimes_R^2 (M \oplus N) \simeq (\otimes_R^2 M) \oplus (M \otimes_R N) \oplus (N \otimes_R M) \oplus (\otimes_R^2 N)$$

and more generally $\otimes_R^n (M \oplus N)$ can be written as a direct sum of 2^n modules, each an n-fold tensor product of M's and N's in some order.

We wish to prove a corresponding result for exterior powers. We cannot expect merely to replace $\otimes$ by $\wedge$ in the above, as $M \wedge_R N$ is undefined. In fact we shall obtain

$$\wedge_R^2 (M \oplus N) \simeq (\wedge_R^2 M) \oplus (M \otimes_R N) \oplus (\wedge_R^2 N)$$

the only surprise, perhaps, being that the term $N \otimes_R M$ is missing.

First, some more definitions. A ring A (not necessarily commutative) is an *R-algebra* if R is a subring of the centre of A. Then A is a *graded* R-algebra if, as an additive group, $A = A_0 \oplus A_1 \oplus A_2 \oplus \ldots$, with $A_0 = R$ and $A_n A_m \subset A_{n+m}$, all n, m. Note that this makes A_n into an R-module, all n. If $x \in A_n$, we say x is of *degree n*. A homomorphism $f : A \to B$ of graded R-algebras is both a ring homomorphism and an R-module homomorphism, with $f(A_n) \subset B_n$, all n.

For example, if M is an R-module, we define the *tensor algebra* $T = T(M)$ by putting $T_n = \otimes_R^n M$; here the multiplication is just $\otimes$. More explicitly, for each n, m there is a bilinear map $T_n \times T_m \to T_{n+m}$ induced by

$$(x_1 \otimes \ldots \otimes x_n, y_1 \otimes \ldots \otimes y_m)$$
$$\mapsto x_1 \otimes \ldots \otimes x_n \otimes y_1 \otimes \ldots \otimes y_m$$

and these maps induce a bilinear map $T \times T \to T$, making T into a graded R-algebra. Similarly, we define the *exterior algebra* $E = E(M)$ by putting $E_n = \wedge_R^n M$; here the multiplication is given by $\wedge$.

Let A, B be graded R-algebras. We can make the module $C =$

$A \otimes_R B$ into a graded R-algebra (the *anticommutative tensor product* of A and B) as follows. We put

$$C_n = \bigoplus_{r=0}^{n} (A_r \otimes_R B_{n-r})$$

so that $C = C_0 \oplus C_1 \oplus C_2 \oplus \ldots$. For each p, q, r, s, there is a bilinear map

$$(A_p \otimes_R B_q) \times (A_r \otimes_R B_s) \to A_{p+r} \otimes_R B_{q+s}$$

induced by

$$(x \otimes y,\ x' \otimes y') \mapsto (-1)^{qr}(xx' \otimes yy').$$

These maps induce bilinear maps $C_n \times C_m \to C_{n+m}$, which in turn induce the multiplication $C \times C \to C$.

Proposition 23. Let M, N be R-modules. Then

$$E(M \oplus N) \simeq E(M) \otimes_R E(N)$$

and comparing parts of degree n,

$$\Lambda_R^n (M \oplus N) \simeq \bigoplus_{r=0}^{n} \{(\Lambda_R^r M) \otimes_R (\Lambda_R^{n-r} N)\}.$$

Proof. There is a homomorphism $T(M) \twoheadrightarrow E(M)$ whose kernel is the ideal generated by all $x \otimes x$ ($x \in M$); similarly there is a homomorphism $T(N) \twoheadrightarrow E(N)$, and these induce a homomorphism $T(M) \otimes_R T(N) \twoheadrightarrow E(M) \otimes_R E(N)$.

Next, there is a homomorphism $T(M \oplus N) \twoheadrightarrow E(M \oplus N)$ whose kernel is the ideal generated by all $(x + y) \otimes (x + y)$ ($x \in M$, $y \in N$); but

$$(x + y) \otimes (x + y) = (x \otimes x) + (y \otimes y) + (x \otimes y + y \otimes x)$$

and $x \otimes x$, $y \otimes y$ also lie in the kernel. Further, there is a homomorphism $T(M \oplus N) \twoheadrightarrow T(M) \otimes_R T(N)$ whose kernel is the ideal generated by all $(x \otimes y + y \otimes x)$ ($x \in M$, $y \in Y$), and this therefore induces an isomorphism $E(M \oplus N) \to E(M) \otimes_R E(N)$. (For a more detailed proof see N. Bourbaki, *Elements of Mathematics, Algebra I*, Chapter 3, section 7.7, Hermann/Addison-Wesley, 1974.)

Exercise: Obtain the direct sum decomposition of $\Lambda_R^n (M \oplus N)$ directly, by constructing suitable maps

$$(\Lambda_R^r M) \otimes_R (\Lambda_R^{n-r} N) \overset{\hookrightarrow}{\twoheadleftarrow} \Lambda_R^n (M \oplus N).$$

Let $\binom{m}{n}$ denote the coefficient of t^n in the expansion of $(1 + t)^m$.

Corollary (i). $\Lambda_R^n R^m \simeq R^{\binom{m}{n}}$. (Conventionally, $R^0 = 0$.)
Proof. The result is true for $m = 1$, since $\Lambda_R^0 R \simeq R$, $\Lambda_R^1 R \simeq R$, and $\Lambda_R^n R = 0$ for $n \geq 2$. We use induction on m. We have

$$\Lambda_R^n R^m = \Lambda_R^n (R \oplus R^{m-1})$$

$$\simeq (\Lambda_R^0 R) \otimes_R (\Lambda_R^n R^{m-1}) \oplus (\Lambda_R^1 R) \otimes_R (\Lambda_R^{n-1} R^{m-1})$$

(by the proposition, other terms vanishing)

$$\simeq R \otimes_R (\Lambda_R^n R^{m-1}) \oplus R \otimes_R (\Lambda_R^{n-1} R^{m-1})$$

$$\simeq R^{\binom{m-1}{n}+\binom{m-1}{n-1}} \quad \text{(by the inductive hypothesis)}$$

$$\simeq R^{\binom{m}{n}}, \text{ as required.}$$

Corollary (ii). If M is a finitely generated projective R-module, so is $\Lambda_R^n M$. Further, if P is a prime ideal of R, then $\operatorname{rank}_P(\Lambda_R^n M) = \binom{m}{n}$, where $m = \operatorname{rank}_P(M)$.
Proof. If $M \oplus N$ is free of finite rank, then the proposition says $\Lambda_R^n M$ is a direct summand of $\Lambda_R^n (M \oplus N)$, which is free of finite rank by Corollary (i). So $\Lambda_R^n M$ is finitely generated and projective; the expression for its rank at P is immediate from Proposition 22, together with Corollary (i), using the fact that $M_P \simeq R_P^{\ m}$.

Corollary (iii). If M is a finitely generated projective R-module of constant rank m, then $\Lambda_R^n M$ is a finitely generated projective R-module of constant rank $\binom{m}{n}$.

Proof. Immediate.

Corollary (iv). If M is an invertible R-module, then $\wedge_R^n M = 0$ for $n \geq 2$.

Proof. We have $\operatorname{rank}(M) = 1$, so $\operatorname{rank}(\wedge_R^n M) = 0$ for $n \geq 2$, by Corollary (iii). The result follows by Proposition 19.

Corollary (v). If $M_1, M_2, \ldots, M_m, N_1, N_2, \ldots, N_n$ are invertible R-modules such that

$$M_1 \oplus M_2 \oplus \ldots \oplus M_m \simeq N_1 \oplus N_2 \oplus \ldots \oplus N_n$$

then $m = n$, and

$$M_1 \otimes_R M_2 \otimes_R \ldots \otimes_R M_m \simeq N_1 \otimes_R N_2 \otimes_R \ldots \otimes_R N_n.$$

Proof. Since each M_i, N_j has rank 1, $\operatorname{rank}(M_1 \oplus \ldots \oplus M_m) = m$ and $\operatorname{rank}(N_1 \oplus \ldots \oplus N_n) = n$, so $m = n$. The result follows by applying $\wedge_R^m$ to each side of the given isomorphism. For $\wedge_R^k M_1 = 0$ for $k \geq 2$, by Corollary (iv), and by a similar argument $\wedge_R^m (M_2 \oplus \ldots \oplus M_m) = 0$, since $\operatorname{rank}(M_2 \oplus \ldots \oplus M_m) = m - 1$. By the proposition,

$$\wedge_R^m (M_1 \oplus (M_2 \oplus \ldots \oplus M_m)) \simeq M_1 \otimes_R (\wedge_R^{m-1} (M_2 \oplus \ldots \oplus M_m))$$

the other terms vanishing, and repeating the argument we obtain

$$\wedge_R^m (M_1 \oplus M_2 \oplus \ldots \oplus M_m) \simeq M_1 \otimes_R M_2 \otimes_R \ldots \otimes_R M_m$$

and similarly

$$\wedge_R^m (N_1 \oplus N_2 \oplus \ldots \oplus N_m) \simeq N_1 \otimes_R N_2 \otimes_R \ldots \otimes_R N_m.$$

Corollary (vi). The natural homomorphism $\operatorname{Pic} R \to (K_0R)^{\bullet}$ is injective.

Proof. Let $\langle M\rangle, \langle N\rangle \in \operatorname{Pic} R$, and suppose $[M] = [N]$ in K_0R. Then $M \oplus R^n \simeq N \oplus R^n$, for some n, by Proposition 2, and applying Corollary (v) gives

$$M \otimes_R R \otimes_R \ldots \otimes_R R \simeq N \otimes_R R \otimes_R \ldots \otimes_R R$$

or $M \simeq N$ and $\langle M\rangle = \langle N\rangle$.

3.6 *The prime spectrum and Zariski topology*

Let R be a (commutative) ring, and write Spec R for the set of all prime ideals of R, the *prime spectrum* of R. We have already mentioned the geometrical significance of this set. When R is the coordinate ring of a variety V, then there is a natural map $V \to \text{Spec } R$; since V is a geometrical object, we expect it to carry a topology, and by means of the above map this topology can be transferred to Spec R. For our purposes, we introduce the topology directly in Spec R; the reader who wishes to see the geometrical constructions and motivation is referred to [8] in the Bibliography.

If $I \lhd R$ (I not necessarily prime) put

$$V(I) = \{P \in \text{Spec } R : I \subset P\}.$$

Clearly $V(0) = \text{Spec } R$, and $V(R) = \emptyset$ (by definition prime ideals are proper ideals, so $R \notin \text{Spec } R$). If $P \in \text{Spec } R$ and $I, J \lhd R$ with $IJ \subset P$, then $I \subset P$ or $J \subset P$. So $V(I) \cup V(J) = V(I \cap J) = V(IJ)$. Also, if $I_\lambda \lhd R$, $\lambda \in \Lambda$, then $\Sigma_{\lambda\in\Lambda} I_\lambda \lhd R$ and $\cap_{\lambda\in\Lambda} V(I_\lambda) = V(\Sigma_{\lambda\in\Lambda} I_\lambda)$. So the sets $V(I)$, all $I \lhd R$, form the closed sets for a topology on Spec R, the *Zariski topology*. The open sets are

$$D(I) = \text{Spec } R - V(I) = \{P \in \text{Spec } R : I \not\subset P\}.$$

Note that $D(R) = \text{Spec } R$, $D(0) = \emptyset$, $D(I) \cap D(J) = D(I \cap J) = D(IJ)$, and $\cup_{\lambda\in\Lambda} D(I_\lambda) = D(\Sigma_{\lambda\in\Lambda} I_\lambda)$. Also, if $D(I) = \text{Spec } R$, then $I = R$, for otherwise $I \subset P$ for some maximal ideal P, and then $P \in \text{Spec } R = D(I)$, so $I \not\subset P$, which is absurd.

The space Spec R is always compact. For suppose we have an open cover, $\text{Spec } R = \cup_{\lambda\in\Lambda} D(I_\lambda) = D(\Sigma_{\lambda\in\Lambda} I_\lambda)$. So $\Sigma_{\lambda\in\Lambda} I_\lambda = R$ and we can find $\lambda_1, \lambda_2, \ldots, \lambda_n \in \Lambda$, and $r_i \in I_{\lambda_i}$, each i, such that $1 = r_1 + r_2 + \ldots + r_n$. But now $\Sigma_{i=1}^n I_{\lambda_i} = R$, so $\text{Spec } R = \cup_{i=1}^n D(I_{\lambda_i})$, and we have a finite sub-cover. Note that in the definition of compactness some authors require the space to be Hausdorff, which fails here. (They would call this space quasi-compact.)

Let $f : R \to S$ be a homomorphism of rings. If $P \in \text{Spec } S$, it is clear that $f^{-1}(P) \in \text{Spec } R$. So we can define a map $f_* :$ $\text{Spec } S \to \text{Spec } R$ (note the direction) by $f_*(P) = f^{-1}(P)$. If $I \lhd R$, then $f_*^{-1}(V(I)) = V(f(I)S)$ (note that $f(I)$ need not be an ideal of S) and so f_* is continuous. It is clear that, if $g :$ $S \to T$ is another ring homomorphism, $(gf)_* = f_*g_*$; also if $1 :$ $R \to R$ is the identity homomorphism, then $1_* : \text{Spec } R \to \text{Spec } R$ is the identity map.

We write H_0R for the set of all continuous maps $\text{Spec } R \to \mathbb{Z}$, where $\mathbb{Z}$ is given the discrete topology. H_0R is a ring under pointwise addition and multiplication; that is, for $f, g \in H_0R$, we define $f + g$, $fg \in H_0R$ by $(f + g)(P) = f(P) + g(P)$, $(fg)(P) = f(P)g(P)$, all $P \in \text{Spec } R$. If $f : R \to S$ is a homomorphism of rings, we define a homomorphism of rings $H_0f : H_0R \to H_0S$ by $(H_0f)(g) = gf_*$, all $g \in H_0R$:

$$\begin{array}{ccc} \text{Spec } R & \xleftarrow{f_*} & \text{Spec } S \\ & g \searrow \quad \swarrow (H_0f)(g) & \\ & \mathbb{Z} & \end{array}$$

Clearly, if $f : R \to S$ and $g : S \to T$ are homomorphisms of rings, then $H_0(gf) = (H_0g)(H_0f)$; also $H_01 = 1$, since $1_* = 1$.

Proposition 24. Let M be a finitely generated R-module. Then

$$\{P \in \text{Spec } R : \text{rank}_P(M) = 0\} = D(\text{ann}(M)).$$

Proof. Let $m_1, m_2, \ldots, m_n$ generate M. If $P \in \text{Spec } R$ and $M_P =$ 0, then $m_i/1 = 0$ in M_P, so $s_im_i = 0$ for some $s_i \in R - P$, each i. Put $s = s_1s_2\ldots s_n$; so $sm_i = 0$, all i, and thus $s \in \text{ann}(M)$. But $s \in R - P$, so $\text{ann}(M) \not\subset P$, or $P \in D(\text{ann}(M))$. Conversely, if $P \in D(\text{ann}(M))$, then $\text{ann}(M) \not\subset P$, so we can choose $s \in \text{ann}(M) - P$. Let $m \in M$ and $t \in R - P$; in M_P, $m/t = sm/st = 0/st = 0$, so $M_P = 0$.

The above result enables us to show that, for a fixed finitely generated projective R-module M, the rank function is continuous

on Spec R, as follows.

Corollary. Let M be a finitely generated projective R-module. Define f_M : Spec $R \to \mathbb{Z}$ by $f_M(P) = \text{rank}_P(M)$. Then $f_M \in H_0R$.

Proof. Since $\mathbb{Z}$ is discrete, we must show that $f_M^{-1}(m)$ is an open set in Spec R, for each $m \in \mathbb{Z}$.

Let $M \oplus N \cong R^n$. For each $P \in$ Spec R, $f_M(P) + f_N(P) = n$. Then $f_M^{-1}(m) = \emptyset$ if $m < 0$ or $m > n$, and this is open, So suppose $0 \le m \le n$. We have

$$f_M^{-1}(m)$$

$$= \{P : \text{rank}_P(M) \le m\} \cap \{P : \text{rank}_P(M) \ge m\}$$

$$= \{P : \text{rank}_P(M) \le m\} \cap \{P : \text{rank}_P(N) \le n - m\}$$

$$= \{P : \text{rank}_P(\wedge_R^{m+1} M) = 0\} \cap \{P : \text{rank}_P(\wedge_R^{n-m+1} N) = 0\}$$

(by Corollary (ii) to Proposition 23)

$$= D(\text{ann}(\wedge_R^{m+1} M)) \cap D(\text{ann}(\wedge_R^{n-m+1} N)$$

by the proposition, and this is an open set.

Corollary. There is a natural ring homomorphism $K_0R \xrightarrow{\text{rank}} H_0R$ given by $[M] \mapsto f_M$.

Proof. If $[M] = [N]$, then M, N are stably isomorphic and so $f_M = f_N$. So our map is well defined on the generators of K_0R. We leave it to the reader to show that it extends to a ring homomorphism.

The kernel of the above homomorphism $K_0R \to H_0R$ is denoted rk_0R; it is the ideal of K_0R consisting of all $[M] - [N]$ with $\text{rank}_P(M) = \text{rank}_P(N)$, all $P \in$ Spec R. Alternatively, since any element of K_0R can be put in the form $[M] - [R^n]$, for some M, n, we see that rk_0R is the set of all $[M] - [R^n]$ with M of constant rank n and $n \ge 0$.

3.7 *Connectedness in Spec R*

Recall that $R = R_1 \times R_2$ if and only if $R_1 = Re_1$, $R_2 = Re_2$ for some $e_1, e_2 \in R$ with $e_1 + e_2 = 1$, $e_1{}^2 = e_1$, $e_2{}^2 = e_2$, and $e_1e_2 = 0$.

Given an idempotent $e \in R$, that is, $e^2 = e$, we have $1 = e + (1 - e)$, and $(1 - e)^2 = 1 - e$. Also $e(1 - e) = 0$, and so if $P \in \text{Spec } R$, either $e \in P$ or $1 - e \in P$, but not both. So $\text{Spec } R = D(Re) \mathbin{\dot{\cup}} D(R(1 - e))$, the dot indicating that we have a disjoint union. Thus if $Re, R(1 - e) \neq R$, that is, if $e \neq 0,1$, we have written Spec R as a disjoint union of two non-empty open sets, and so Spec R is not connected. In fact, if $R = R_1 \times R_2$, then the natural homomorphism $\pi_i : R \twoheadrightarrow R_i$ induces an embedding $\pi_{i*} : \text{Spec } R_i \hookrightarrow \text{Spec } R$ with image $D(R_i)$; identifying, Spec R_i is an open subset of Spec R, $i = 1,2$, and then $\text{Spec } R = \text{Spec } R_1 \mathbin{\dot{\cup}} \text{Spec } R_2$. (Here we are identifying $P_1 \in \text{Spec } R_1$ with $P_1 \oplus R_2 \in \text{Spec } R$, and $P_2 \in \text{Spec } R_2$ with $R_1 \oplus P_2 \in \text{Spec } R$.)

We now establish the converse, namely that every disconnection of Spec R, that is, every way of writing Spec R as a disjoint union of two non-empty open sets, corresponds to a decomposition of R as a product $R_1 \times R_2$. First recall that the element $x \in R$ is said to be *nilpotent* if $x^n = 0$, some n. Write nil(R) for the set of all nilpotent elements of R, the *nil radical* of R. We claim $\text{nil}(R) = \cap\{P : P \in \text{Spec } R\}$. For if x is nilpotent, $x^n = 0$, say, then $x.x^{n-1} = 0$, so for any $P \in \text{Spec } R$, $x \in P$ or $x^{n-1} \in P$; in the latter case, $x.x^{n-2} \in P$ so $x \in P$ or $x^{n-2} \in P$; and so on: we deduce $x \in P$. Conversely, if $x \in R$ and x is not nilpotent, then put $S = \{1, x, x^2, x^3, \ldots\}$. By Zorn's lemma, there is an ideal P of R, maximal subject to $S \cap P = \emptyset$. Suppose $a, b \in R$ with $ab \in P$. If $a, b \notin P$, then by the construction of P we can find n, m with $x^n \in P + aR$ and $x^m \in P + bR$, so $x^{n+m} \in (P + aR)(P + bR) \subset P$, since P is an ideal and $ab \in P$. But this is absurd, and hence P is a prime ideal, and $x \notin P$. So $\text{nil}(R) = \cap\{P : P \in \text{Spec } R\}$; in particular,

$\text{nil}(R) \lhd R$.

Suppose $I, J \lhd R$ with Spec $R = D(I) \dot{\cup} D(J)$. So $D(I + J) =$ Spec R, and $I + J = R$, and also $D(I \cap J) = \emptyset$, whence $IJ \subset I \cap J \subset P$, all $P \in$ Spec R, or $IJ \subset I \cap J \subset \text{nil}(R)$. Let $1 = a_1 + a_2$, with $a_1 \in I$, $a_2 \in J$. So $a_1a_2 = a_1(1 - a_1) \in IJ$, and therefore $(a_1a_2)^n = 0$, some n, or $a_1{}^n(1 - a_1)^n = 0$. Now, working in the polynomial ring $\mathbb{Z}[x]$, we have

$$1 = (x + (1 - x))^{2n} = \Sigma_{r=0}^{2n} \binom{2n}{r} x^{2n-r}(1 - x)^r.$$

Put $f(x) = \Sigma_{r=0}^{n} \binom{2n}{r} x^{2n-r}(1 - x)^r$: so $f(x) \equiv 0 \bmod x^n$, and $f(x) \equiv 1 \bmod (1 - x)^n$. Hence $f(x)^2 \equiv f(x) \bmod x^n(1 - x)^n$; also note $f(x) \equiv x^{2n} \bmod x(1 - x)$. Going back to R, put $e_1 = f(a_1)$, and $e_2 = 1 - e_1$. So $e_1{}^2 \equiv e_1 \bmod a_1{}^n(1 - a_1)^n$, that is, mod 0, so $e_1{}^2 = e_1$. It follows that $e_2{}^2 = e_2$ and $e_1e_2 = 0$, and $R = R_1 \times R_2$, where $R_1 = Re_1$, $R_2 = Re_2$. Also, since $a_1(1 - a_1) \in IJ$, we have $e_1 \equiv a_1{}^{2n} \bmod IJ$. But $1 = a_1 + a_2$, so $a_1{}^r = a_1{}^{r+1} + a_1{}^r a_2$, or $a_1{}^r \equiv a_1{}^{r+1} \bmod IJ$, all r. Therefore $e_1 \equiv a_1 \bmod IJ$, and also $e_2 \equiv a_2 \bmod IJ$ (since $a_1 + a_2 = e_1 + e_2$). So $e_1 \in I$, and $e_2 \in J$, and therefore $R_1 \subset I$ and $R_2 \subset J$. Thus $D(R_1) \subset D(I)$ and $D(R_2) \subset D(J)$; but Spec $R = D(I) \dot{\cup} D(J) = D(R_1) \dot{\cup} D(R_2)$, so we conclude $D(R_1) = D(I)$ and $D(R_2) = D(J)$. So the given disconnection of Spec R corresponds to the decomposition $R = R_1 \times R_2$.

We say that the ring R is *connected* if the topological space Spec R is connected. By the above argument, this is the same as saying that the only idempotents in R are 0 and 1. For instance, if R is an integral domain, and e is an idempotent of R, then $e(1 - e) = 0$, so $e = 0$ or $1 - e = 0$. Again, if R is a local ring and e is an idempotent of R, then $e(1 - e) = 0$; if $e \in R^{\bullet}$, we deduce $1 - e = 0$, and if $e \notin R^{\bullet}$, then $1 - e \in R^{\bullet}$, so we deduce $e = 0$. In either case, we have a connected ring. Here is an alternative, somewhat whimsical argument that avoids mention of idempotents. If [0, 1] stands, as usual, for the unit real closed interval, and if $P_1, P_2 \in \operatorname{Spec} R$ with $P_1 \subset P_2$, define $f : [0, 1] \to \operatorname{Spec} R$ by $f(x) = P_1$ for $x < 1$ and $f(1) = P_2$. Then f is continuous, so it is a path from P_1 to P_2, in the ordinary sense of topology. Thus, if R is an integral domain, $0 \in \operatorname{Spec} R$, and $0 \subset P$, any $P \in \operatorname{Spec} R$, so Spec R is path-connected. Similarly, if R is a local ring, $R - R^{\bullet} \in \operatorname{Spec} R$, and $P \subset R - R^{\bullet}$, any $P \in \operatorname{Spec} R$, and again Spec R is path-connected. In either case, we deduce that Spec R is connected.

The ring R is *Noetherian* if it satisfies the maximal condition on ideals.

Exercise: (i) If R is a Noetherian connected ring, show that Spec R is path-connected.

(ii) If R is an integral domain or a local ring, show that Spec R is n-connected, for all n.

There seem to be a lot of unanswered questions in this area. If Spec R is connected, is it path-connected? If it is path-connected, is it simply connected? In general, is the homotopy of Spec R of any interest? We shall pursue the matter no further here.

Let us recall from topology that the *connected components* of a topological space X are its maximal connected subspaces. Since the closure of a connected subspace is connected, and the union of two connected subspaces is connected if their intersection is non-empty, it follows that the connected components partition X into closed subsets. If these are finite in number, they are also open. On the other hand, if they are all open, and if X is compact, they must be finite in number. But even if X is compact, the connected components may be infinite in number, for instance if $X = \{1, \frac{1}{2}, \frac{1}{4}, \frac{1}{8}, \ldots , 0\}$, with topology as a subspace of the real numbers $\mathbb{R}$.

Given an open and closed subset $D(I)$ $(I \lhd R)$ of Spec R, that is, $\operatorname{Spec} R = D(I) \mathbin{\dot\cup} D(J)$ for some $J \lhd R$, we have seen that there is an idempotent e such that $D(I) = D(Re)$. Suppose $\bar{e}$ is another idempotent such that $D(Re) = D(R\bar{e})$. Taking complements, $V(Re) = V(R\bar{e})$, and $V(R(1 - e)) = V(R(1 - \bar{e}))$. So, for each $P \in \text{Spec } R$, either $e, \bar{e} \in P$ or $1 - e, 1 - \bar{e} \in P$, and in either case $e - \bar{e} \in P$. Since P was arbitrary, $e - \bar{e} \in \text{nil}(R)$; suppose $(e - \bar{e})^n = 0$, where without loss of generality we may assume n is odd. Expanding, and using the fact that $e, \bar{e}$ are idempotent, we obtain $e = \bar{e}$.

Suppose $f \in H_0R$. Since f is continuous and Spec R is compact, image(f) is a compact subset of the discrete space $\mathbb{Z}$, so it is finite, say image$(f) = \{n_1, n_2, \ldots, n_r\}$. Now $\{n_i\}$ is open and closed in $\mathbb{Z}$, so $f^{-1}\{n_i\}$ is open and closed in Spec R, so there is a unique idempotent $e_i \in R$ with $f^{-1}\{n_i\} = D(Re_i)$, each i. If $i \neq j$, then $D(Re_i) \cap D(Re_j) = \emptyset$, so $D(Re_ie_j) = \emptyset$, and $e_ie_j \subset \text{nil}(R)$. But e_ie_j is idempotent, so $e_ie_j = 0$. (Such e_i, e_j are called *orthogonal* idempotents.) Next, $\Sigma_{i=1}^{r} D(Re_i) = \text{Spec } R$, so $R = Re_1 + Re_2 + \ldots + Re_r$. Thus

$$1 = s_1e_1 + s_2e_2 + \ldots + s_re_r$$

some $s_i \in R$, and on multiplying each side by e_i we see $e_i = s_ie_i$, so $1 = e_1 + e_2 + \ldots + e_r$.

We have found a decomposition of 1 as a sum $e_1 + \ldots + e_r$ of orthogonal idempotents $e_1, \ldots, e_r$ such that f is constant on $D(Re_i)$, each i. Such a decomposition we call a decomposition *corresponding to f*. Any finer decomposition is another such; by a *finer* decomposition, or a *refinement*, we mean an expression $1 = \Sigma_{i,j}\, e_{ij}$ where the e_{ij} are orthogonal idempotents, and $e_i = \Sigma_j\, e_{ij}$, each i. Notice that any two decompositions $1 = \Sigma_i\, e'_i = \Sigma_j\, e''_j$ of 1 as a sum of orthogonal idempotents have a common refinement given by $1 = \Sigma_{i,j}\, e_{ij}$, where $e_{ij} = e'_ie''_j$, each i, j.

We now use the above to construct a ring homomorphism θ : $H_0R \to K_0R$. Given $f \in H_0R$, let $1 = \Sigma_i\, e_i$ be a decomposition of 1 as a sum of orthogonal idempotents, corresponding to f. Let

$f(D(Re_i)) = n_i$, each i; put $\theta(f) = \Sigma_i n_i[Re_i] \in K_0R$. If we pass to a finer decomposition by $e_i = \Sigma_j e_{ij}$, each i, then clearly $D(Re_i) = \dot{\cup}_j D(Re_{ij})$ and $Re_i = \oplus_j Re_{ij}$, each i, so θ is well defined. Given $f, g \in H_0R$, by passing to a common refinement of decompositions corresponding to f, g, we can find a decomposition $1 = \Sigma_i e_i$ of 1 as a sum of orthogonal idempotents that corresponds to both f and g. Then $\theta(f) = \Sigma_i n_i[Re_i]$ and $\theta(g) = \Sigma_i m_i[Re_i]$, say, and it is now obvious that $\theta(f + g) = \theta(f) + \theta(g)$. The fact that $\theta(fg) = \theta(f)\theta(g)$ follows from

$$Re_i \otimes_R Re_j \simeq \begin{cases} Re_i & (i = j) \\ 0 & (i \neq j) \end{cases}$$

the proof of which is left to the reader. Finally, $1 \in H_0R$ is defined by $1 : P \mapsto 1$, all $P \in \operatorname{Spec} R$, so $\theta(1) = [R]$, which is the multiplicative identity of K_0R. So we have a homomorphism of rings.

Proposition 25. For any (commutative) ring R, $K_0R \simeq H_0R \oplus rk_0R$.

Proof. We already have constructed homomorphisms

$$H_0R \xrightarrow{\theta} K_0R \xrightarrow{\text{rank}} H_0R$$

with $rk_0R = \ker(\text{rank})$, so it remains to show that $(\text{rank})(\theta) = 1 : H_0R \to H_0R$.

Let $f \in H_0R$, and let $1 = e_1 + e_2 + \dots + e_n$ be an orthogonal idempotent decomposition of 1, corresponding to f. Suppose $f(P) = n_i$, all $P \in D(Re_i)$, each i. We must show $\text{rank}(\theta(f)) = f$, or $\text{rank}(\theta(f)) : P \mapsto n_i$, all $P \in D(Re_i)$, each i.

Consider Re_1 as an R-module. We have $\wedge_R^2 Re_1 = 0$, since $re_1 \wedge se_1 = (rs)(e_1 \wedge e_1) = 0$, all $r, s \in R$. So $\text{rank}_P(Re_1) \leq 1$, all $P \in \operatorname{Spec} R$. But by Proposition 24,

$$\begin{aligned} \{P : \text{rank}_P(Re_1) = 0\} &= D(\text{ann}(Re_1)) \\ &= D(Re_2 + Re_3 + \dots + Re_n) \\ &= \cup_{i=2}^n D(Re_i). \end{aligned}$$

Since Spec $R = \cup_{i=1}^{n} D(Re_i)$, we have the following complete description of $\text{rank}_P(Re_1)$:

$$\text{rank}_P(Re_1) = \begin{cases} 1 & (P \in D(Re_1)) \\ 0 & (P \in D(Re_i),\ i \geq 2). \end{cases}$$

Similarly we obtain, for each i,

$$\text{rank}_P(Re_i) = \begin{cases} 1 & (P \in D(Re_i)) \\ 0 & (\text{otherwise}). \end{cases}$$

But

$$\theta(f) = \Sigma_i\, n_i[Re_i] = [(Re_1)^{n_1} \oplus \ldots \oplus (Re_n)^{n_n}]$$

and by the above results,

$$\text{rank}_P((Re_1)^{n_1} \oplus \ldots \oplus (Re_n)^{n_n}) = n_i \text{ if } P \in D(Re_i)$$

so $\text{rank}(\theta(f)) : P \mapsto n_i$ for $P \in D(Re_i)$, and the proof is complete.

Let us identify H_0R with its image in K_0R. Recall that, since R is commutative, the homomorphism $\mathbb{Z} \to R$ induces a homomorphism $\mathbb{Z}\ (= K_0\mathbb{Z}) \hookrightarrow K_0R$. Since this is a ring homomorphism, and H_0R is a subring of K_0R, the image of $\mathbb{Z} \hookrightarrow K_0R$ lies inside H_0R. Indeed, this image is the subring of H_0R consisting of all constant functions Spec $R \to \mathbb{Z}$. Identifying $\mathbb{Z}$ with its image in K_0R, we have

$$\tilde{K}_0R = (K_0R)/\mathbb{Z} \simeq ((H_0R)/\mathbb{Z}) \oplus rk_0R \text{ (as abelian groups).}$$

If R is connected, then $H_0R = \mathbb{Z}$, $\tilde{K}_0R \simeq rk_0R$, and so in this case we obtain a canonical decomposition $K_0R = \mathbb{Z} \oplus \tilde{K}_0R$ (cf. section 2.7).

3.8 *The determinant map det : $K_0R \to$ Pic R*

Any element of rk_0R is of the form $[M] - [R^m]$, where M is a finitely generated projective R-module of constant rank m. For such M, $\wedge_R^m M$ is of constant rank 1, by Proposition 23, Corollary (iii), so it is invertible. If $[M] - [R^m] = [N] - [R^n]$, then

$[M \oplus R^n] = [N \oplus R^m]$, so $M \oplus R^{n+r} \simeq N \oplus R^{m+r}$, some r. Thus

$$\wedge_R^{m+n+r} (M \oplus R^{n+r}) \simeq \wedge_R^{m+n+r} (N \oplus R^{m+r}).$$

But by Proposition 23,

$$\wedge_R^{m+n+r} (M \oplus R^{n+r}) \simeq (\wedge_R^m M) \otimes_R (\wedge_R^{n+r} R^{n+r})$$

(the other terms vanishing)

$$\simeq (\wedge_R^m M) \otimes_R R \simeq \wedge_R^m M$$

and similarly

$$\wedge_R^{m+n+r} (N \oplus R^{m+r}) \simeq \wedge_R^n N$$

so we deduce $\wedge_R^m M \simeq \wedge_R^n N$, and we have a well defined map $rk_0 R \to$ Pic R given by $[M] - [R^m] \mapsto \langle \wedge_R^m M \rangle$.

Now suppose $[M] - [R^m]$, $[N] - [R^n] \in rk_0 R$: so

$$([M] - [R^m]) + ([N] - [R^n]) = [M \oplus N] - [R^{m+n}]$$

and by proposition 23,

$$\wedge_R^{m+n} (M \oplus N) \simeq (\wedge_R^m M) \otimes_R (\wedge_R^n N)$$

the other terms vanishing. So our map $rk_0 R \to$ Pic R is a homomorphism from the additive group of $rk_0 R$ to the multiplicative group Pic R. It is surjective, since for any invertible module M, $[M] - [R] \in rk_0 R$, and then $\wedge_R^1 M = M$.

We now define the *determinant map*, det : $K_0 R \twoheadrightarrow$ Pic R, to be the composite of the natural map $K_0 R \twoheadrightarrow rk_0 R$ (using Proposition 25) and the map $rk_0 R \twoheadrightarrow$ Pic R constructed above. So det is an epimorphism from the additive group $K_0 R$ to the multiplicative group Pic R. The reason for using the word 'determinant' here will emerge later when we discuss matrices and $K_1 R$ (Chapter 6).

Given rings R, S and a ring homomorphism $f : R \to S$, there is a group homomorphism Pic f : Pic $R \to$ Pic S, given by $\langle M \rangle \mapsto \langle S \otimes_f M \rangle$. It is the restriction to Pic R, regarded as a subgroup of $(K_0 R)^\bullet$ (Proposition 23, Corollary (vi)), of the ring homomorphism $K_0 f$.

Exercise: Show that Pic f is well defined, that is, if M is an invertible R-module, then $S \otimes_f M$ is an invertible S-module.

Exercise: In the above situation, show that the diagram

$$\begin{array}{ccc} K_0R & \xrightarrow{K_0 f} & K_0S \\ \det \downarrow & & \downarrow \det \\ \text{Pic } R & \xrightarrow{\text{Pic } f} & \text{Pic } S \end{array}$$

commutes.

We now give a more explicit description of the determinant map. If $[M]$ is a generator of K_0R, the natural map $K_0R \to rk_0R$ gives $[M] \mapsto [M] - \theta(f)$, where θ is the homomorphism $H_0R \to K_0R$ constructed just before Proposition 25 and $f \in H_0R$ is defined by $f : P \mapsto \text{rank}_P(M)$, all $P \in \text{Spec } R$. If $1 = \Sigma_i e_i$ is an orthogonal idempotent decomposition of 1 corresponding to f, then $\theta(f) = \Sigma_i n_i[Re_i]$, say. Put $n = \max_i n_i$ and let $m_i + n_i = n$, all i. So

$$\theta(f) = \Sigma_i n[Re_i] - \Sigma_i m_i[Re_i] = n[\oplus_i Re_i] - \Sigma_i m_i[Re_i]$$
$$= n[R] - \Sigma_i m_i[Re_i] = [R^n] - \Sigma_i m_i[Re_i].$$

Therefore the map $K_0R \to rk_0R$ gives

$$[M] \mapsto [M] + \Sigma_i m_i[Re_i] - [R^n] = [N] - [R^n]$$

where

$$N = M \oplus (\oplus_i (Re_i)^{m_i}).$$

By construction, $[N] - [R^n] \in rk_0R$, so N is of constant rank n, and therefore $\det[M] = \langle\wedge_R^n N\rangle$. We sometimes write, loosely, $\det M = \wedge_R^n N$; in this notation, $\det[M] = \langle\det M\rangle$.

Now $R = Re_1 \oplus Re_2 \oplus \ldots \oplus Re_r = Re_1 \times Re_2 \times \ldots \times Re_r$. Each Re_i is a ring, with multiplicative identity e_i, and the natural ring homomorphism $\pi_i : R \twoheadrightarrow Re_i$ is given by $r \mapsto re_i$. Of course, Re_i is also an R-module, and π_i is also an R-module homomorphism. Next, $M = e_1M \oplus e_2M \oplus \ldots \oplus e_rM$, and for each i, $e_iM \simeq M/(\ker \pi_i)M \simeq Re_i \otimes_{\pi_i} M$, as in Lemma 6. Applying this to

$\wedge_R^s M$, we have

$$e_i(\wedge_R^s M) \simeq Re_i \otimes_{\pi_i} (\wedge_R^s M) \simeq \wedge_{Re_i}^s (Re_i \otimes_{\pi_i} M) \simeq \wedge_{Re_i}^s (e_i M)$$

for any non-negative integer s, the middle isomorphism coming from an argument similar to that in the proof of Proposition 22. Applying the above to det M, we have

$$\det M = \wedge_R^n N \simeq \oplus_i e_i(\wedge_R^n N) \simeq \oplus_i (\wedge_{Re_i}^n e_i N)$$

$$\simeq \oplus_i \left(\wedge_{Re_i}^n (e_i M \oplus (Re_i)^{m_i})\right)$$

$$\simeq \oplus_i \left((\wedge_{Re_i}^{n_i} e_i M) \otimes_{Re_i} (\wedge_{Re_i}^{m_i} (Re_i)^{m_i})\right)$$

by Proposition 23, the other terms vanishing because $e_i M$ has constant rank n_i as an Re_i-module. So

$$\det M \simeq \oplus_i \left((\wedge_{Re_i}^{n_i} e_i M) \otimes_{Re_i} Re_i\right)$$

$$\simeq \oplus_i (\wedge_{Re_i}^{n_i} e_i M) \simeq \oplus_i e_i(\wedge_R^{n_i} M).$$

Exercise: Given $f \in H_0 R$, with $f(P) \geq 0$, all $P \in \operatorname{Spec} R$, let $1 = \Sigma_i e_i$ be an orthogonal idempotent decomposition corresponding to f, and let $f(P) = n_i$ for $P \in D(Re_i)$, each i. Define $\wedge_R^f M = \oplus_i e_i(\wedge_R^{n_i} M)$, where M is an R-module. Show that this is well defined, up to isomorphism, and that it corresponds to the previous definition if $f \in \mathbb{Z}$, where $\mathbb{Z}$ is embedded in $H_0 R$ as the constant functions Spec $R \to \mathbb{Z}$. Show also that, if $g : R \to S$ is a ring homomorphism, and $H_0 g : f \mapsto f' \in H_0 S$, then $S \otimes_g (\wedge_R^f M) \simeq \wedge_S^{f'} (S \otimes_g M)$. State and prove an appropriate generalization of Proposition 23. (First extend the definition of $\wedge_R^f M$ to arbitrary $f \in H_0 R$ by defining $\wedge_R^n M = 0$, all $n < 0$.)

In the above notation, det $M = \wedge_R^f M$, or more correctly $\det[M] = \langle \wedge_R^f M \rangle$, where $f \in H_0 R$ is the rank map of M, that is, $f(P) = \operatorname{rank}_P(M)$, all $P \in \operatorname{Spec} R$.

CHAPTER FOUR

K_0 of Integral Domains and Dedekind Domains

All rings in this chapter are commutative. Recall that the (commutative) ring R is an *integral domain* if 0 is a prime ideal of R, and then $F = R_0$ is a field, the *field of fractions* of R. Also, the natural map $R \to R_0 = F$ is injective; identifying, R is embedded as a subring of F. Of course, F is an R-module, the scalar multiplication being just multiplication in F. If M is an invertible R-module, then $M_0 \simeq R_0$, so the natural map $M \to M_0$, which we leave the reader to show is injective, gives an isomorphism between M and some R-submodule of F. Conversely, if M is a non-zero R-submodule of F, then $M_0 \simeq F$ as R-modules. Explicitly, choose $r/s \in M - \{0\}$ and define a map $F \to M_0$ by $t/u \mapsto (tr/s)/u$; we leave the reader to show that this is a well defined isomorphism. If M is finitely generated and projective, this shows $\mathrm{rank}_0(M) = 1$, and so M is invertible, since R is connected. Thus we may regard Pic R as the isomorphism classes of non-zero finitely generated projective R-submodules of F.

Let M be an R-submodule of F, the field of fractions of the integral domain R. M is called a *fractional ideal* of R if it satisfies the following equivalent conditions:

(i) $aR \subset M \subset bR$ for some $a, b \in F^\bullet$

(ii) $M = cI$ for some $I \lhd R$, $I \neq 0$, and some $c \in F^\bullet$.

The equivalence of (i) and (ii) is immediate: (i) => (ii) by putting $I = b^{-1}M$, $c = b$, and (ii) => (i) by putting $b = c$ and $a = cd$ for any $d \in I - \{0\}$.

Of course, every non-zero ideal of R is a fractional ideal. If M, M_1 are fractional ideals of R, put MM_1 = all finite sums $\Sigma_i m_i m_i'$ ($m_i \in M$, $m_i' \in M_1$, all i). Then MM_1 is another fractional ideal, for if $M = cI$, $M_1 = c_1I_1$, where I, I_1 are non-zero ideals of R and c, $c_1 \in F^\bullet$, then $MM_1 = cc_1II_1$; II_1 is a non-zero ideal of R and $cc_1 \in F^\bullet$.

Let M be a fractional ideal of R. Put $\bar{M} = \{x \in F : xM \subset R\}$. Clearly $\bar{M}$ is an R-submodule of F; then if a, $b \in F^\bullet$ with $aR \subset M \subset bR$, then $b^{-1}M \subset R$, so $b^{-1} \in \bar{M}$ and therefore $b^{-1}R \subset \bar{M}$. Also $a \in M$ since $aR \subset M$, and so $a\bar{M} \subset R$, or in other words $\bar{M} \subset a^{-1}R$. Of course, a^{-1}, $b^{-1} \in F^\bullet$, and so $\bar{M}$ is a fractional ideal of R. Note that $M\bar{M} \subset R$, any fractional ideal M.

Proposition 26. Let M be a non-zero R-submodule of F, the field of fractions of the integral domain R. The following conditions are equivalent:

(i) M is a fractional ideal of R, and $M\bar{M} = R$

(ii) M is finitely generated and projective as an R-module.

Proof. (i) => (ii). Write $1 = m_1n_1 + m_2n_2 + \ldots + m_rn_r$, where $m_i \in M$, $n_i \in \bar{M}$, all i. Define a homomorphism $i : M \to R^r$ by $m \mapsto (mn_1, mn_2, \ldots, mn_r)$, and a homomorphism $\pi : R^r \to M$ by $(x_1, x_2, \ldots, x_r) \mapsto x_1m_1 + x_2m_2 + \ldots + x_rm_r$. Then $\pi i = 1 : M \to M$, so $R^r = \text{image}(i) \oplus \ker(\pi) \simeq M \oplus \ker(\pi)$.

(ii) => (i). Since $M \neq 0$, we can choose $a \in M - \{0\}$, and then $a \in F^\bullet$ and $aR \subset M$. If r_i, $s_i \in R$ are such that r_i/s_i, $1 \leq i \leq n$, generate M as an R-module, put $b = 1/s_1s_2\ldots s_n \in F^\bullet$, and it is clear that $M \subset bR$. So M is a fractional ideal.

Now suppose M is isomorphic to a direct summand of R^r; so there are homomorphisms $i : M \to R^r$ and $\pi : R^r \to M$ with $\pi i = 1 : M \to M$. Suppose i is given by $i(m) = (f_1(m), f_2(m), \ldots, f_r(m))$, and π is given by $\pi(x_1, x_2, \ldots, x_r) = x_1m_1 + x_2m_2 + \ldots + x_rm_r$, where $m_1 = \pi(1,0,\ldots,0)$, etc. Then

$$m = f_1(m)m_1 + f_2(m)m_2 + \ldots + f_r(m)m_r, \text{ all } m \in M.$$

Now $f_i : M \to R$ is an R-module homomorphism, each i. If $m, m' \in M$, then $m = r/s$, $m' = r'/s$ for some $r, r', s \in R$, and then $(r/s)f_i(r'/s) = (1/s)f_i(rr'/s) = (r'/s)f_i(r/s)$, or $mf_i(m') = m'f_i(m)$. Therefore if we choose $m \in M - \{0\}$ and put $n_i = m^{-1}f_i(m) \in F$, each i, then $f_i(m) = mn_i$ for *all* $m \in M$; since $mn_i \in R$, all $m \in M$, then $n_i \in \bar{M}$, all i, and furthermore $m = mn_1m_1 + mn_2m_2 + \ldots + mn_rm_r$, all $m \in M$, and since $M \neq 0$ and $M \subset F$, we deduce $n_1m_1 + n_2m_2 + \ldots + n_rm_r = 1$, whence $M\bar{M} = R$.

Note that in the above situation, $\bar{\bar{M}} = M$. For $\bar{\bar{M}} = \{x \in F : x\bar{M} \subset R\}$, whence $M \subset \bar{\bar{M}}$. Thus $R = M\bar{M} \subset \bar{\bar{M}}\bar{M} \subset R$, and so $\bar{\bar{M}}\bar{M} = R$. Then $M = MR = M(\bar{M}\bar{\bar{M}}) = (M\bar{M})\bar{\bar{M}} = R\bar{\bar{M}} = \bar{\bar{M}}$. Note also that, if M is as above, and $MN = R$ for some R-submodule N of F, then $N = RN = (\bar{M}M)N = \bar{M}(MN) = \bar{M}R = \bar{M}$. Now suppose M, N are fractional ideals of R with $M\bar{M} = R = N\bar{N}$. We claim that $\overline{MN} = \bar{M}\bar{N}$, and $(MN)(\overline{MN}) = R$. For firstly $(MN)(\bar{M}\bar{N}) = (M\bar{M})(N\bar{N}) = RR = R$, and it follows that $\bar{M}\bar{N} \subset \overline{MN}$. But now $R = (MN)(\bar{M}\bar{N}) \subset (MN)(\overline{MN}) \subset R$, so $(MN)(\overline{MN}) = R$. Since $(MN)(\bar{M}\bar{N}) = R$, it follows by the preceding remarks that $\bar{M}\bar{N} = \overline{MN}$.

Let M, N be R-submodules of F. The map $M \times N \to MN$ given by $(m, n) \mapsto mn$ is bilinear, and so induces an R-module homomorphism $M \otimes_R N \to MN$. Explicitly, this homomorphism is given by $\Sigma_i\, m_i \otimes n_i \mapsto \Sigma_i\, m_in_i$, so it is an epimorphism.

Corollary. Let M, N be fractional ideals of R with $M\bar{M} = R = N\bar{N}$. Then the natural epimorphism $M \otimes_R N \twoheadrightarrow MN$ is an isomorphism.

Proof. By the preceding remarks, $(MN)(\overline{MN}) = R$, and so by the proposition, MN, M, N, and hence also $M \otimes_R N$, are finitely generated projective R-modules of constant rank 1. The result follows by the second corollary to Proposition 19.

Corollary. Let M be a fractional ideal of R with $M\bar{M} = R$. Then $\bar{M} \simeq M^*$.

Proof. By the first corollary, $M \otimes_R \bar{M} \simeq M\bar{M} = R$. The result follows by Proposition 21.

If M is a fractional ideal of R with $M\bar{M} = R$, we shall call M an *invertible* fractional ideal. By the above results, this is the same as saying that the fractional ideal M is invertible as an R-module. Since, by the remarks at the beginning of this chapter, every invertible R-module is isomorphic to an invertible fractional ideal, we may think of Pic R as the set of all isomorphism classes $\langle M \rangle$ of invertible fractional ideals M of R. By the corollaries above, the group structure of Pic R is then given by $\langle M \rangle\langle N \rangle = \langle MN \rangle$, and $\langle M \rangle^{-1} = \langle \bar{M} \rangle$, all $\langle M \rangle, \langle N \rangle \in$ Pic R.

Note that, if $\alpha : M \to N$ is an isomorphism of R-submodules of F, then there is a fixed $a \in F^{\bullet}$ such that $\alpha(m) = am$, all $m \in M$. For if $m, m' \in M$, then $m = r/s$, $m' = r'/s$ for some $r, r', s \in R$, and then

$$(r/s)\alpha(r'/s) = (1/s)\alpha(rr'/s) = (r'/s)\alpha(r/s)$$

or $m\alpha(m') = m'\alpha(m)$. Thus for instance, if M is a fractional ideal of R, then $M \simeq R$ as R-modules if and only if M is a *principal* fractional ideal, that is $M = aR$, some $a \in F^{\bullet}$. In particular, if $I \lhd R$, then $I \simeq R$ as R-modules if and only if $I = aR$ for some $a \in R - \{0\}$, that is, if and only if I is a principal ideal of R.

Write $\mathrm{Cl}(R)$ for the set of isomorphism classes $\langle I \rangle$ of non-zero ideals I of R which are finitely generated and projective as R-modules, that is, which are invertible. Putting $\langle I \rangle\langle J \rangle = \langle IJ \rangle$, all $\langle I \rangle, \langle J \rangle \in \mathrm{Cl}(R)$, makes $\mathrm{Cl}(R)$ into a group, called the *ideal class group*, or *class group*, of R. The fact that $\mathrm{Cl}(R)$ is a group follows from the fact that every fractional ideal M of R is of the form cI, some $I \lhd R$ and $c \in F^{\bullet}$, whence $M \simeq I$ as R-modules. This also shows that $\mathrm{Cl}(R) \simeq$ Pic R, the isomorphism being the obvious one, $\langle I\rangle \mapsto \langle I\rangle$, implied by the notation. Note carefully that, if $I \lhd R$ and $\langle I\rangle \in \mathrm{Cl}(R)$, then in general $\bar{I}$ is not an ideal of R; indeed, since $I \subset R$, we have $R \subset \bar{I}$. But $\bar{I} = cJ$ for some $J \lhd R$ and $c \in F^{\bullet}$, and then $\langle I\rangle^{-1} = \langle J\rangle$; put another way, if $\langle I\rangle \in \mathrm{Cl}(R)$, then $\langle I\rangle^{-1} = \langle J\rangle$, where $J \lhd R$ is chosen so that IJ is a principal ideal of R.

The order of the group $\mathrm{Cl}(R)$ (or Pic R) is called the *class number* of R. Note that if R is a principal ideal domain (that is, every ideal of R is a principal ideal) then the class number of R is 1. The converse is false in general: if the class number of R is 1, all we know is that every *invertible* ideal is principal.

Consider the map i : Pic $R \to K_0R$ given by $\langle M\rangle \mapsto [M] - [R]$. In general this is not a homomorphism of any sort, but it is always injective, since $\det[M] = \langle M\rangle$ (M is invertible) and $\det[R] = \langle R\rangle$, so $\det([M] - [R]) = \langle M\rangle\langle R\rangle^{-1} = \langle M\rangle$. The image of i lies inside rk_0R, which is isomorphic to $\widetilde{K}_0R$ since R is an integral domain, and hence connected. We shall see that, under suitable conditions on R, i is a homomorphism from the multiplicative group Pic R to the additive group K_0R, and its image is rk_0R; under these conditions, if we write the abelian group Pic R additively instead of multiplicatively, we shall have $K_0R \simeq \mathbb{Z} \oplus \text{Pic } R \simeq \mathbb{Z} \oplus \mathrm{Cl}(R)$.

Since R is connected, every finitely generated projective R-module is of constant rank, and the image of i will generate rk_0R if every finitely generated projective R-module is stably isomorphic to a direct sum of invertible R-modules, or to a direct sum of ideals of R. The condition for i to be a homomorphism is

$$i\langle M\rangle + i\langle N\rangle = i(\langle M\rangle\langle N\rangle)$$

or

$$([M] - [R]) + ([N] - [R]) = [M \otimes_R N] - [R]$$

or

$$[M \oplus N] = [(M \otimes_R N) \oplus R]$$

or that $M \oplus N$, $(M \otimes_R N) \oplus R$ are stably isomorphic, for all invertible R-modules M, N. A sufficient condition therefore is $M \oplus N \simeq (M \otimes_R N) \oplus R$, all invertible M, N.

The integral domain R is called a *Dedekind domain* if the following equivalent conditions are satisfied:

(i) Every ideal of R is finitely generated and projective as an R-module, i.e., it is zero or invertible

(ii) Given ideals I, J of R with $I \subset J$, there is an ideal I_1 of R with $I = JI_1$.

The equivalence is shown thus: (i) => (ii): Given $I, J \lhd R$ with $I \subset J$, suppose $J \neq 0$ (otherwise put $I_1 = 0$, done!), so by hypothesis J is invertible and we can find $J_1 \lhd R$ and $b \in R - \{0\}$ such that $JJ_1 = bR$. Then $J_1I \subset J_1J = bR$, so $b^{-1}J_1I \subset R$ (working in F). Put $I_1 = b^{-1}J_1I$; I_1 is a fractional ideal of R and $I_1 \subset R$, so $I_1 \lhd R$. Then $JI_1 = b^{-1}JJ_1I = b^{-1}bRI = RI = I$, as required. (ii) => (i): Given $I \lhd R$, $I \neq 0$, choose $a \in I - \{0\}$, and we have $aR \subset I$. Applying (ii), $aR = II_1$ for some $I_1 \lhd R$, that is, II_1 is principal, and so I is invertible. (There are many equivalent ways of defining a Dedekind domain; the two definitions we have given are perhaps not the most usual definitions, but they are the most suitable to our present purposes.)

Note that a Dedekind domain is always Noetherian, since by (i) every ideal is finitely generated. Explicitly, if $I_1 \subset I_2 \subset I_3 \subset \ldots$ are ideals of the Dedekind domain R, put $I = \cup_i I_i$, and we have $I \lhd R$, so we can find $a_1, a_2, \ldots, a_n \in I$ such that $I = Ra_1 + Ra_2 + \ldots + Ra_n$. For each r, $a_r \in I$, and so $a_r \in I_{i_r}$, some i_r. If $m = \max_r i_r$, then $a_1, a_2, \ldots, a_n \in I_m \subset I$, so $I = I_m = I_{m+1} = I_{m+2} = \ldots$.

Lemma 27. Let R be a Dedekind domain. Then every non-zero finitely generated projective R-module is isomorphic to a direct sum of invertible R-modules.

Proof. Suppose $M \oplus N \simeq R^n$, where $M \neq 0$. If $\pi_i : R^n \twoheadrightarrow R$ is the natural epimorphism onto the ith component, and $\alpha : M \hookrightarrow R^n$ is the natural monomorphism, then the homomorphisms $\pi_i\alpha : M \to R$ cannot all be zero. If i is chosen so that $\pi_i\alpha \neq 0$, and $I =$ image$(\pi_i\alpha)$, then $I \lhd R$ and $I \neq 0$. Since R is Dedekind, I is invertible. We have an epimorphism $\pi_i\alpha : M \twoheadrightarrow I$, and since I is projective, the epimorphism splits and $M \simeq I \oplus M_1$, say. Counting ranks, rank(M_1) = rank(M) - 1, so the proof follows by induction on rank(M).

Corollary. If R is a Dedekind domain, the image of i : Pic $R \to K_0R$ generates rk_0R, as additive group.

Proof. Immediate.

We proceed to show that, when R is Dedekind, i is a homomorphism. We have seen that it is sufficient to show $M \oplus N \simeq (M \otimes_R N) \oplus R$ for all invertible R-modules M, N; since R is Dedekind, this amounts to showing that $I \oplus J \simeq IJ \oplus R$, for all non-zero I, $J \lhd R$. We need two lemmas first:

Lemma 28. Let R be a Dedekind domain. Every proper non-zero ideal of R can be expressed uniquely as a product of maximal ideals of R.

Proof. Suppose $I \lhd R$, $I \neq R$, $I \neq 0$. Choose a maximal $M_1 \lhd R$ with $I \subset M_1$, and find $I_1 \lhd R$ such that $I = M_1I_1$. Clearly $I \subset I_1$, and we cannot have $I = I_1$, otherwise $I = M_1I$ gives (working in the field of fractions) $I\bar{I} = M_1I\bar{I}$, or $R = M_1R = M_1$, which is absurd. If $I_1 = R$, then $I = M_1$; otherwise we can choose a maximal $M_2 \lhd R$ with $I_1 \subset M_2$, and find $I_2 \lhd R$ such that $I_1 = M_2I_2$. Again, $I_1 \subset I_2$ and $I_1 \neq I_2$. If $I_2 = R$ then $I_1 = M_2$ and $I = M_1M_2$. Otherwise, we can choose a maximal $M_3 \lhd R$ with $I_2 \subset M_3$, and so on. The ascending chain $I \subsetneqq I_1 \subsetneqq I_2 \subsetneqq I_3 \subsetneqq \ldots$ must terminate, since R is Noetherian, and so at some stage we have $I_k = R$, and $I = M_1M_2\ldots M_k$. If also $I = M_1'M_2'\ldots M_\ell'$, with each M_i'

maximal, then $M_1' \supset M_1M_2\ldots M_k$; but M_1' is maximal, hence prime, and so $M_1' \supset M_i$ for some i. But M_i is maximal, so $M_1' = M_i$. We can cancel M_1' (strictly, work in the field of fractions and multiply each side of the equation $M_1M_2\ldots M_k = M_1'M_2'\ldots M_\ell'$ by $\overline{M_1'}$) and the result now follows by induction on $\max(k, \ell)$.

Corollary. If R is a Dedekind domain, and $I = M_1M_2\ldots M_k$, where each M_i is a maximal ideal of R, then the *distinct* M_i are precisely those maximal ideals of R that contain I.

Proof. Clearly $I \subset M_i$, all i. If M is a maximal ideal of R and $I \subset M$, then in the proof of Lemma 28 we can take $M_1 = M$, and the result follows.

Lemma 29. Let R be a Dedekind domain, and let $I \lhd R$, $I \neq 0$. Then R/I is a principal ideal ring (that is, it is a ring in which every ideal is principal; it is not necessarily an integral domain).

The proof depends on the Chinese remainder theorem, which we state and prove first.

Chinese remainder theorem. Let R be a commutative ring, and let $I_1, I_2, \ldots, I_k$ be ideals of R which are pairwise comaximal, that is, $I_i + I_j = R$ whenever $i \neq j$. Then, given $a_i \in R$, $1 \leq i \leq k$, there exists $a \in R$ such that $a \equiv a_i \bmod I_i$, all i.

Proof. If $k = 1$, put $a = a_1$. If $k = 2$, then $I_1 + I_2 = R$, so $x_1 + x_2 = 1$ for some $x_1 \in I_1$, $x_2 \in I_2$. Put $a = x_2a_1 + x_1a_2$, and the result follows, since $x_1 \equiv 0$ and $x_2 \equiv 1 \bmod I_1$, and $x_1 \equiv 1$ and $x_2 \equiv 0 \bmod I_2$. Now use induction: if $k > 2$, find, by the inductive hypothesis, $a_0 \in R$ such that $a_0 \equiv a_i \bmod I_i$ for all $i > 1$. Then $I_1 + I_2I_3\ldots I_k = R$, for if $x_i \in I_1$, $y_i \in I_i$, $2 \leq i \leq k$, with $x_i + y_i = 1$, all i, then

$$(x_2 + y_2)(x_3 + y_3)\ldots(x_k + y_k) = 1$$

and on expanding we see every term is in I_1 except $y_2 y_3 \ldots y_k$, which is in $I_2 I_3 \ldots I_k$. So by the case $k = 2$, we can find $a \in R$ such that $a \equiv a_1 \bmod I_1$ and $a \equiv a_0 \bmod I_2 I_3 \ldots I_k$; the result follows.

Proof of Lemma 29. Let $M_1, M_2, \ldots, M_k$ be the distinct maximal ideals of R which contain I; they are finite in number, by the corollary to Lemma 28. So R/I has only finitely many distinct maximal ideals, namely $M_1/I, M_2/I, \ldots, M_k/I$. Further, every non-zero proper ideal of R is a product of maximal ideals, so the same is true in R/I. Since any product of principal ideals is another principal ideal, it suffices to show that M_i/I is a principal ideal of R/I, each i, or in other words that $M_i = I + y_i R$, some $y_i \in R$, each i. Now $M_1^2 \subset M_1$, but $M_1^2 \neq M_1$, by Lemma 28. Choose $x_1 \in M_1 - M_1^2$. Also $M_1^2 \not\subset M_i$, all $i > 1$, by Lemma 28, and since M_i is maximal, all i, it follows that the ideals $M_1^2, M_2, M_3, \ldots, M_k$ are pairwise comaximal. By the Chinese remainder theorem, we can find $y_1 \in R$ with $y_1 \equiv x_1 \bmod M_1^2$ and $y_1 \equiv 1 \bmod M_i$, all $i > 1$. We claim $I + y_1 R = M_1$. For $I \subset M_1$, and $x_1 \in M_1$, and $y_1 - x_1 \in M_1^2 \subset M_1$, so $y_1 \in M_1$ and therefore $I + y_1 R \subset M_1$. If $I + y_1 R \subset M$ for some maximal ideal M of R, then $I \subset M$, and so $M = M_i$, some i, $1 \leq i \leq k$. If $i > 1$, then $y_1 \in M_i$; but $y_1 - 1 \in M_i$ also, and so $1 \in M_i$, a contradiction. So $i = 1$, and M_1 is the unique maximal ideal of R containing $I + y_1 R$. By Lemma 28, we must have $I + y_1 R = M_1^r$, some r. If $r > 1$, then $I + y_1 R \subset M_1^2$, so $y_1 \in M_1^2$; but $y_1 - x_1 \in M_1^2$, so $x_1 \in M_1^2$, another contradiction. So $r = 1$, and $I + y_1 R = M_1$. Similar arguments give $I + y_i R = M_i$, suitable y_i, $i > 1$, and the result follows.

Exercise: Recall that a commutative ring is local if and only if it has a unique maximal ideal. Prove that a commutative ring is semi-local if and only if it has only finitely many maximal ideals.

Proposition 30. Let R be a Dedekind domain. Then the map i : Pic $R \to rk_0R$, $\langle M\rangle \mapsto [M] - [R]$, is a group isomorphism, and if we write Pic R, Cl(R) additively, then

$$K_0R \simeq \mathbb{Z} \oplus \text{Pic } R \simeq \mathbb{Z} \oplus \text{Cl}(R).$$

Proof. We have already seen that i is injective, since $\det([M] - [R]) = \langle M\rangle$, and so if i is a homomorphism, it is an isomorphism by the corollary to Lemma 27. To show that i is a homomorphism, we have seen that it is sufficient to show that $I \oplus J \simeq IJ \oplus R$, all non-zero $I, J \lhd R$. So let $I, J \lhd R$, $I, J \neq 0$, and suppose first that I, J are comaximal, that is, $I + J = R$. There is an epimorphism $I \oplus J \twoheadrightarrow R$ given by $(a, b) \mapsto a + b$, and its kernel is $\{(a, -a) : a \in I \cap J\} \simeq I \cap J$. But R is projective, so the epimorphism splits, giving $I \oplus J \simeq (I \cap J) \oplus R$. Then $IJ \subset I \cap J$, and if $x \in I \cap J$ and $1 = a + b$ ($a \in I$, $b \in J$) then $x = xa + xb \in IJ$, so $IJ = I \cap J$, and $I \oplus J \simeq IJ \oplus R$.

Now suppose I, J are not comaximal. Choose $x \in I$, $x \neq 0$; so $xR \subset I$, and we can find $I_1 \lhd R$ with $xR = II_1$. By Lemma 29, I_1/I_1J is a principal ideal of R/I_1J, so there exists $y \in R$ such that $I_1 = I_1J + yR$. Multiplying through by I, $xR = II_1 = II_1J + yI = xJ + yI$. Working in the field of fractions, put $I' = x^{-1}yI$. I' is a fractional ideal of R, and $xR = xJ + xI'$; multiplying through by x^{-1}, $R = J + I'$, so $I' \subset R$ and therefore $I' \lhd R$. Further, $I' = x^{-1}yI$, so $I' \simeq I$ as R-modules, and I', J are comaximal. The result follows by the first part.

Let us examine the above result more closely. It says that any element of K_0R can be written uniquely in the form $n + i\langle I\rangle$, some $n \in \mathbb{Z}$ and $I \lhd R$, or, more explicitly, in the form $n[R] + ([I] - [R])$, and that the additive structure of K_0R is then given by

$$\begin{aligned}(n + i\langle I\rangle) + (m + i\langle J\rangle) &= (n + m) + (i\langle I\rangle + i\langle J\rangle)\\ &= (n + m) + i\langle IJ\rangle.\end{aligned}$$

Let us look at the multiplicative structure of K_0R. We have of course $[R] = 1$, and then for any $I, J \lhd R$, $I, J \neq 0$,

$$\begin{aligned} i\langle I\rangle i\langle J\rangle &= ([I] - [R])([J] - [R]) \\ &= [I][J] - [I] - [J] + [R] \\ &= [IJ \oplus R] - [I \oplus J] = 0. \end{aligned}$$

So the multiplication in K_0R is given by

$$\begin{aligned} (n + i\langle I\rangle)(m + i\langle J\rangle) &= nm + (n.i\langle J\rangle + m.i\langle I\rangle) \\ &= nm + i\langle J^n I^m\rangle \end{aligned}$$

where J^n means $JJJ\ldots J$ (n terms) if $n > 0$, R if $n = 0$, and $\bar{J}\bar{J}\bar{J}\ldots\bar{J}$ ($|n|$ terms) if $n < 0$, and similarly for I^m. In particular, if $n + i\langle I\rangle \in (K_0R)^{\bullet}$, then $n = \pm 1$. But then

$$(1 + i\langle I\rangle)(1 - i\langle I\rangle) = 1 = (-1 + i\langle I\rangle)(-1 - i\langle I\rangle)$$

any I, so $n + i\langle I\rangle \in (K_0R)^{\bullet}$ if and only if $n = \pm 1$. Now

$$1 + i\langle I\rangle = [R] + [I] - [R] = [I]$$

and

$$\begin{aligned} -1 + i\langle I\rangle &= -[R] + [I] - [R] = [I] - [R^2] \\ &= [I \oplus J] - [J \oplus R^2] \end{aligned}$$

(where $J \lhd R$ is chosen so that $IJ \simeq R$)

$$\begin{aligned} &= [IJ \oplus R] - [J \oplus R^2] = [R \oplus R] - [J \oplus R^2] \\ &= -[J]. \end{aligned}$$

Recall from Proposition 23, Corollary (vi), that there is a monomorphism of multiplicative groups $\mathrm{Pic}\, R \to (K_0R)^{\bullet}$ given by $\langle I\rangle \mapsto [I]$. By the above remarks, if R is a Dedekind domain, this monomorphism embeds $\mathrm{Pic}\, R$ as a subgroup of $(K_0R)^{\bullet}$ of index 2; identifying, we have in fact that $(K_0R)^{\bullet}$ is the direct sum of $\mathrm{Pic}\, R$ and the cyclic group of order 2, $\{\pm 1\}$.

4.1 *Examples: the calculation of $K_0\mathbb{Z}[\sqrt{-d}]$ for small d*

We shall now construct some Dedekind domains, and calculate their Grothendieck groups. Let

$$R = \mathbb{Z}[\sqrt{-d}] = \{n + m\sqrt{-d} : n, m \in \mathbb{Z}\}$$

where d is a fixed positive integer. R is a subring of the field of complex numbers, $\mathbb{C}$; it is an integral domain, but not in general a Dedekind domain. We shall look at small values of d to see when R is a Dedekind domain, and then calculate K_0R. We need to examine the ideals of R; the method we shall use is an attempt to generalize the familiar process for finding the ideals of $\mathbb{Z}$ or of $\mathbb{Z}[\sqrt{-1}]$. As we proceed, we place restrictions on the value of d as necessary to make the method work. Note that $|n + m\sqrt{-d}|^2 = n^2 + m^2d$ is a non-negative integer for $n, m \in \mathbb{Z}$, so if $X \subset R$, $X \neq \emptyset$, we can choose $x \in X$ with $|x|$ minimal.

Let $I \lhd R$, $I \neq 0$. Choose $a \in I$ with $|a|$ minimal, $a \neq 0$. Then $\mathbb{Z}a \subset I$ but $\mathbb{Z}a \neq I$ since $a\sqrt{-d} \in I - \mathbb{Z}a$. Thus we can choose $b \in I - \mathbb{Z}a$ with $|b|$ minimal. So $|b| \geq |a|$ and $|b \pm a| \geq |b|$.

For any $r \in R$, we have $r, r + 1, r + \sqrt{-d}, r + 1 + \sqrt{-d} \in R$, and these four points in the Argand diagram lie at the vertices of a rectangle whose diagonal has half-length $\frac{1}{2}\sqrt{(d + 1)}$. Every point of $\mathbb{C}$ lies in or on some such rectangle, so given $x \in \mathbb{C}$ we can find $r \in R$ with $|x - r| \leq \frac{1}{2}\sqrt{(d + 1)}$.

Case (i). $|b| > \frac{1}{2}|a|\sqrt{(d + 1)}$. Given $c \in I$ we can find $r \in R$ such that $|\frac{c}{a} - r| \leq \frac{1}{2}\sqrt{(d + 1)}$, or $|c - ar| \leq \frac{1}{2}|a|\sqrt{(d + 1)} < |b|$. But $c - ar \in I$, so by choice of b, $c - ar \in \mathbb{Z}a$, and so $c \in Ra$, and so I is principal, $I = Ra$.

Case (ii). $|b| \leq \frac{1}{2}|a|\sqrt{(d + 1)}$. If $d = 1$ or 2, this gives $|b| < |a|$, a contradiction. So if $d = 1$ or 2, R is a principal ideal domain, hence a Dedekind domain, and $K_0R \simeq \mathbb{Z}$. So now we assume $d \geq 3$. Put $z = \frac{b}{a} = x + iy$ $(x, y \in \mathbb{R})$. We have

$$|a| \leq |b| \leq \tfrac{1}{2}|a|\sqrt{(d + 1)}$$

whence

$$1 \leq x^2 + y^2 \leq \tfrac{1}{4}(d + 1). \qquad \ldots (1)$$

Then $|b \pm a| \geq |b|$, so $|z \pm 1| \geq |z|$, whence $|x| \leq \frac{1}{2}$. Without loss of generality, we may assume $x \geq 0$ (replace b by $-b$ if necessary), so we then have

$$0 \leq x \leq \tfrac{1}{2}. \qquad \ldots (2)$$

Now $2b - (1 \pm \sqrt{-d})a \in I$, so either $2b = (1 \pm \sqrt{-d})a$ or else

$|2b - (1 \pm \sqrt{-d})a| \geq |a|$, that is, $|2z - (1 \pm \sqrt{-d})| \geq 1$, or

$$(x - \tfrac{1}{2})^2 + (y \pm \tfrac{1}{2}\sqrt{d})^2 \geq \tfrac{1}{4}. \qquad \ldots (3)$$

If d is even and $2b = (1 \pm \sqrt{-d})a$, then $2(1 \mp \sqrt{-d})b = (1 - d^2)a$, so $(1 \mp \sqrt{-d})b = \frac{1}{2}(1 - d^2)a \in I$; but $a \in I$ and $1 - d^2$ is odd, so $\frac{1}{2}a \in I$, which is a contradiction, by the choice of a. So (3) holds when d is even. Next, $2b \pm (\sqrt{-d})a \in I$, so either $2b = \mp(\sqrt{-d})a$ or else $|2b \pm (\sqrt{-d})a| \geq |a|$, that is, $|2z \pm \sqrt{-d}| \geq 1$, or

$$x^2 + (y \pm \tfrac{1}{2}\sqrt{d})^2 \geq \tfrac{1}{4}. \qquad \ldots (4)$$

If d is odd and $2b = \mp(\sqrt{-d})a$, then $\pm(\sqrt{-d})b = \frac{1}{2}da \in I$; but $a \in I$, so $\frac{1}{2}a \in I$, which is a contradiction, by the choice of a. So (4) holds when d is odd.

We shall show that (1),(2),(3),(4) cannot hold simultaneously when d is small; it then follows that, for d even, $2b = \pm(\sqrt{-d})a$, and for d odd, $2b = (1 \pm \sqrt{-d})a$.

Note that, since $d \geq 1$, we have $4\sqrt{(3d)} > 1$, whence

$$\frac{d+1}{4} < \frac{3 + 4\sqrt{(3d)} + 4d}{16}$$

or

$$\tfrac{1}{2}\sqrt{(d + 1)} < \tfrac{1}{4}\sqrt{3} + \tfrac{1}{2}\sqrt{d}. \qquad \ldots (5)$$

Now $10 < 6\sqrt{5}$, so $28 < 18 + 6\sqrt{5} = (\sqrt{3} + \sqrt{15})^2$, whence

$$2\sqrt{d} < \sqrt{3} + \sqrt{15} \text{ if } d \leq 7. \qquad \ldots (6)$$

Also $3 < 2\sqrt{3}$, so $7 < 4 + 2\sqrt{3} = (1 + \sqrt{3})^2$, whence

$$\sqrt{d} < 1 + \sqrt{3} \text{ if } d \leq 7. \qquad \ldots (7)$$

(a) Suppose $y \geq \frac{1}{2}\sqrt{d}$. By (1),

$$y^2 \leq \tfrac{1}{4}(d + 1) - x^2 \leq \tfrac{1}{4}(d + 1)$$

so

$$\begin{aligned}\tfrac{1}{2}\sqrt{d} \leq y &\leq \tfrac{1}{2}\sqrt{(d + 1)} \\ &< \tfrac{1}{4}\sqrt{3} + \tfrac{1}{2}\sqrt{d} \text{ by (5)}\end{aligned}$$

and thus $0 \leq y - \frac{1}{2}\sqrt{d} < \frac{1}{4}\sqrt{3}$, or $(y - \frac{1}{2}\sqrt{d})^2 < 3/16$. Therefore:

(i) if $0 \leq x \leq \frac{1}{4}$, then $x^2 + (y - \frac{1}{2}\sqrt{d})^2 < \frac{1}{4}$, so (4) fails;

(ii) if $\frac{1}{4} \leq x \leq \frac{1}{2}$, then $(x - \frac{1}{2})^2 + (y - \frac{1}{2}\sqrt{d})^2 < \frac{1}{4}$, so (3) fails.

(b) Suppose $0 \leq y \leq \frac{1}{2}\sqrt{d}$, and suppose $d \leq 7$.

(i) If $0 \leq x \leq \frac{1}{4}$, then by (1), $y^2 \geq 1 - x^2 \geq 15/16$. So

$\frac{1}{2}\sqrt{d} \geq y \geq \frac{1}{4}\sqrt{15} > \frac{1}{2}\sqrt{d} - \frac{1}{4}\sqrt{3}$, by (6), and thus $0 \geq y - \frac{1}{2}\sqrt{d} > -\frac{1}{4}\sqrt{3}$, or $(y - \frac{1}{2}\sqrt{d})^2 < 3/16$, whence $x^2 + (y - \frac{1}{2}\sqrt{d})^2 < \frac{1}{4}$, and (4) fails.

(ii) For $\frac{1}{4} \leq x \leq \frac{1}{2}$, the curve $y = \sqrt{(1 - x^2)}$ lies strictly above the curve $y = \frac{1}{2}\sqrt{d} - \sqrt{(x - x^2)}$, since

$$F(x) = \sqrt{(1 - x^2)} + \sqrt{(x - x^2)}$$

is increasing on $[\frac{1}{4}, \frac{1}{3}]$ and decreasing on $[\frac{1}{3}, \frac{1}{2}]$, and furthermore $F(\frac{1}{4}) = \frac{1}{4}(\sqrt{15} + \sqrt{3}) > \frac{1}{2}\sqrt{d}$, by (6), and $F(\frac{1}{2}) = \frac{1}{2}(\sqrt{3} + 1) > \frac{1}{2}\sqrt{d}$, by (7). Since, by (1), $x^2 + y^2 \geq 1$, we have

$$\begin{aligned}\tfrac{1}{2}\sqrt{d} \geq y \geq \sqrt{(1 - x^2)} &> \tfrac{1}{2}\sqrt{d} - \sqrt{(x - x^2)} \\ &= \tfrac{1}{2}\sqrt{d} - \sqrt{(\tfrac{1}{4} - (x - \tfrac{1}{2})^2)}\end{aligned}$$

so

$$0 \geq y - \tfrac{1}{2}\sqrt{d} > -\sqrt{(\tfrac{1}{4} - (x - \tfrac{1}{2})^2)}$$

or

$$(x - \tfrac{1}{2})^2 + (y - \tfrac{1}{2}\sqrt{d})^2 < \tfrac{1}{4}$$

and (3) fails.

(c) If $y \leq 0$, similar arguments to (a) and (b) apply, using the other sign in conditions (3) and (4).

Thus when $d \leq 7$, conditions (1),(2),(3), and (4) cannot all hold, and we deduce

$$2b = \begin{cases} (1 \pm \sqrt{-d})a & (d = 3, 5, 7) \\ (\pm\sqrt{-d})a & (d = 4, 6). \end{cases}$$

Now suppose $c \in I$; then we can find $r \in R$ with $|\frac{c}{a} - r| \leq \frac{1}{2}\sqrt{(d + 1)}$. Then $c - ar \in I$, and $|c - ar| \leq \frac{1}{2}|a|\sqrt{(d + 1)}$. Repeating the above argument with $c - ar$ in place of b gives

$$2(c - ar) = \begin{cases} (1 \pm \sqrt{-d})a & (d = 3, 5, 7) \\ (\pm\sqrt{-d})a & (d = 4, 6) \end{cases}$$

so $c - ar = b$ or $b \pm (\sqrt{-d})a$, and in each case $c - ar \in aR + bR$, so $c \in aR + bR$. Thus $I = aR + bR$, and since

$$aR + \tfrac{1}{2}(1 + \sqrt{-d})aR = aR + \tfrac{1}{2}(1 - \sqrt{-d})aR$$

and

$$aR + \tfrac{1}{2}(\sqrt{-d})aR = aR - \tfrac{1}{2}(\sqrt{-d})aR$$

we deduce, dividing through by $\frac{1}{2}a$, that every non-principal ideal of R is isomorphic, as an R-module, to $2R + (1 + \sqrt{-d})R$ if $d = 3$, 5, or 7, or to $2R + (\sqrt{-d})R$ if $d = 4$ or 6.

Now put

$$I = \begin{cases} 2R + (1 + \sqrt{-d})R & (d = 3, 5, 7) \\ 2R + (\sqrt{-d})R & (d = 4, 6) \end{cases}$$

and note that

$$II = \begin{cases} 4R + 2(1 + \sqrt{-d})R + (1 - d + 2(\sqrt{-d}))R & (d = 3, 5, 7) \\ 4R + 2(\sqrt{-d})R + (-d)R & (d = 4, 6) \end{cases}$$

$$= \begin{cases} 4R + 2(1 + \sqrt{-d})R + (1 + d)R & (d = 3, 5, 7) \\ 4R + 2(\sqrt{-d})R + dR & (d = 4, 6) \end{cases}$$

$$= \begin{cases} 2I & (d = 3, 4, 7) \\ 2R & (d = 5, 6). \end{cases}$$

If $II = 2R$, then $I(\frac{1}{2}I) = R$, and it follows easily that I is invertible, with $\bar{I} = \frac{1}{2}I$. Since, in this case, I is not principal, we have shown that for $d = 5$ or 6, R is a Dedekind domain, with class number 2. So if $d = 5$ or 6, it follows that $K_0R \approx \mathbb{Z} \oplus (\mathbb{Z}/2\mathbb{Z})$, as an additive group, by Proposition 30. These are the first examples we have seen of rings whose Grothendieck groups are not free abelian groups (cf. Proposition 14).

If $d = 3$, 4, or 7, we have $II = 2I$; if I were invertible, we should have, in Pic R, $\langle I\rangle\langle I\rangle = \langle 2I\rangle = \langle I\rangle$, so $\langle I\rangle = 1$, or $\langle I\rangle = \langle R\rangle$, $I \approx R$. But I is not principal, so it cannot be invertible, and R is not a Dedekind domain. It is straightforward nonetheless to calculate K_0R in these cases. For if $a, b \in R$ with $2b = (1 + \sqrt{-d})a$ ($d = 3$ or 7) or $2b = (\sqrt{-d})a$ ($d = 4$), then it is easy to see that $a, b \in I$; details are left to the reader. So in each case the ideal I contains every non-principal ideal of R.

Now let M be a finitely generated projective R-module. We shall show that M is free. First we show that, if $M \neq 0$, then $M \neq IM$. For if $M = IM$, then $M = IM = IIM = 2IM = 2M$, so $M = 2^nM$, all n. Suppose $M \oplus N = R^r$, and $m \in M \subset R^r$; if $m = (a_1, \ldots, a_r)$ ($a_i \in R$, all i), then $m = 2^n m'$, some $m' \in M$. If $m' = (b_1, \ldots, b_r)$ ($b_i \in R$, all i), then $a_i = 2^n b_i$, each i, so if $a_i \neq 0$, then $b_i \neq 0$, so $|b_i| \geq 1$ and $|a_i| \geq 2^n$. But n is arbitrary, so this is impossible, and so $a_i = 0$, all i, and $m = 0$, and hence $M = 0$. So if $M \neq 0$, then $M \neq IM$, and $M/IM \neq 0$. Now M/IM is an (R/I)-module, and $R/I \simeq \mathbb{Z}/2\mathbb{Z}$ is a field. So M/IM is a non-zero free (R/I)-module, and so there is an (R/I)-module epimorphism $M/IM \twoheadrightarrow R/I$. It is easy to see that this map is in fact an R-module epimorphism, so combining with the natural R-module epimorphism $M \twoheadrightarrow M/IM$ we obtain an R-module epimorphism $M \twoheadrightarrow R/I$. Since M is projective, this epimorphism factors through the natural epimorphism $R \twoheadrightarrow R/I$ to give a homomorphism $M \to R$:

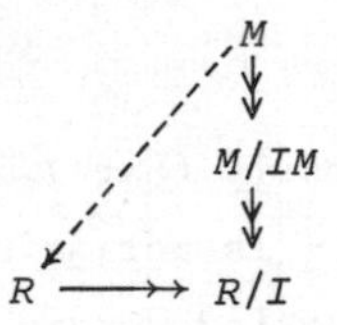

The image of this homomorphism $M \to R$ is an ideal J of R. Now $J \not\subset I$, otherwise the composite $M \to R \twoheadrightarrow R/I$ is zero, whereas we know that the composite is an epimorphism. Since $J \not\subset I$, J must be a principal ideal, so $J \simeq R$, and we obtain an epimorphism $M \twoheadrightarrow R$. This splits, since R is projective, and so $M \simeq R \oplus M'$, say, with $\text{rank}(M') = \text{rank}(M) - 1$. So we deduce that M is free, by induction on its rank. Therefore $K_0 R \simeq \mathbb{Z}$. Later, when we come to discuss exact sequences, we shall have an alternative method of making this calculation.

Now let d be a positive integer with $d \equiv 3 \bmod 4$. Then if $\alpha = \frac{1}{2}(1 + \sqrt{-d})$, we have $\alpha^2 = \alpha - \frac{1}{4}(1 + d)$, and $\frac{1}{4}(1 + d) \in \mathbb{Z}$, so $R = \{n + m\alpha : n, m \in \mathbb{Z}\}$ is a ring, $R = \mathbb{Z}[\alpha]$.

Exercise: With the above notation, show that for any $x \in \mathbb{C}$ there exists $r \in R$ with $|x - r| \leq (1 + d)/4\sqrt{d}$. Deduce that R is a principal ideal domain for $d = 3$, 7, or 11, and show that for $d = 15$, 19, or 23 every ideal of R is either principal or is isomorphic to $I = 2R + \alpha R$ or to $J = 2R + \bar{\alpha}R$, where $\bar{\alpha}$ is the complex conjugate of α.

In the above exercise, when $d = 19$, we have $\alpha - 5 = \alpha^2 \in I$, so $I = R$, and similarly $J = R$. So R is a principal ideal domain in this case also. Thus R is a Dedekind domain, with $K_0R \simeq \mathbb{Z}$, when $d = 3, 7, 11$, or 19.

Now let $d = 15$ or 23. It is easy to show that I, J are not principal in these cases; details are left to the reader. Then

$$\begin{aligned} IJ &= 4R + 2\alpha R + 2\bar{\alpha}R + \alpha\bar{\alpha}R \\ &= 4R + (1 + \sqrt{-d})R + (1 - \sqrt{-d})R + \tfrac{1}{4}(1 + d)R \\ &= 2R, \text{ since } \tfrac{1}{4}(1 + d) = 4 \text{ or } 6. \end{aligned}$$

Thus I, J are invertible, and once again R is a Dedekind domain.

Let $d = 15$. Then $\bar{\alpha}I = 2\bar{\alpha}R + \alpha\bar{\alpha}R = 2\bar{\alpha}R + 4R = 2J$, so $I \simeq J$, and the class number of R is 2. So $K_0R \simeq \mathbb{Z} \oplus (\mathbb{Z}/2\mathbb{Z})$.

Let $d = 23$. Then

$$II = 4R + 2\alpha R + \alpha^2 R = 4R + 2\alpha R + (\alpha - 6)R = 4R + (\alpha - 2)R$$

so

$$\begin{aligned} (\bar{\alpha} - 2)II &= (4\bar{\alpha} - 8)R + (\alpha\bar{\alpha} - 2(\alpha + \bar{\alpha}) + 4)R \\ &= (4\bar{\alpha} - 8)R + 8R = 4\bar{\alpha}R + 8R = 4J. \end{aligned}$$

Thus $II \simeq J$. We cannot have $I \simeq J$, otherwise in Pic R, $\langle I\rangle\langle I\rangle = \langle J\rangle = \langle I\rangle$, or $\langle I\rangle = 1$; but I is not principal. So the class number of R is 3, and $K_0R \simeq \mathbb{Z} \oplus (\mathbb{Z}/3\mathbb{Z})$.

In general, the ring $\mathbb{Z}[\sqrt{-d}]$ is a Dedekind domain provided d is square-free (that is, provided d is not divisible by the square of any prime number) and $d \not\equiv 3 \bmod 4$. If $d = n^2d_1$, where d_1 is square-free, then $\mathbb{Z}[\sqrt{-d}]$ is a proper subring of $\mathbb{Z}[\sqrt{-d_1}]$, which is a Dedekind domain if $d_1 \not\equiv 3 \bmod 4$. If $d \equiv 3 \bmod 4$ and d is square-free, then $\mathbb{Z}[\sqrt{-d}]$ is a proper subring of the ring $\mathbb{Z}[\frac{1}{2}(1 + \sqrt{-d})]$, which is a Dedekind domain. So $\mathbb{Z}[\sqrt{-d}]$ is always contained in a Dedekind domain R, which is itself a subring of the field $\mathbb{Q}[\sqrt{-d}]$, and R is in fact the ring of (algebraic) integers in $\mathbb{Q}[\sqrt{-d}]$, that is, the zeros of monic polynomials in $\mathbb{Z}[x]$ (x an indeterminate) which lie in $\mathbb{Q}[\sqrt{-d}]$. The proofs of these statements, and the calculation of the class numbers and class groups of the rings obtained, belongs to number theory: see, for example, H.Cohn, *A Second Course in Number Theory*, Wiley (1962).

CHAPTER FIVE

Categories

We introduce categories for two reasons. Firstly, categories provide a convenient language in which to summarize a lot of information: for example, the statement that, if $R \xrightarrow{f} S \xrightarrow{g} T$ are homomorphisms of rings, then $K_0R \xrightarrow{K_0f} K_0S \xrightarrow{K_0g} K_0T$ are homomorphisms of groups, with $(K_0g)(K_0f) = K_0(gf)$ and $K_01 = 1$, can be summarized by saying that K_0 is a functor from rings to groups. Secondly, and more importantly, we shall generalize the construction of the Grothendieck group K_0R to construct the Grothendieck group of certain categories (categories with product), in such a way that K_0R, where R is a ring, is just K_0 of the category of finitely generated projective R-modules, the product being the direct sum operation. We can then apply a loop construction to this category, and calculate K_0 of the new category (in an even more general sense) to obtain a group which we call K_1R, the Whitehead group of R. This makes the passage from K_0 to K_1 a natural process, and aids the construction (in Chapter 7) of certain exact sequences containing K_0 and K_1 terms. In Chapter 6, we obtain a more elementary construction of K_1R in terms of matrices; in the commutative case it becomes clear that the group of units $R^{\bullet}$ replaces the Picard group, and the determinant map becomes a map det : $K_1R \to R^{\bullet}$ which is related in an obvious way to the usual determinant map for matrices.

A *category* $\underline{\underline{C}}$ consists of elements called *objects* and elements called *morphisms*, satisfying certain axioms. We write $A \in$ ob $\underline{\underline{C}}$ to mean A is an object of $\underline{\underline{C}}$. For each A, $B \in$ ob $\underline{\underline{C}}$ there is a set $\underline{\underline{C}}(A, B)$ of morphisms *from* A *to* B; we write $f \in \underline{\underline{C}}(A, B)$, or $f : A \to B$ in $\underline{\underline{C}}$, to mean f is a morphism from A to B. Given $f : A \to B$ and $g : B \to C$ in $\underline{\underline{C}}$, there is a *composite* morphism $gf : A \to C$ in $\underline{\underline{C}}$. Composition of morphisms is associative, where it is defined, that is, if $f : A \to B$, $g : B \to C$, and $h : C \to D$ in $\underline{\underline{C}}$, then $(hg)f = h(gf)$. Finally, for each $A \in$ ob $\underline{\underline{C}}$, there is an *identity* morphism $1_A : A \to A$, such that for any $f : A \to B$ and any $g : C \to A$ in $\underline{\underline{C}}$ we have $f1_A = f$ and $1_A g = g$.

Notice that in the above definition we have avoided saying that ob $\underline{\underline{C}}$ is a set; often, ob $\underline{\underline{C}}$ will be too large to be considered a set in any formal sense. For instance, in Example 1 below, ob $\underline{\underline{C}}$ is the 'set' of all sets, a concept which, as is well known, leads to contradictions and paradoxes. However, we shall adopt a naive approach, and we assure the reader that this will not lead to logical difficulties in the sequel.

We give below examples of categories, and notation for some standard categories. In most of the examples, the objects are sets, usually with some additional structure, and the morphisms are structure-preserving set-theoretic maps. However, the objects and morphisms can be purely abstract entities, as in Example 6.

Examples. 1. $\underline{\underline{\text{Set}}}$, the category of sets and set-theoretic maps. Here A, $B \in$ ob $\underline{\underline{\text{Set}}}$ just means A, B are sets, and $f \in \underline{\underline{\text{Set}}}(A, B)$ means f is a map from A to B. Composition of morphisms is ordinary composition of maps.

2. $\underline{\underline{\text{Gp}}}$, the category of groups and group homomorphisms. Here G, $H \in$ ob $\underline{\underline{\text{Gp}}}$ means G, H are groups, and $f \in \underline{\underline{\text{Gp}}}(G, H)$ means f is a group homomorphism from G to H. Composition is as before.

3. $\underline{\underline{\text{Ab}}}$, the category of abelian groups and group homomorph-

isms.

4. $\underline{\underline{\text{Rg}}}$, the category of rings (with 1, of course) and ring homomorphisms (preserving 1).

5. $\underline{\underline{\text{CR}}}$, the category of commutative rings and ring homomorphisms.

6. Let G be a multiplicative group. Let $\underline{\underline{C}}$ be the category with just one object, denoted $*$, and with $\underline{\underline{C}}(*, *) = G$. That is, elements of G are morphisms from $*$ to $*$, composition being multiplication in G.

7. Let R be a ring. $\underline{\underline{M}}(R)$, or $\underline{\underline{M}}_\ell(R)$, is the category of (left) R-modules and R-module homomorphisms, and $\underline{\underline{P}}(R)$, or $\underline{\underline{P}}_\ell(R)$, is the category of finitely generated projective (left) R-modules and R-module homomorphisms. Similarly, $\underline{\underline{M}}_r(R)$, $\underline{\underline{P}}_r(R)$ are the corresponding categories of right R-modules.

8. Let R be a commutative ring. $\underline{\underline{\text{Pic}}}(R)$ is the category of invertible R-modules and R-module homomorphisms.

9. $\underline{\underline{\text{Top}}}$ is the category of topological spaces and continuous maps.

Let $\underline{\underline{C}}$, $\underline{\underline{D}}$ be categories. A *functor*, or *covariant functor*, $F : \underline{\underline{C}} \to \underline{\underline{D}}$, consists of a map $F : \text{ob } \underline{\underline{C}} \to \text{ob } \underline{\underline{D}}$, and maps $F : \underline{\underline{C}}(A, B) \to \underline{\underline{D}}(F(A), F(B))$, all $A, B \in \text{ob } \underline{\underline{C}}$, such that $F(1_A) = 1_{F(A)}$, all $A \in \text{ob } \underline{\underline{C}}$, and if $f : A \to B$, $g : B \to C$ in $\underline{\underline{C}}$, then $F(gf) = F(g)F(f)$. A *contravariant functor* $F : \underline{\underline{C}} \to \underline{\underline{D}}$ is defined similarly, except that F reverses the direction of morphisms, that is, $F : \underline{\underline{C}}(A, B) \to \underline{\underline{D}}(F(B), F(A))$, all $A, B \in \text{ob } \underline{\underline{C}}$, and if $f : A \to B$, $g : B \to C$ in $\underline{\underline{C}}$, then $F(gf) = F(f)F(g)$.

Examples. 1. The *forgetful functor* $F : \underline{\underline{\text{Gp}}} \to \underline{\underline{\text{Set}}}$, as its name implies, forgets the algebraic structure: for $G \in \text{ob } \underline{\underline{\text{Gp}}}$, $F(G) = G$ (as a set), and for $f \in \underline{\underline{\text{Gp}}}(G, H)$, $F(f) = f$ (as a set-theoretic map). Forgetful functors can also be defined on $\underline{\underline{\text{Ab}}}$, $\underline{\underline{\text{Rg}}}$, etc., and indeed on any category in which the objects are sets and the morphisms are maps with ordinary map composition.

2. The *abelianizing functor ab* : $\underline{\underline{Gp}} \to \underline{\underline{Ab}}$, given by $G \mapsto G^{ab} = G/G'$, where G' is the derived, or commutator, subgroup of G, and if $f : G \to H$ in $\underline{\underline{Gp}}$, then $f \mapsto f^{ab}$, where $f^{ab} : G^{ab} \to H^{ab}$ is the homomorphism induced by f.

3. Let G, H be groups, and let $\underline{\underline{C}}$, $\underline{\underline{D}}$ be the corresponding one-object categories, as in Example 6 above. If $f : G \to H$ is a homomorphism of groups, then it is a functor $\underline{\underline{C}} \to \underline{\underline{D}}$ in an obvious way.

4. Let R, S be rings, and let $f : R \to S$ be a ring homomorphism. There is a functor $F : \underline{\underline{M}}(R) \to \underline{\underline{M}}(S)$ given by $F(M) = S \otimes_f M$, $M \in \text{ob } \underline{\underline{M}}(R)$, and $F(g) = 1 \otimes g : S \otimes_f M \to S \otimes_f N$, where $g : M \to N$ in $\underline{\underline{M}}(R)$. By restriction, F is also a functor $\underline{\underline{P}}(R) \to \underline{\underline{P}}(S)$, and if R, S are commutative then by restricting again, F is a functor $\underline{\underline{Pic}}(R) \to \underline{\underline{Pic}}(S)$.

5. $K_0 : \underline{\underline{Rg}} \to \underline{\underline{Ab}}$ is a functor; that is, $K_0 : R \mapsto K_0R$, $R \in \text{ob } \underline{\underline{Rg}}$, and $K_0 : f \mapsto K_0f : K_0R \to K_0S$, where $f : R \to S$ in $\underline{\underline{Rg}}$, and K_0 respects composition of homomorphisms and preserves identity maps. K_0 is also a functor $\underline{\underline{CR}} \to \underline{\underline{CR}}$, using the fact that K_0R can be given the structure of a commutative ring when R is commutative.

6. Pic : $\underline{\underline{CR}} \to \underline{\underline{Ab}}$ is a functor, with Pic : $R \mapsto \text{Pic } R$, etc.

7. Spec : $\underline{\underline{CR}} \to \underline{\underline{Top}}$ is a contravariant functor, with Spec : $R \mapsto \text{Spec } R$, $R \in \text{ob } \underline{\underline{CR}}$, and Spec : $f \mapsto f_*$ (recall that $f : R \to S$ induces $f_* : \text{Spec } S \to \text{Spec } R$).

8. Let R be a ring. If M is a left R-module, then the dual $M^* = \text{Hom}_R(M, R)$ is a right R-module, and if $f : M \to N$ is a homomorphism of left R-modules, there is an induced homomorphism $f^* : N^* \to M^*$ given by $g \mapsto gf$ ($g \in N^*$). It is easy to see that $* : \underline{\underline{M}}_\ell(R) \to \underline{\underline{M}}_r(R)$ is a contravariant functor. By restriction, $*$ is a contravariant functor $\underline{\underline{P}}_\ell(R) \to \underline{\underline{P}}_r(R)$. In a similar way, if M is a right R-module, the dual $M^* = \text{Hom}_R(M, R)$ is a left R-module, and we obtain a contravariant functor $* : \underline{\underline{M}}_r(R) \to \underline{\underline{M}}_\ell(R)$. It is clear that the composite functor $** : \underline{\underline{M}}_\ell(R) \to \underline{\underline{M}}_\ell(R)$ is

covariant.

We can now give a further example of a category: $\underline{\underline{\text{Cat}}}$ is the category whose objects are categories and whose morphisms are (covariant) functors.

Let $\underline{\underline{C}}$, $\underline{\underline{D}}$ be categories, and let F, $G : \underline{\underline{C}} \to \underline{\underline{D}}$ be functors. A *natural transformation* $\alpha : F \to G$ is a collection of morphisms $\alpha : F(A) \to G(A)$ in $\underline{\underline{D}}$, one for each $A \in \text{ob } \underline{\underline{C}}$, such that for any $f : A \to B$ in $\underline{\underline{C}}$, the diagram

$$\begin{array}{ccc} F(A) & \xrightarrow{F(f)} & F(B) \\ \alpha \downarrow & & \downarrow \alpha \\ G(A) & \xrightarrow{G(f)} & G(B) \end{array}$$

commutes, that is, $(G(f))(\alpha) = (\alpha)(F(f))$.

Examples. 1. K_0 and Pic can both be regarded as functors $\underline{\underline{\text{CR}}} \to \underline{\underline{\text{Ab}}}$, and then det : $K_0 \to$ Pic is a natural transformation (see the second exercise on p. 59).

2. Let R be a ring, and let M be an R-module. If $m \in M$, define $\bar{m} \in M^{**}$ by $\bar{m}(\theta) = \theta(m)$, all $\theta \in M^*$. Then if $\alpha : M \to M^{**}$ is defined by $\alpha(m) = \bar{m}$, α is a homomorphism, and we leave it to the reader to show that α, thus defined, is a natural transformation from the identity functor $1 : \underline{\underline{M}}_\ell(R) \to \underline{\underline{M}}_\ell(R)$ to the functor $** :$ $\underline{\underline{M}}_\ell(R) \to \underline{\underline{M}}_\ell(R)$.

Let $\underline{\underline{C}}$ be a category. A morphism $f : A \to B$ in $\underline{\underline{C}}$ is an *isomorphism* if there exists a morphism $g : B \to A$ in $\underline{\underline{C}}$ such that $gf = 1_A$ and $fg = 1_B$; we then say that the objects A, B are *isomorphic*, and write $A \simeq B$.

Let $\underline{\underline{C}}$, $\underline{\underline{D}}$ be categories, let F, $G : \underline{\underline{C}} \to \underline{\underline{D}}$ be functors, and let $\alpha : F \to G$ be a natural transformation. Then α is a *natural equivalence* if $\alpha : F(A) \to G(A)$ is an isomorphism for every $A \in$ ob $\underline{\underline{C}}$. We then say that the functors F, G are *naturally equivalent*.

Exercise: Let R be a ring. Use the homomorphisms constructed in Example 2 above to show that the functors 1, $** : \underline{\underline{P}}_\ell(R) \to \underline{\underline{P}}_\ell(R)$ are naturally equivalent.

Let $\underline{\underline{C}}$, $\underline{\underline{D}}$ be categories, and let $F : \underline{\underline{C}} \to \underline{\underline{D}}$ be a functor. We say F is an *equivalence* if there is a functor $G : \underline{\underline{D}} \to \underline{\underline{C}}$ such that GF, $1 : \underline{\underline{C}} \to \underline{\underline{C}}$ are naturally equivalent and FG, $1 : \underline{\underline{D}} \to \underline{\underline{D}}$ are naturally equivalent. We then say that $\underline{\underline{C}}$, $\underline{\underline{D}}$ are *equivalent*. It seems at first sight that the above exercise, together with the corresponding result for $\underline{\underline{P}}_r(R)$, shows that $\underline{\underline{P}}_\ell(R)$ and $\underline{\underline{P}}_r(R)$ are equivalent, but this is only true in a generalized sense, since the functors $*$ are contravariant, not covariant.

Let $\underline{\underline{C}}$, $\underline{\underline{D}}$ be categories. We can define the *cartesian product category* $\underline{\underline{C}} \times \underline{\underline{D}}$ in the obvious way: the objects are pairs (A, B), $A \in \text{ob } \underline{\underline{C}}$, $B \in \text{ob } \underline{\underline{D}}$, and then $(\underline{\underline{C}} \times \underline{\underline{D}})((A, B), (A', B')) = \underline{\underline{C}}(A, A') \times \underline{\underline{D}}(B, B')$, with componentwise composition.

5.1 *Categories with product*

We now wish to introduce into a category a binary operation which generalizes the notions of direct sum $\oplus$ and tensor product $\otimes$ for modules, which were used to construct the groups K_0R and Pic R. Now it is not true that the modules $(M_1 \oplus M_2) \oplus M_3$ and $M_1 \oplus (M_2 \oplus M_3)$ are equal, but there is a natural isomorphism between them, and for most practical purposes one does not need to distinguish between them; similar remarks apply to $M_1 \oplus M_2$ and $M_2 \oplus M_1$, and to $M_1 \oplus 0$, $0 \oplus M_1$, and M_1. Recall that homomorphisms can also be combined by $\oplus$ and $\otimes$, and in fact these operations are functors from $\underline{\underline{P}}(R) \times \underline{\underline{P}}(R)$ to $\underline{\underline{P}}(R)$ (in the latter case, provided R is commutative). We generalize all this as follows.

A *category with product* $(\underline{\underline{C}}, \perp, \theta, \phi, 0, \lambda, \mu)$, or $(\underline{\underline{C}}, \perp)$ for short, consists of a category $\underline{\underline{C}}$ together with a functor $\perp : \underline{\underline{C}} \times \underline{\underline{C}} \to \underline{\underline{C}}$ which is *coherently associative and commutative*, that is:

(i) The functors $\perp(1 \times \perp)$, $\perp(\perp \times 1) : \underline{\underline{C}} \times \underline{\underline{C}} \times \underline{\underline{C}} \to \underline{\underline{C}}$ are naturally equivalent, that is, for any A, B, $C \in$ ob $\underline{\underline{C}}$ there is an isomorphism $\theta : A \perp (B \perp C) \to (A \perp B) \perp C$ such that, if $\alpha : A \to A'$, $\beta : B \to B'$, $\gamma : C \to C'$ in $\underline{\underline{C}}$, then the diagram

$$\begin{array}{ccc} A \perp (B \perp C) & \xrightarrow{\alpha \perp (\beta \perp \gamma)} & A' \perp (B' \perp C') \\ \theta \downarrow & & \downarrow \theta \\ (A \perp B) \perp C & \xrightarrow{(\alpha \perp \beta) \perp \gamma} & (A' \perp B') \perp C' \end{array}$$

commutes.

(ii) If $T : \underline{\underline{C}} \times \underline{\underline{C}} \to \underline{\underline{C}} \times \underline{\underline{C}}$ is given by $(A, B) \mapsto (B, A)$, etc., then the functors $\perp T$, $\perp : \underline{\underline{C}} \times \underline{\underline{C}} \to \underline{\underline{C}}$ are naturally equivalent, that is, for any A, $B \in$ ob $\underline{\underline{C}}$ there is an isomorphism $\phi : B \perp A \to A \perp B$ such that, if $\alpha : A \to A'$ and $\beta : B \to B'$ in $\underline{\underline{C}}$, then the diagram

$$\begin{array}{ccc} B \perp A & \xrightarrow{\beta \perp \alpha} & B' \perp A' \\ \phi \downarrow & & \downarrow \phi \\ A \perp B & \xrightarrow{\alpha \perp \beta} & A' \perp B' \end{array}$$

commutes.

(iii) θ and ϕ are compatible, in the sense that the isomorphisms of two products of several factors obtained one from the other by successive reassociations and permutations are all the same. Explicitly, we require that the diagrams

$$\begin{array}{ccc} A \perp (B \perp (C \perp D)) & \xrightarrow{1 \perp \theta} & A \perp ((B \perp C) \perp D) \\ \theta \downarrow & & \downarrow \theta \\ (A \perp B) \perp (C \perp D) & & (A \perp (B \perp C)) \perp D \\ \theta \downarrow & & \downarrow \theta \perp 1 \\ ((A \perp B) \perp C) \perp D & \xrightarrow{1} & ((A \perp B) \perp C) \perp D \end{array}$$

and

$$\begin{array}{ccc} A \perp B & \xrightarrow{\phi} & B \perp A \\ & {\scriptstyle 1} \searrow & \downarrow \phi \\ & & A \perp B \end{array}$$

and

$$\begin{array}{ccccc} A \perp (B \perp C) & \xrightarrow{1 \perp \phi} & A \perp (C \perp B) & \xrightarrow{\phi} & (C \perp B) \perp A \\ \theta \downarrow & & & & \uparrow \theta \\ (A \perp B) \perp C & \xrightarrow{\phi} & C \perp (A \perp B) & \xrightarrow{1 \perp \phi} & C \perp (B \perp A) \end{array}$$

should commute for all A, B, C, $D \in$ ob $\underline{\underline{C}}$.

In order to avoid having an infinite list of types of diagram to check for commutativity, it is convenient technically to require our category to have a neutral object 0 with the following properties: firstly, the functors $\cdot \perp 0 : A \mapsto A \perp 0$, $\alpha \mapsto \alpha \perp 1_0$, and $0 \perp \cdot : A \mapsto 0 \perp A$, $\alpha \mapsto 1_0 \perp \alpha$ are both naturally equivalent to the identity functor on $\underline{\underline{C}}$, that is, there are isomorphisms $\lambda : A \perp 0 \to A$ and $\mu : 0 \perp A \to A$ such that if $\alpha : A \to B$ in $\underline{\underline{C}}$, then the diagrams

$$\begin{array}{ccc} A \perp 0 & \xrightarrow{\alpha \perp 1} & B \perp 0 \\ \lambda \downarrow & & \downarrow \lambda \\ A & \xrightarrow{\alpha} & B \end{array} \quad \text{and} \quad \begin{array}{ccc} 0 \perp A & \xrightarrow{1 \perp \alpha} & 0 \perp B \\ \mu \downarrow & & \downarrow \mu \\ A & \xrightarrow{\alpha} & B \end{array}$$

commute. Then, when $A = 0$, $\lambda = \mu : 0 \perp 0 \to 0$. Further, λ and μ are compatible with θ and ϕ, that is, the diagrams

$$\begin{array}{ccc} A \perp (0 \perp B) & \xrightarrow{\theta} & (A \perp 0) \perp B \\ 1 \perp \lambda \downarrow & & \downarrow \mu \perp 1 \\ A \perp B & \xrightarrow{1} & A \perp B \end{array} \quad \text{and} \quad \begin{array}{ccc} 0 \perp A & \xrightarrow{\phi} & A \perp 0 \\ \lambda \downarrow & & \downarrow \mu \\ A & \xrightarrow{1} & A \end{array}$$

must commute for all A, $B \in$ ob $\underline{\underline{C}}$. It can be shown that the commutativity of the diagrams we have given is sufficient to ensure that all such diagrams commute. That is, we can write a multiple product as $\perp_{i=1}^{n} A_i$, without specifying the order of the terms or the bracketing, unambiguously up to a uniquely determined isomorphism. We shall not prove this here, but instead refer the reader to [9], Chapter VII, where such categories are called *symmetric monoidal categories*.

Let $(\underline{\underline{C}}, \perp)$, $(\underline{\underline{C}}', \perp')$ be categories with product. A *product-preserving functor* $F : (\underline{\underline{C}}, \perp) \to (\underline{\underline{C}}', \perp')$ is a functor $F : \underline{\underline{C}} \to \underline{\underline{C}}'$

together with a natural equivalence $\psi : F\perp \to \perp'(F \times F)$. That is, for each $A, B \in \text{ob } \underline{\underline{C}}$ there is an isomorphism $\psi : F(A \perp B) \to F(A) \perp' F(B)$ such that if $\alpha : A \to A_0$ and $\beta : B \to B_0$ in $\underline{\underline{C}}$, the diagram

$$\begin{array}{ccc} F(A \perp B) & \xrightarrow{F(\alpha \perp \beta)} & F(A_0 \perp B_0) \\ \psi \downarrow & & \downarrow \psi \\ F(A) \perp' F(B) & \xrightarrow{F(\alpha) \perp' F(\beta)} & F(A_0) \perp' F(B_0) \end{array}$$

commutes. Further, F is required to be compatible with θ, ϕ in the sense that the diagrams

$$\begin{array}{ccc} F(A \perp (B \perp C)) & \xrightarrow{F(\theta)} & F((A \perp B) \perp C) \\ \psi \downarrow & & \downarrow \psi \\ F(A) \perp' F(B \perp C) & & F(A \perp B) \perp' F(C) \\ 1 \perp' \psi \downarrow & & \downarrow \psi \perp' 1 \\ F(A) \perp' (F(B) \perp' F(C)) & \xrightarrow{\theta'} & (F(A) \perp' F(B)) \perp' F(C) \end{array}$$

and

$$\begin{array}{ccc} F(B \perp A) & \xrightarrow{F(\phi)} & F(A \perp B) \\ \psi \downarrow & & \downarrow \psi \\ F(B) \perp' F(A) & \xrightarrow{\phi'} & F(A) \perp' F(B) \end{array}$$

both commute for all $A, B, C \in \text{ob } \underline{\underline{C}}$. We also require that $F(0) = 0'$, and that the diagrams

$$\begin{array}{ccc} F(0 \perp A) & \xrightarrow{F(\lambda)} & F(A) \\ \psi \downarrow & & \downarrow 1 \\ 0' \perp' F(A) & \xrightarrow{\lambda'} & F(A) \end{array} \quad \text{and} \quad \begin{array}{ccc} F(A \perp 0) & \xrightarrow{F(\mu)} & F(A) \\ \psi \downarrow & & \downarrow 1 \\ F(A) \perp' 0' & \xrightarrow{\mu'} & F(A) \end{array}$$

commute for all $A \in \text{ob } \underline{\underline{C}}$.

We leave it to the reader to draw up a suitable definition of a *contravariant* product-preserving functor.

Examples. 1. Let R be a ring. Then we may take $\underline{\underline{C}} = \underline{\underline{P}}(R)$ and $\perp = \oplus$; that is, $(\underline{\underline{P}}(R), \oplus)$ is a category with product; the neutral object 0 is the zero module. Verification that all the conditions are satisfied is straightforward but tedious, and we leave the details to the reader.

2. Let R be a commutative ring. Then $(\underline{\underline{\mathrm{Pic}}}(R), \otimes_R)$ is a category with product; this time, the neutral object is R. Again, details are left to the reader.

3. Let R, S be rings, and let $f : R \to S$ be a ring homomorphism. Then as in Example 4, p. 81, we can use the tensor product to construct a functor $F : \underline{\underline{P}}(R) \to \underline{\underline{P}}(S)$. It is easy to see that F is a product-preserving functor $(\underline{\underline{P}}(R), \oplus) \to (\underline{\underline{P}}(S), \oplus)$, and by restriction F is also a product-preserving functor $(\underline{\underline{\mathrm{Pic}}}(R), \otimes_R) \to (\underline{\underline{\mathrm{Pic}}}(S), \otimes_S)$, when R, S are both commutative.

4. Let R be a commutative ring. Then the functor det : $\underline{\underline{P}}(R) \to \underline{\underline{\mathrm{Pic}}}(R)$ is a product-preserving functor $(\underline{\underline{P}}(R), \oplus) \to (\underline{\underline{\mathrm{Pic}}}(R), \otimes_R)$.

Let $\underline{\underline{C}}$, $\underline{\underline{D}}$ be categories. $\underline{\underline{D}}$ is a *subcategory* of $\underline{\underline{C}}$ if $A \in \mathrm{ob}\, \underline{\underline{D}} \Rightarrow A \in \mathrm{ob}\, \underline{\underline{C}}$, $\underline{\underline{D}}(A, B) \subset \underline{\underline{C}}(A, B)$, all $A, B \in \mathrm{ob}\, \underline{\underline{D}}$, and if the composition of morphisms in $\underline{\underline{D}}$ is the same as in $\underline{\underline{C}}$. The subcategory $\underline{\underline{D}}$ of $\underline{\underline{C}}$ is *full* if $\underline{\underline{D}}(A, B) = \underline{\underline{C}}(A, B)$, all $A, B \in \mathrm{ob}\, \underline{\underline{D}}$. $\underline{\underline{D}}$ is a *skeletal* subcategory of $\underline{\underline{C}}$ if it is full and also, for each $A \in \mathrm{ob}\, \underline{\underline{C}}$, there is a unique $B \in \mathrm{ob}\, \underline{\underline{D}}$ with $A \cong B$ in $\underline{\underline{C}}$ (that is, $\underline{\underline{D}}$ contains precisely one object from each isomorphism class of objects in $\underline{\underline{C}}$). It is clear that, by the axiom of choice, every category has a skeletal subcategory. The category $\underline{\underline{C}}$ is *small* if $\mathrm{ob}\, \underline{\underline{C}}$ is a set.

Let $(\underline{\underline{C}}, \perp)$ be a category with product such that $\underline{\underline{C}}$ has a small skeletal subcategory. We define the *Grothendieck group* $K_0(\underline{\underline{C}}, \perp)$, or $K_0\underline{\underline{C}}$ for short, as the abelian group given by the following generators and relations: we take one generator $[A]$ for each isomorphism class of objects A in $\underline{\underline{C}}$ (so $[A] = [B]$ if $A \cong B$) and one relation $[A] + [B] = [A \perp B]$ for each pair $A, B \in \mathrm{ob}\, \underline{\underline{C}}$.

Examples. 1. Let R be a ring. If the cardinal of the underlying set of R is λ, then R^n has cardinal λ^n, and so the number of direct summands of R^n cannot exceed 2^{λ^n}. It is now clear that $\underline{\underline{P}}(R)$ has a small skeletal subcategory, and then $K_0(\underline{\underline{P}}(R), \oplus) \simeq K_0 R$.

2. Let R be a commutative ring. Then $K_0(\underline{\underline{Pic}}(R), \otimes_R) \simeq$ Pic R.

Any product-preserving functor $f : (\underline{\underline{C}}, \perp) \to (\underline{\underline{C}}', \perp')$ induces a group homomorphism $K_0 f : K_0\underline{\underline{C}} \to K_0\underline{\underline{C}}'$ by the rule $[A] \mapsto [f(A)]$. It is left to the reader to show that this is well defined and induces a homomorphism, making K_0 into a functor from the category of categories with product and small skeletal subcategory, and product-preserving functors, to $\underline{\underline{Ab}}$.

Examples. 1. Let R, S be rings, and let $\alpha : R \to S$ be a ring homomorphism. Then $f = S \otimes_\alpha \cdot : \underline{\underline{P}}(R) \to \underline{\underline{P}}(S)$ induces $K_0 f :$ $K_0\underline{\underline{P}}(R) \to K_0\underline{\underline{P}}(S)$, which is the map we have previously called $K_0\alpha : K_0 R \to K_0 S$. When R, S are commutative, the restriction $f :$ $\underline{\underline{Pic}}(R) \to \underline{\underline{Pic}}(S)$ induces $K_0 f : K_0\underline{\underline{Pic}}(R) \to K_0\underline{\underline{Pic}}(S)$, which is the map we have previously called Pic α : Pic $R \to$ Pic S.

2. Let R be a commutative ring. Then det : $\underline{\underline{P}}(R) \to \underline{\underline{Pic}}(R)$ induces $K_0\det : K_0\underline{\underline{P}}(R) \to K_0\underline{\underline{Pic}}(R)$, which is just the map det : $K_0 R \to$ Pic R.

3. Let R be a ring. Then $M \mapsto M^*$ is a product-preserving contravariant functor $\underline{\underline{P}}_\ell(R) \xrightarrow{*} \underline{\underline{P}}_r(R)$ and so induces a homomorphism $K_0{*} : K_0\underline{\underline{P}}_\ell(R) \to K_0\underline{\underline{P}}_r(R)$. Similarly, $\underline{\underline{P}}_r(R) \xrightarrow{*} \underline{\underline{P}}_\ell(R)$ induces $K_0{*} : K_0\underline{\underline{P}}_r(R) \to K_0\underline{\underline{P}}_\ell(R)$. The composite $(K_0{*})(K_0{*}) =$ $K_0(**) : K_0\underline{\underline{P}}_\ell(R) \to K_0\underline{\underline{P}}_\ell(R)$ is given by $[M] \mapsto [M^{**}]$. But $1, ** :$ $\underline{\underline{P}}_\ell(R) \to \underline{\underline{P}}_\ell(R)$ are naturally equivalent (exercise, p. 83), so $M \simeq M^{**}$, or $[M] = [M^{**}]$, and so $K_0(**)$ is the identity homomorphism on $K_0\underline{\underline{P}}_\ell(R)$; similarly on $K_0\underline{\underline{P}}_r(R)$. Thus $K_0{*} : K_0\underline{\underline{P}}_\ell(R) \to$ $K_0\underline{\underline{P}}_r(R)$ is an isomorphism, and so it does not matter whether we set up our K-theory in terms of left modules or in terms of right modules; the groups obtained will be the same, up to isomorphism.

4. (Sketch of connection with topological K-theory.) Let X be a compact Hausdorff topological space. The set of all vector bundles $p : E \to X$ forms a category with product, $(\underline{\underline{\text{Vect}}}(X), \oplus)$, and topological K-theory is concerned with the group $K_0\underline{\underline{\text{Vect}}}(X)$. If the vector spaces used are complex, the group obtained is called $K(X)$, or $K_{\mathbb{C}}(X)$, and if they are real, the group is called $KO(X)$, or $K_{\mathbb{R}}(X)$. For definitions and general details, see [7]; we shall give one concrete example.

The circle, or 1-sphere, S^1, is obtained from the real unit interval $[0, 1]$ by glueing the ends together, that is, by identifying the points 0 and 1. Then the strip $[0, 1] \times \mathbb{R}$, with its edges glued by identifying $(0, x)$ with $(1, x)$, all $x \in \mathbb{R}$, becomes a cylinder, $S^1 \times \mathbb{R}$, and this is a (trivial) real vector bundle over S^1. A more interesting example of a real vector bundle over S^1 is obtained by glueing the edges of $[0, 1] \times \mathbb{R}$ with a twist, by identifying $(0, x)$ with $(1, -x)$, all $x \in \mathbb{R}$; we then obtain a Möbius band. In each case note that the bundle projection $p : E \to S^1$ is induced by $p : [0, 1] \times \mathbb{R} \to [0, 1]$ (projection to the first component), and each fibre $p^{-1}(x)$ ($x \in [0, 1]$) is a vector space (in fact, just $\mathbb{R}$, or rather $\{x\} \times \mathbb{R}$).

Let R be the ring of all continuous functions $S^1 \to \mathbb{R}$, that is, all continuous functions $f : [0, 1] \to \mathbb{R}$ with $f(0) = f(1)$, with addition and multiplication defined pointwise. Now given a vector bundle $p : E \to S^1$, a *section* of E is a continuous function $s : S^1 \to E$ with $ps = 1_{S^1}$. The set $\Gamma(E)$ of all sections of E has the structure of an R-module if we define addition and scalar multiplication pointwise, that is, $(s + t)(x) = s(x) + t(x)$ and $(fs)(x) = f(x)s(x)$, all $s, t \in \Gamma(E)$, $f \in R$, $x \in S^1$. Note that $s(x), t(x) \in p^{-1}(x)$, which is a real vector space, and $f(x) \in \mathbb{R}$, so the definitions make sense. For example, a section of the cylinder $S^1 \times \mathbb{R}$ is a continuous map $s : S^1 \to S^1 \times \mathbb{R}$, $x \mapsto (x', x'')$ say, and since $ps = 1$, we have $x' = x$. The section thus determines and is determined by a map $S^1 \to \mathbb{R}$, $x \mapsto x''$, i.e., an element of R, and we deduce that

$\Gamma(S^1 \times \mathrm{R}) \simeq R$ (as R-modules).

Now let E be the Möbius band, with bundle projection $p : E \to S^1$. A similar argument to the above shows that $M = \Gamma(E)$ may be regarded as the set of all continuous functions $s : [0, 1] \to \mathrm{R}$ with $s(0) = -s(1)$. Note that every such function must take the value zero somewhere, by the intermediate value theorem. We shall show that the module M is projective of rank 1, but not free.

Let $q : F \to S^1$ be a second Möbius band. The product bundle $E \oplus F$ is defined by $E \oplus F = \{(e, f) \in E \times F : p(e) = q(f)\}$ with the obvious projection to S^1, and this can be regarded as the bundle obtained from $[0, 1] \times \mathrm{R}^2$ by identifying $(0, y, z)$ with $(1, -y, -z)$, all $(y, z) \in \mathrm{R}^2$, the two Möbius bands being embedded as $z = 0$ and $y = 0$, respectively. Now it is easily seen that the two sections $x \mapsto (x, \cos(x/\pi), \sin(x/\pi))$ and $x \mapsto (x, -\sin(x/\pi), \cos(x/\pi))$ form a free basis for $\Gamma(E \oplus F)$ (for each x, the second and third entries give a basis of R^2, so any section can be expressed uniquely in terms of this basis at each x), so that $M \oplus M \simeq \Gamma(E) \oplus \Gamma(F) \simeq \Gamma(E \oplus F) \simeq R^2$. So M is projective of rank 1. If M is free, then $M \simeq R$, since $\mathrm{rank}(M) = 1$, but this cannot happen, since every $s \in M$ has a zero somewhere, and thus every scalar multiple fs $(f \in R)$ vanishes at the same point. So $\{s\}$ cannot be a basis, and M is not free.

In fact, every real vector bundle E over S^1 gives rise to a finitely generated projective R-module $\Gamma(E)$, and every finitely generated projective R-module arises in this way; indeed, Γ induces an isomorphism $K_{\mathrm{R}}(S^1) \simeq K_0R$. There is even a tensor product construction for vector bundles which makes $K_{\mathrm{R}}(S^1)$ into a ring isomorphic to the ring K_0R. All this works similarly for real vector bundles over any compact Hausdorff topological space X, and it works for complex bundles also, giving $K_{\mathrm{C}}(X) \simeq K_0R$, where of course the corresponding ring R is the ring of all continuous functions $X \to \mathrm{C}$. For details, see [7], [15].

Exercise: Show that any real vector bundle E over S^1 may be regarded as $[0, 1] \times \mathbb{R}^n$, with $(0, x)$ identified with $(1, Ax)$, all $x \in \mathbb{R}^n$, where $A \in GL_n(\mathbb{R})$ is fixed. Deduce that, if $\det A > 0$, then E is isomorphic to the (bundle) product of n cylinders, and $\Gamma(E) \simeq R^n$ (where R = real continuous functions on S^1), and if $\det A < 0$, then E is isomorphic to the (bundle) product of a Möbius band and $n - 1$ cylinders, and $\Gamma(E) \simeq M \oplus R^{n-1}$ (where M = sections of a Möbius band). (Hint: if $\det A > 0$, there is a path in $GL_n(\mathbb{R})$ from the identity matrix to A; this path is a matrix of functions $[0, 1] \to \mathbb{R}$, and its columns are paths in $\mathbb{R}^n$, in fact, sections of E, and are a free basis of $\Gamma(E)$. If $\det A < 0$, and B is the matrix obtained from A by multiplying the first column by -1, then $\det B > 0$.) Deduce that $K_0R \simeq \mathbb{Z} \oplus \mathbb{Z}/2\mathbb{Z}$, generated by $[R]$ and $[R] - [M]$ respectively. What is the multiplicative structure of K_0R? (In terms of topology, we are asking whether the tensor product of two Möbius bands is a cylinder or another Möbius band.)

Definition: Let $(\underline{\underline{C}}, \perp)$ be a category with product. A subcategory $\underline{\underline{D}}$ of $\underline{\underline{C}}$ (as category with product, that is, we are assuming $A \perp B \in \text{ob } \underline{\underline{D}}$ if $A, B \in \text{ob } \underline{\underline{D}}$, etc.) is *cofinal* if, for each $A \in \text{ob } \underline{\underline{C}}$ there exists $A' \in \text{ob } \underline{\underline{C}}$ and $B \in \text{ob } \underline{\underline{D}}$ with $A \perp A' \simeq B$.

Examples. 1. Let R be a ring, $\underline{\underline{C}} = \underline{\underline{P}}(R)$, and let $\underline{\underline{D}}$ be the full subcategory whose objects are the free R-modules of finite rank.

2. Let R be a commutative ring, $\underline{\underline{C}} = \underline{\underline{\text{Pic}}}(R)$, and let $\underline{\underline{D}}$ be the full subcategory with just one object, R itself.

Exercise: Let $\underline{\underline{C}}$ be a category with product, and let $\underline{\underline{D}}$ be a cofinal subcategory. Show that any element of $K_0\underline{\underline{C}}$ can be written in the form $[A] - [B]$ for some $A \in \text{ob } \underline{\underline{C}}$, $B \in \text{ob } \underline{\underline{D}}$. Show also that $[A] - [B] = [A'] - [B']$ if and only if there exists $C \in \text{ob } \underline{\underline{D}}$ with $A \perp B' \perp C \simeq A' \perp B \perp C$. (cf. Lemma 1 and Proposition 2.)

5.2 *Categories with product and composition*

We now start to work our way towards a definition of K_1R. In linear algebra, after one has shown that finite-dimensional vector spaces are classified up to isomorphism by their dimension, and matrices up to equivalence by their rank, one then starts to look at maps from a space to itself, leading to the study of square matrices and the various canonical forms under similarity; here the determinant appears as a test for whether such a map is an automorphism.

Following our construction of the group K_0R, we now wish to make a group out of the *automorphisms* of finitely generated projective R-modules, that is, not just the automorphisms of one such module (which in the classical case when R is a field would give the general linear group $\mathrm{GL}_n(R)$), but the automorphisms of all such modules.

Imitating the procedure for K_0R, we wish to construct a group K_1R with one generator $[\alpha]$ for each such automorphism α. Now first we require $[\alpha] = [\beta]$ if $\alpha \simeq \beta$: so what is this to mean? It is convenient technically to write (M, α) in place of α, where α is an automorphism of M; the M is there just as a label, so that we can tell at a glance on which module α acts. The definition we shall then adopt is that $(M, \alpha) \simeq (N, \beta)$ if there is an isomorphism $f : M \to N$ such that $f\alpha = \beta f$, that is, such that the diagram

$$\begin{array}{ccc} M & \xrightarrow{f} & N \\ \alpha \downarrow & & \downarrow \beta \\ M & \xrightarrow{f} & N \end{array}$$

commutes. (In the classical case, if $N = R^n$, then β is an $n \times n$ matrix representing α with repect to a choice of basis of M given by f, and if $M = N$ then α, β represent the same map with respect to different bases, the change of basis being given by f.) We can then form a direct sum of automorphisms in the obvious way, by $(M, \alpha) \oplus (N, \beta) = (M \oplus N, \alpha \oplus \beta)$, and this enables us to make the set of all automorphisms of finitely generated projective R-modules into a category with product.

However, there is a second, rather more obvious, way of combining automorphisms of which we wish to take account: if α, β are automorphisms of the *same* module M, then they have a composite $\alpha\beta$, also an automorphism of M. We are led to define a *composition* on our category of automorphisms; the composite of (M, α) with (N, β) is not always defined, but only when $M = N$, and then it is defined by $(M, \alpha) \circ (M, \beta) = (M, \alpha\beta)$. The product and the composition are related by the fact that, if α, α' are automorphisms of M and β, β' are automorphisms of N, then $\alpha \oplus \beta$, $\alpha' \oplus \beta'$ are automorphisms of $M \oplus N$, and

$$(\alpha \oplus \beta)(\alpha' \oplus \beta') = (\alpha\alpha' \oplus \beta\beta').$$

Both the product and the composition will give rise to relations in the group K_1R.

Definition: Let $(\underline{C}, \perp)$ be a category with product. A *composition* on $\underline{C}$ is a sometimes-defined binary composition $\circ$ of objects of $\underline{C}$ such that, if $A \circ B$ and $C \circ D$ are defined, so is $(A \perp C) \circ (B \perp D)$, and furthermore

$$(A \circ B) \perp (C \circ D) = (A \perp C) \circ (B \perp D).$$

(Note that we insist on equality here, not just isomorphism.) We refer to $\underline{C} = (\underline{C}, \perp, \circ)$ as a *category with product and composition.*

Let $(\underline{C}, \perp, \circ)$ be a category with product and composition, such that $\underline{C}$ has a small skeletal subcategory. We define the *Grothendieck group* $K_0(\underline{C}, \perp, \circ)$, or $K_0\underline{C}$ for short, as the abelian group given by the following generators and relations: we take one generator $[A]$ for each isomorphism class of objects A in $\underline{C}$ (so $[A] = [B]$ if $A \simeq B$), one relation $[A] + [B] = [A \perp B]$ for each pair $A, B \in \text{ob } \underline{C}$, and one relation $[A] + [B] = [A \circ B]$ for each pair $A, B \in \text{ob } \underline{C}$ for which $A \circ B$ is defined. Of course, if $\circ$ is nowhere defined, we recover the previous definition of $K_0\underline{C}$, and in general there is a group epimorphism $K_0(\underline{\underline{C}}, \perp) \to K_0(\underline{\underline{C}}, \perp, \circ)$ given by $[A] \mapsto [A]$. Exactly as before, any element of $K_0(\underline{\underline{C}}, \perp, \circ)$ can be written as $[A] - [B]$ for some $A, B \in \text{ob } \underline{\underline{C}}$, and if $\underline{\underline{D}}$ is a cofinal subcategory of $\underline{\underline{C}}$, we can insist $B \in \text{ob } \underline{\underline{D}}$. Also any functor $f : (\underline{\underline{C}}, \perp, \circ) \to (\underline{\underline{C}}', \perp', \circ')$ preserving both product and composition (where the latter means that whenever $A \circ B$ is defined, so is $f(A) \circ' f(B)$, and $f(A \circ B) = f(A) \circ' f(B)$) induces a group homomorphism $K_0 f : K_0\underline{\underline{C}} \to K_0\underline{\underline{C}}'$ by the rule $[A] \mapsto [f(A)]$.

Proposition 31. Let $\underline{\underline{C}}$ be a category with product and composition, and let $A, B \in \text{ob } \underline{\underline{C}}$. The following are equivalent:

(i) $[A] = [B]$ in $K_0\underline{\underline{C}}$;

(ii) There exist $C, D, E, D', E' \in \text{ob } \underline{\underline{C}}$ with

$$A \perp C \perp (D \circ E) \perp D' \perp E' \simeq B \perp C \perp D \perp E \perp (D' \circ E').$$

(Note that $\circ$ must be defined somewhere for this to make sense; alternatively, D, E, D', and E' may be omitted if $\circ$ is nowhere defined.)

Proof. Let $\underline{\underline{C}}'$ be the same category as $\underline{\underline{C}}$, but forgetting the composition. As remarked above, there is an epimorphism $K_0\underline{\underline{C}} \to K_0\underline{\underline{C}}'$; and its kernel is generated by all expressions

$$[D] + [E] - [D \circ E]$$

where $D, E \in \text{ob } \underline{\underline{C}}$ and $D \circ E$ is defined. Let $[A] = [B]$ in $K_0\underline{\underline{C}}$. Working in $K_0\underline{\underline{C}}'$, we have

$$[A] - [B] = \Sigma_i ([D_i] + [E_i] - [D_i \circ E_i]) + \Sigma_j ([D_j' \circ E_j'] - [D_j'] - [E_j'])$$

for some $D_i, E_i, D_j', E_j' \in \text{ob } \underline{\underline{C}}$, and so

$$[A] + \Sigma_i [D_i \circ E_i] + \Sigma_j [D_j'] + \Sigma_j [E_j'] = [B] + \Sigma_i [D_i] + \Sigma_i [E_i] + \Sigma_j [D_j' \circ E_j']$$

or

$$[A \perp (D \circ E) \perp D' \perp E'] = [B \perp D \perp E \perp (D' \circ E')]$$

where $D = \perp_i D_i$, $E = \perp_i E_i$, $D' = \perp_j D_j'$, and $E' = \perp_j E_j'$ (note that $\perp_i (D_i \circ E_i) = D \circ E$ and $\perp_j (D'_j \circ E'_j) = D' \circ E')$. By the second exercise on p. 91 it now follows that there exists $C \in$ ob $\underline{C}$ such that

$$A \perp C \perp (D \circ E) \perp D' \perp E' \simeq B \perp C \perp D \perp E \perp (D' \circ E')$$

as required. The converse is immediate.

Definition: Let $\underline{C}$ be a category with product. We construct a new category $\Omega\underline{C}$, the *loop category* of $\underline{C}$, as follows. The objects of $\Omega\underline{C}$ are all pairs (A, α) with $A \in$ ob $\underline{C}$ and $\alpha \in \mathrm{aut}(A)$ (that is, α is an automorphism of A, i.e., an isomorphism from A to A). A morphism $f : (A, \alpha) \to (B, \beta)$ in $\Omega\underline{C}$ is a morphism $f \in \underline{C}(A, D)$ such that the diagram

$$\begin{array}{ccc} A & \xrightarrow{f} & B \\ \alpha \downarrow & & \downarrow \beta \\ A & \xrightarrow{f} & B \end{array}$$

commutes. $\Omega\underline{C}$ becomes a category with product and composition if we put $(A, \alpha) \perp (B, \beta) = (A \perp B, \alpha \perp \beta)$ and $(A, \alpha) \circ (A, \beta) = (A, \alpha\beta)$ (the composite of (A, α) and (B, β) only being defined when $A = B$). The neutral object is $(0, 1_0)$; we leave it to the reader to show that the remaining conditions for a category with product are satisfied. Note that

$$\begin{aligned} &((A, \alpha) \circ (A, \beta)) \perp ((B, \gamma) \circ (B, \delta)) \\ &= (A, \alpha\beta) \perp (B, \gamma\delta) \\ &= (A \perp B, \alpha\beta \perp \gamma\delta) \\ &= (A \perp B, (\alpha \perp \gamma)(\beta \perp \delta)) \\ &= (A \perp B, \alpha \perp \gamma) \circ (A \perp B, \beta \perp \delta) \\ &= ((A, \alpha) \perp (B, \gamma)) \circ ((A, \beta) \perp (B, \delta)). \end{aligned}$$

Definition: $K_1\underline{C} = K_0\Omega\underline{C}$, the *Whitehead group* of $\underline{C}$. More directly, $K_1\underline{C}$ is an abelian group with generators $[A, \alpha]$ ($A \in$ ob $\underline{C}$, $\alpha \in \mathrm{aut}(A)$) and relations:

(i) $[A, \alpha] = [B, \beta]$ if there is an isomorphism $f : A \to B$ with $\beta f = f\alpha$;

(ii) $[A, \alpha] + [B, \beta] = [A \perp B, \alpha \perp \beta]$;

(iii) $[A, \alpha] + [A, \beta] = [A, \alpha\beta]$.

Let $\underline{\underline{C}}$, $\underline{\underline{D}}$ be categories with product and let $F : \underline{\underline{C}} \to \underline{\underline{D}}$ be a product-preserving functor. We define a functor $\Omega F : \Omega\underline{\underline{C}} \to \Omega\underline{\underline{D}}$ by $\Omega F(A, \alpha) = (FA, F\alpha)$ and, if $f \in \Omega\underline{\underline{C}}((A, \alpha), (B, \beta))$, $\Omega Ff = Ff$ (note that $Ff \in \underline{\underline{D}}(FA, FB)$ and $(F\beta)(Ff) = F(\beta f) = F(f\alpha) = (Ff)(F\alpha)$, so $\Omega Ff \in \Omega\underline{\underline{D}}(F(A, \alpha), F(B, \beta))$. Further,

$$\begin{aligned}\Omega F((A, \alpha) \perp (B, \beta)) &= \Omega F(A \perp B, \alpha \perp \beta)\\ &= (F(A \perp B), F(\alpha \perp \beta))\\ &= (FA \perp FB, F\alpha \perp F\beta)\\ &= (FA, F\alpha) \perp (FB, F\beta)\\ &= \Omega F(A, \alpha) \perp \Omega F(B, \beta)\end{aligned}$$

and also

$$\begin{aligned}\Omega F((A, \alpha) \circ (A, \beta)) &= \Omega F(A, \alpha\beta)\\ &= (FA, F(\alpha\beta))\\ &= (FA, (F\alpha)(F\beta))\\ &= (FA, F\alpha) \circ (FA, F\beta)\\ &= \Omega F(A, \alpha) \circ \Omega F(A, \beta)\end{aligned}$$

where we are using $\perp$ and $\circ$ to denote the product and composition in both $\underline{\underline{C}}$ and $\underline{\underline{D}}$, and hence in $\Omega\underline{\underline{C}}$ and $\Omega\underline{\underline{D}}$ as well. Thus ΩF preserves product and composition (remaining details being left to the reader), and we can define $K_1F : K_1\underline{\underline{C}} \to K_1\underline{\underline{D}}$ by $K_1F = K_0\Omega F$. This makes K_1 into a functor from the category of categories with product and small skeletal subcategory, and product-preserving functors, to $\underline{\underline{\text{Ab}}}$.

CHAPTER SIX

Whitehead Groups

Recall that at the end of the last chapter we defined the Whitehead group $K_1\underline{\underline{C}}$, where $\underline{\underline{C}}$ is a category with product, to be the Grothendieck group $K_0\Omega\underline{\underline{C}}$ of the loop category $\Omega\underline{\underline{C}}$. As a special case of this, if R is a ring, we define the *Whitehead group* K_1R to be $K_1\underline{\underline{P}}(R)$. An equivalent but more straightforward definition of K_1R will be given later; the present approach is chosen for its generality, and because it emphasizes the similarity of construction of K_1R and K_0R, and also because it leads to a natural construction of the K-theory long exact sequence (Chapter 7).

Example. Let F be a field; we shall compute K_1F. Put $\underline{\underline{P}} = \underline{\underline{P}}(R)$. If $V \in \text{ob } \underline{\underline{P}}$, that is, V is a finite-dimensional vector space over F, and $\alpha \in \text{aut}(V)$, then α has a well-defined determinant, $\det \alpha$. (Either choose a basis of V and take the determinant of the matrix corresponding to α, observing that this is independent of the choice of basis, or note that, if $\dim V = n$, then $\dim(\wedge^n_F V) = 1$, so $\text{aut}(\wedge^n_F V)$ can be identified with $F^\bullet$ in a natural way, and then $\det \alpha = \wedge^n \alpha$.) Next, if $f : (V, \alpha) \to (W, \beta)$ is an isomorphism in $\Omega\underline{\underline{P}}$, then $\beta f = f\alpha$, and $f : V \to W$ is an isomorphism, so choosing bases for V, W, we have $\det(\beta f) = \det(f\alpha)$, or $(\det \beta)(\det f) = (\det f)(\det \alpha)$, or $\det \beta = \det \alpha$, since

$\det f \neq 0$. Moreover, $(V, \alpha) \circ (V, \beta) = (V, \alpha\beta)$, and $\det \alpha\beta = (\det \alpha)(\det \beta)$; also $(V, \alpha) \oplus (W, \beta) = (V \oplus W, \alpha \oplus \beta)$, and $\det \alpha \oplus \beta = (\det \alpha)(\det \beta)$. It follows that there is a homomorphism from the additive group K_1F to the multiplicative group $F^{\bullet}$, $\det : K_1F \to F^{\bullet}$, given by $\det[V, \alpha] = \det \alpha$. (*Note:* we have a functor $\det : \underline{\underline{P}} \to \underline{\underline{\text{Pic}}}(R)$; we shall see later that $K_1\underline{\underline{\text{Pic}}}(F) \simeq F^{\bullet}$, and the map that we have just constructed is really $K_1(\det) : K_1F \to F^{\bullet}$.) Now $\text{aut}(F) = \text{GL}_1(F) = F^{\bullet}$, and $\det : \text{aut}(F) \to F^{\bullet}$ is an isomorphism, so $\det : K_1F \to F^{\bullet}$ is surjective. (Indeed, $\det : \text{aut}(V) \to F^{\bullet}$ is surjective if $V \neq 0$.) Next, let $[V, \alpha]$ be a generator of K_1F, choose a basis for V, and let α have matrix A with respect to this basis. We can reduce A to diagonal form by means of elementary row operations, that is, we can write $A = E_1E_2\ldots E_rD$, where the E_i are elementary matrices (matrices differing from the identity matrix in at most one entry, not on the diagonal) and D is a diagonal matrix. Now if $\lambda \in F^{\bullet}$, we have

$$\begin{pmatrix} \lambda & 0 \\ 0 & \lambda^{-1} \end{pmatrix} = \begin{pmatrix} 1 & \lambda-1 \\ 0 & 1 \end{pmatrix}\begin{pmatrix} 1 & 0 \\ 1 & 1 \end{pmatrix}\begin{pmatrix} 1 & \lambda^{-1}-1 \\ 0 & 1 \end{pmatrix}\begin{pmatrix} 1 & 0 \\ -\lambda & 1 \end{pmatrix}$$

and if $\lambda, \mu \in F^{\bullet}$, then

$$\begin{pmatrix} \lambda & 0 \\ 0 & \mu \end{pmatrix} = \begin{pmatrix} \mu^{-1} & 0 \\ 0 & \mu \end{pmatrix}\begin{pmatrix} \lambda\mu & 0 \\ 0 & 1 \end{pmatrix}$$

so $\begin{pmatrix} \lambda & 0 \\ 0 & \mu \end{pmatrix}$ can be reduced to $\begin{pmatrix} \lambda\mu & 0 \\ 0 & 1 \end{pmatrix}$ by elementary row operations. Similarly,

$$\begin{pmatrix} \lambda_1 & & 0 \\ & \lambda_2 & \\ 0 & & \ddots \\ & & & \lambda_n \end{pmatrix} \text{ can be reduced to } \begin{pmatrix} \mu & & 0 \\ & 1 & \\ 0 & & \ddots \\ & & & 1 \end{pmatrix}$$

by elementary row operations, where $\mu = \lambda_1\lambda_2\ldots\lambda_n$. Thus we may assume that D is already of the same form. Then $\det D = \mu$, and $\det E_i = 1$, all i, so $\det A = \mu$. So in K_1F,

$$[V, \alpha] = [F^n, E_1E_2\ldots E_rD]$$
$$= \Sigma_i [F^n, E_i] + [F^n, D]$$
$$= \Sigma_i [F^n, E_i] + [F, \mu] + [F^{n-1}, I_{n-1}]$$

(identifying aut(F) with $F^{\bullet}$). Now $[F^{n-1}, I_{n-1}] = 0$; indeed, $[W, 1] = 0$, any W, since

$$[W, 1] + [W, 1] = [(W, 1) \circ (W, 1)] = [W, 1].$$

Next, since

$$[F^n, E_i] = [F^n, E_i] + [F, I_1] = [F^{n+1}, E_i \oplus I_1]$$

we may assume $n \geq 3$. But now each E_i is a commutator; explicitly, if $B_{ij}(x)$ is the elementary matrix with x in the i, j position ($i \neq j$), then if i, j, k are distinct,

$$B_{ij}(x) = B_{ik}(x)B_{kj}(1)B_{ik}(x)^{-1}B_{kj}(1)^{-1}.$$

However, K_1F is abelian, and so $[F^n, E_i] = 0$, each i, and we conclude $[V, \alpha] = [F, \mu]$, where $\mu = \det \alpha$. Now suppose

$$[V, \alpha] - [W, \beta] \in \ker(\det : K_1F \to F^{\bullet}).$$

So $\det \alpha = \mu = \det \beta$, say, and $[V, \alpha] = [F, \mu] = [W, \beta]$, as above. Thus $[V, \alpha] - [W, \beta] = 0$, and $\det : K_1F \to F^{\bullet}$ is a monomorphism, hence an isomorphism. So $K_1F \simeq F^{\bullet}$.

Proposition 32. Let $(\underline{\underline{C}}, \perp)$ be a category with product and let $(\underline{\underline{C}}', \perp)$ be a full cofinal subcategory. Then

$$K_1(\underline{\underline{C}}' \hookrightarrow \underline{\underline{C}}) : K_1\underline{\underline{C}}' \to K_1\underline{\underline{C}}$$

is an isomorphism.

Proof. (i) Surjectivity: let $[A, \alpha]$ be a generator of $K_1\underline{\underline{C}}$. Thus $A \in$ ob $\underline{\underline{C}}$, and since $\underline{\underline{C}}'$ is cofinal there exist $A_1 \in$ ob $\underline{\underline{C}}$ and $A' \in$ ob $\underline{\underline{C}}'$ with an isomorphism $f : A \perp A_1 \to A'$. Then $(A \perp A_1, \alpha \perp 1) \simeq (A', \beta)$ in $\Omega\underline{\underline{C}}$, where $\beta = f(\alpha \perp 1)f^{-1}$, and since $A' \in$ ob $\underline{\underline{C}}'$ and $\underline{\underline{C}}'$ is full, $(A', \beta) \in$ ob $\Omega\underline{\underline{C}}'$. So in $K_1\underline{\underline{C}}$,

$$[A, \alpha] = [A, \alpha] + [A_1, 1] = [A \perp A_1, \alpha \perp 1] = [A', \beta]$$

and this is in the image of $K_1\underline{\underline{C}}'$.

(ii) Injectivity: suppose $[A, \alpha] - [B, \beta] \in K_1\underline{\underline{C}}'$ is such that

$[A, \alpha] = [B, \beta]$ in $K_1\underline{\underline{C}}$. We must show that $[A, \alpha] = [B, \beta]$ in $K_1\underline{\underline{C}}'$. Note first that $A, B \in \text{ob } \underline{\underline{C}}'$. Next, by Proposition 31, there exist $(C, \gamma), (D, \delta_1), (D, \delta_2), (E, \varepsilon_1), (E, \varepsilon_2) \in \text{ob } \Omega\underline{\underline{C}}$ such that

$$(A, \alpha) \perp (C, \gamma) \perp ((D, \delta_1) \circ (D, \delta_2)) \perp (E, \varepsilon_1) \perp (E, \varepsilon_2)$$
$$\simeq (B, \beta) \perp (C, \gamma) \perp (D, \delta_1) \perp (D, \delta_2)$$
$$\perp ((E, \varepsilon_1) \circ (E, \varepsilon_2))$$

in $\Omega\underline{\underline{C}}$, that is,

$$(A \perp C \perp D \perp E^2, \alpha \perp \gamma \perp \delta_1\delta_2 \perp \varepsilon_1 \perp \varepsilon_2)$$
$$\simeq (B \perp C \perp D^2 \perp E, \beta \perp \gamma \perp \delta_1 \perp \delta_2 \perp \varepsilon_1\varepsilon_2) \qquad \ldots (1)$$

where we are writing D^2 for $D \perp D$, etc. It follows that $A \perp C \perp D \perp E^2 \simeq B \perp C \perp D^2 \perp E$ in $\underline{\underline{C}}$, and so

$$(A \perp C \perp D \perp E^2, 1) \simeq (B \perp C \perp D^2 \perp E, 1) \qquad \ldots (2)$$

in $\Omega\underline{\underline{C}}$. Put $F = A \perp B \perp C^2 \perp D^3 \perp E^3$, so

$$F \simeq (A \perp C \perp D \perp E^2) \perp (B \perp C \perp D^2 \perp E)$$

the isomorphism being obtained by changing the order of the terms. So, taking the product of the left-hand side of (1) with the right-hand side of (2), and vice versa, we obtain, after permuting the terms,

$$(F, \alpha \perp 1 \perp (\gamma \perp 1) \perp (\delta_1\delta_2 \perp 1 \perp 1) \perp (\varepsilon_1 \perp \varepsilon_2 \perp 1))$$
$$\simeq (F, 1 \perp \beta \perp (\gamma \perp 1) \perp (\delta_1 \perp \delta_2 \perp 1) \perp (\varepsilon_1\varepsilon_2 \perp 1 \perp 1))$$
$$\ldots (3).$$

As in (i) above, we can find C_0, C_1, and γ' such that $(C \perp C_0, \gamma \perp 1) \simeq (C_1, \gamma')$ (and of course $(C \perp C_0, 1) \simeq (C_1, 1)$) and $(C_1, \gamma') \in \text{ob } \Omega\underline{\underline{C}}'$; similarly for D and E. Put

$$F_1 = A \perp B \perp C_1{}^2 \perp D_1{}^3 \perp E_1{}^3 \simeq F \perp (C_0{}^2 \perp D_0{}^3 \perp E_0{}^3).$$

By forming the product of each side of (3) with $(C_0{}^2 \perp D_0{}^3 \perp E_0{}^3, 1)$ and permuting the terms, we obtain

$$(F_1, \alpha \perp 1 \perp (\gamma' \perp 1) \perp ((\delta_1\delta_2)' \perp 1 \perp 1) \perp (\varepsilon_1' \perp \varepsilon_2' \perp 1))$$
$$\simeq (F_1, 1 \perp \beta \perp (\gamma' \perp 1) \perp (\delta_1' \perp \delta_2' \perp 1)$$
$$\perp ((\varepsilon_1\varepsilon_2)' \perp 1 \perp 1))$$

and since $A, B, C_1, D_1, E_1 \in \text{ob } \underline{\underline{C}}'$ and $\underline{\underline{C}}'$ is full, this is an isomorphism in $\Omega\underline{\underline{C}}'$. Passing to $K_1\underline{\underline{C}}'$, and remembering that

$[M, 1] = 0$, all $M \in \text{ob}\,\underline{\underline{C}}'$, we have

$$[A, \alpha] + [C_1, \gamma'] + [D_1, (\delta_1\delta_2)'] + [E_1, \varepsilon_1'] + [E_1, \varepsilon_2']$$
$$= [B, \beta] + [C_1, \gamma'] + [D_1, \delta_1'] + [D_1, \delta_2'] + [E_1, (\varepsilon_1\varepsilon_2)'].$$

If $f : D \perp D_0 \to D_1$ is the isomorphism used in the construction of δ_1', δ_2', and $(\delta_1\delta_2)'$, we have $(\delta_1\delta_2)' = f((\delta_1\delta_2) \perp 1)f^{-1} = f(\delta_1 \perp 1)(\delta_2 \perp 1)f^{-1} = (f(\delta_1 \perp 1)f^{-1})(f(\delta_2 \perp 1)f^{-1}) = \delta_1'\delta_2'$. So

$$[D_1, (\delta_1\delta_2)'] = [D_1, \delta_1'\delta_2'] = [D_1, \delta_1'] + [D_1, \delta_2']$$

and similarly

$$[E_1, (\varepsilon_1\varepsilon_2)'] = [E_1, \varepsilon_1'] + [E_1, \varepsilon_2'].$$

Substituting back, we obtain $[A, \alpha] = [B, \beta]$ in $K_1\underline{\underline{C}}'$, and the proof is complete.

Remark: In the above proof, we used the fact that any element of $K_1\underline{\underline{C}}$ can be written in the form $[A, \alpha] - [B, \beta]$, for some (A, α), $(B, \beta) \in \text{ob}\,\Omega\underline{\underline{C}}$. In fact, one can do better than this, for $[B, \beta] + [B, \beta^{-1}] = [B, 1] = 0$, so $[A, \alpha] - [B, \beta] = [A, \alpha] + [B, \beta^{-1}] = [A \perp B, \alpha \perp \beta^{-1}]$. Thus any element of $K_1\underline{\underline{C}}$ can be written in the form $[A, \alpha]$ for some $(A, \alpha) \in \text{ob}\,\Omega\underline{\underline{C}}$. This should be compared with the situation in K_0C, where in general Lemma 1 is the best we can hope for; e.g., in the isomorphism $K_0\mathbb{Z} \simeq \mathbb{Z}$, elements of the form $[M]$ correspond to non-negative integers only.

Proposition 32 can be used to simplify the computation of $K_1\underline{\underline{C}}$, for it allows us to throw away part of $\underline{\underline{C}}$ and work with a suitable subcategory.

Example. Let R be a ring, and let $\underline{\underline{P}}'(R)$ be the full subcategory of $\underline{\underline{P}}(R)$ whose objects are 0 and R^n, $n = 1,2,\ldots$. That this is cofinal is immediate from the definition of projective module, and so $K_1R \simeq K_1\underline{\underline{P}}'(R)$. Note that $\text{aut}(R^n) = \text{GL}_n(R)$ (the group of all $n \times n$ invertible matrices with coefficients in R), so K_1R must be constructible by glueing together the groups $\text{GL}_n(R)$ in some way. We shall now make this more explicit.

6.1 *$K_1\underline{\underline{C}}$ as a direct limit*

Let $(\underline{\underline{C}}, \perp)$ be a category with product. If $f : A \to B$ is an isomorphism in $\underline{\underline{C}}$, it induces an isomorphism $f' : \mathrm{aut}(A) \to \mathrm{aut}(B)$ by $\alpha \mapsto f\alpha f^{-1}$, and this induces an isomorphism $\bar{f} : \mathrm{aut}(A)^{\mathrm{ab}} \to \mathrm{aut}(B)^{\mathrm{ab}}$ (recall $G^{\mathrm{ab}} = G/G'$, where G' is the derived subgroup of the group G). Now suppose $g : A \to B$ is another isomorphism; then $gf^{-1} \in \mathrm{aut}(B)$, and in $\mathrm{aut}(B)$ we have

$$(gf^{-1})f'(\alpha) = (gf^{-1})(f\alpha f^{-1}) = (g\alpha g^{-1})(gf^{-1}) = g'(\alpha)(gf^{-1}).$$

Passing to $\mathrm{aut}(B)^{\mathrm{ab}}$, we see $\bar{f}(\bar{\alpha}) = \bar{g}(\bar{\alpha})$ (where $\bar{\alpha}$ is the image of α in $\mathrm{aut}(A)^{\mathrm{ab}}$), or $\bar{f} = \bar{g}$. Thus if we write $\bar{A}$ for the isomorphism class of A, the group $G(\bar{A}) = \mathrm{aut}(A)^{\mathrm{ab}}$ is determined by $\bar{A}$ up to a unique isomorphism.

Next suppose $h : A \perp A_1 \to B$ is an isomorphism in $\underline{\underline{C}}$. This induces a homomorphism $\hat{h} : \mathrm{aut}(A) \to \mathrm{aut}(B)$ by $\alpha \mapsto h(\alpha \perp 1)h^{-1}$, and this in turn induces $\tilde{h} : \mathrm{aut}(A)^{\mathrm{ab}} \to \mathrm{aut}(B)^{\mathrm{ab}}$. If $k : A \perp A_1 \to B$ is another isomorphism, then $kh^{-1} \in \mathrm{aut}(B)$, and in $\mathrm{aut}(B)$ we have

$$(kh^{-1})(h(\alpha \perp 1)h^{-1}) = (k(\alpha \perp 1)k^{-1})(kh^{-1})$$

whence, passing to $\mathrm{aut}(B)^{\mathrm{ab}}$, $\tilde{h}(\bar{\alpha}) = \tilde{k}(\bar{\alpha})$, or $\tilde{h} = \tilde{k}$. Further, if $f : A \to A'$, $f_1 : A_1 \to A_1'$, $h : A \perp A_1 \to B$ and $k : A' \perp A_1' \to B'$ are isomorphisms, then $g : B \to B'$ is an isomorphism, where $g = k(f \perp f_1)h^{-1}$, and the diagram

$$\begin{array}{ccc} A \perp A_1 & \xrightarrow{h} & B \\ {\scriptstyle f \perp f_1}\downarrow & & \downarrow {\scriptstyle g} \\ A' \perp A_1' & \xrightarrow{k} & B' \end{array}$$

then commutes. Let $\alpha \in \mathrm{aut}(A)$: then

$$\begin{aligned} g'\hat{h}(\alpha) &= g'(h(\alpha \perp 1)h^{-1}) = gh(\alpha \perp 1)h^{-1}g^{-1} \\ &= k(f \perp f_1)(\alpha \perp 1)(f^{-1} \perp f_1^{-1})k^{-1} \\ &= k(f\alpha f^{-1} \perp 1)k^{-1} = \hat{k}(f\alpha f^{-1}) = \hat{k}f'(\alpha). \end{aligned}$$

So $g'\hat{h} = \hat{k}f'$, and hence $\bar{g}\tilde{h} = \tilde{k}\bar{f}$, that is, the diagram

$$\begin{array}{ccc} \mathrm{aut}(A)^{ab} & \xrightarrow{\tilde{h}} & \mathrm{aut}(B)^{ab} \\ \bar{f}\downarrow & & \downarrow \bar{g} \\ \mathrm{aut}(A')^{ab} & \xrightarrow{\tilde{k}} & \mathrm{aut}(B')^{ab} \end{array}$$

commutes. So given A, A_1 with $A \perp A_1 \simeq B$, the homomorphism $\tilde{h}$: $G(\bar{A}) \to G(\bar{B})$ is uniquely determined by $\bar{A}$ and $\bar{A}_1$. We denote it by $g(\bar{A}, \bar{A}_1)$. Now we form a new category $\underline{\underline{G}}$ with ob $\underline{\underline{G}}$ = $\{G(\bar{A}) : A \in \text{ob } \underline{\underline{C}}\}$ and $\underline{\underline{G}}(G(\bar{A}), G(\bar{B})) = \{g(\bar{A}, \bar{A}_1) : A \perp A_1 \simeq B\}$. So the objects of $\underline{\underline{G}}$ are abelian groups, one for each isomorphism class of objects in $\underline{\underline{C}}$, and the morphisms are certain group homomorphisms. Note that if $h_1 : A \perp A_1 \to B$ and $h_2 : B \perp B_1 \to C$ are isomorphisms in $\underline{\underline{C}}$, then, if $\alpha \in \mathrm{aut}(A)$,

$$\begin{aligned} \hat{h}_2\hat{h}_1(\alpha) &= \hat{h}_2(h_1(\alpha \perp 1)h_1^{-1}) = h_2(h_1(\alpha \perp 1)h_1^{-1} \perp 1)h_2^{-1} \\ &= h_2(h_1 \perp 1)(\alpha \perp 1 \perp 1)(h_1^{-1} \perp 1)h_2^{-1} = \hat{k}(\alpha) \end{aligned}$$

(where $k = h_2(h_1 \perp 1)$), whence $g(\bar{B}, \bar{B}_1)g(\bar{A}, \bar{A}_1) = g(\bar{A}, \overline{A_1 \perp B_1})$. So the morphisms of $\underline{\underline{G}}$ are closed under composition. Recall that, if $\lambda : A \perp 0 \to A$ is the natural isomorphism, where 0 is the neutral object of $\underline{\underline{C}}$, then for all $\alpha \in \mathrm{aut}(A)$, the diagram

$$\begin{array}{ccc} A \perp 0 & \xrightarrow{\alpha \perp 1} & A \perp 0 \\ \lambda\downarrow & & \downarrow\lambda \\ A & \xrightarrow{\alpha} & A \end{array}$$

commutes, so that $\alpha = \lambda(\alpha \perp 1)\lambda^{-1}$. It follows that

$$g(\bar{A}, \bar{0}) = 1_{G(\bar{A})} \in \underline{\underline{G}}(G(\bar{A}), G(\bar{A}))$$

so $\underline{\underline{G}}$ is a category.

In fact $\underline{\underline{G}}$ is a *directed* category, that is:
(1) given $G(\bar{A}_1)$ and $G(\bar{A}_2) \in$ ob $\underline{\underline{G}}$ there exists $G(\bar{B}) \in$ ob $\underline{\underline{G}}$ such that $\underline{\underline{G}}(G(\bar{A}_1), G(\bar{B}))$ and $\underline{\underline{G}}(G(\bar{A}_2), G(\bar{B}))$ are both non-empty; and
(2) given $g(\bar{A}, \bar{A}_1)$, $g(\bar{A}, \bar{A}_2) : G(\bar{A}) \to G(\bar{B})$, there exists $g(\bar{B}, \bar{B}_1) : G(\bar{B}) \to G(\bar{C})$ such that $g(\bar{B}, \bar{B}_1)g(\bar{A}, \bar{A}_1) = g(\bar{B}, \bar{B}_1)g(\bar{A}, \bar{A}_2)$, that is, such that $g(\bar{A}, \overline{A_1 \perp B_1}) = g(\bar{A}, \overline{A_2 \perp B_1})$.

For in (1) we may choose $B = A_1 \perp A_2$, and in (2) we may choose $B_1 = A$, noting that $A_1 \perp A \simeq B \simeq A_2 \perp A$. Now for such a category, we can construct the *direct limit* $\varinjlim \underline{\underline{G}}$; it is an abelian group with a homomorphism $h(\bar{A}) : G(\bar{A}) \to \varinjlim \underline{\underline{G}}$ for each $G(\bar{A}) \in$ ob $\underline{\underline{G}}$, such that every diagram

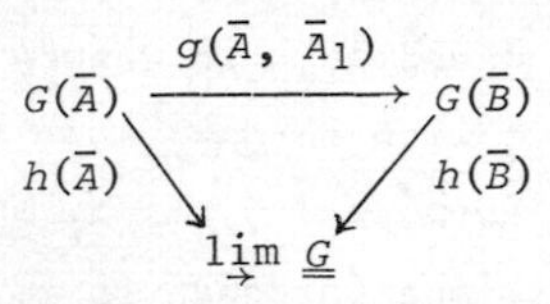

commutes, and is universal for these properties. The existence of this group is easily established: one takes the direct sum of all the groups $G(\bar{A})$ and factors out the smallest subgroup necessary to make the above diagrams commute; details are left to the reader. For our present purposes, we do not need to show that $\varinjlim \underline{\underline{G}}$ can be constructed, for we have constructed it already, as the next proposition shows.

Proposition 33. $K_1\underline{\underline{C}} \simeq \varinjlim \underline{\underline{G}}$.

Proof. All we need to do is show that $K_1\underline{\underline{C}}$ has the required universal property. First, there is a homomorphism $f(\bar{A}) : G(\bar{A}) \to K_1\underline{\underline{C}}$ given by $\bar{\alpha} \mapsto [A, \alpha]$, where $\alpha \in \text{aut}(A)$ and $\bar{\alpha}$ is its image in $G(\bar{A})$. Note that if $\beta \in \text{aut}(A)$ with $\bar{\alpha} = \bar{\beta}$, then $\alpha\beta^{-1}$ belongs to the derived subgroup of aut(A), and hence $[A, \alpha\beta^{-1}] = 0$, since $K_1\underline{\underline{C}}$ is abelian, that is, $[A, \alpha] = [A, \beta]$. Further, if $g : A \to B$ is an isomorphism in $\underline{\underline{C}}$, then in $\Omega\underline{\underline{C}}$, $(A, \alpha) \simeq (B, g\alpha g^{-1}) = (B, \bar{g}(\alpha))$, so in $K_1\underline{\underline{C}}$, $[A, \alpha] = [B, g(\alpha)]$. Thus $f(\bar{A})$ is well defined, and it is a homomorphism since, if $\alpha, \beta \in$ aut(A),

$$f(\bar{A})(\bar{\alpha}\bar{\beta}) = f(\bar{A})(\overline{\alpha\beta}) = [A, \alpha\beta]$$
$$= [A, \alpha] + [A, \beta] = f(\bar{A})(\bar{\alpha}) + f(\bar{A})(\bar{\beta}).$$

Next, let $h : A \perp A_1 \to B$ be an isomorphism in $\underline{\underline{C}}$, and let $\alpha \in$ aut(A). In $\Omega\underline{\underline{C}}$, $(A \perp A_1, \alpha \perp 1) \simeq (B, h(\alpha \perp 1)h^{-1})$, so

$$f(\bar{B})g(\bar{A}, \bar{A}_1)(\bar{\alpha}) = f(\bar{B})(\overline{h(\alpha \perp 1)h^{-1}}) = [B, h(\alpha \perp 1)h^{-1}]$$

$$= [A \perp A_1,\ \alpha \perp 1] = [A,\ \alpha] + [A_1,\ 1] = [A,\ \alpha] = f(\bar{A})(\bar{\alpha}).$$

Thus the diagram

$$\begin{array}{ccc} G(\bar{A}) & \xrightarrow{g(\bar{A},\ \bar{A}_1)} & G(\bar{B}) \\ & {\scriptstyle f(\bar{A})} \searrow \quad \swarrow {\scriptstyle f(\bar{B})} & \\ & K_1\underline{\underline{C}} & \end{array}$$

commutes.

Now let H be an abelian group with a homomorphism $h(\bar{A})$: $G(\bar{A}) \to H$ for each $G(\bar{A}) \in$ ob $\underline{\underline{G}}$, and such that every diagram

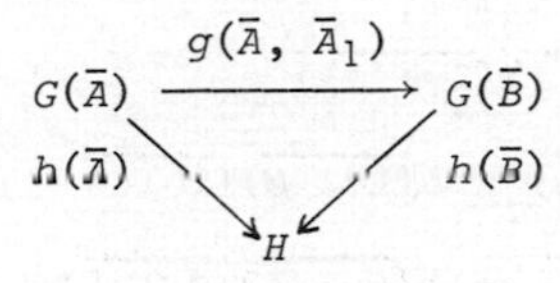

commutes. We must show that there is a unique homomorphism $\theta : K_1\underline{\underline{C}} \to H$ such that every diagram

$$\begin{array}{ccc} G(\bar{A}) & \xrightarrow{f(\bar{A})} & K_1\underline{\underline{C}} \\ & {\scriptstyle h(\bar{A})} \searrow \quad \swarrow {\scriptstyle \theta} & \\ & H & \end{array}$$

commutes. Indeed, since a typical element of $K_1\underline{\underline{C}}$ is of the form $[A,\ \alpha]$ (see the remark on p.101), we must have $\theta([A,\ \alpha]) = h(\bar{A})(\bar{\alpha})$, and it only remains to show that this gives a well defined homomorphism. So suppose $[A,\ \alpha] = [B,\ \beta]$ in $K_1\underline{\underline{C}}$. Then by Proposition 31 there exist $(C,\ \gamma)$, $(D,\ \delta_1)$, $(D,\ \delta_2)$, $(E,\ \varepsilon_1)$, $(E,\ \varepsilon_2) \in$ ob $\Omega\underline{\underline{C}}$ such that, in $\Omega\underline{\underline{C}}$, there is an isomorphism

$$f_1 : (A \perp C \perp D \perp E^2,\ \alpha \perp \gamma \perp \delta_1\delta_2 \perp (\varepsilon_1 \perp \varepsilon_2))$$
$$\to (B \perp C \perp D^2 \perp E,\ \beta \perp \gamma \perp (\delta_1 \perp \delta_2) \perp \varepsilon_1\varepsilon_2).$$

Thus $\overline{A \perp C \perp D \perp E^2} = \overline{B \perp C \perp D^2 \perp E}$, and

$$\bar{f}_1(\overline{\alpha \perp \gamma \perp \delta_1\delta_2 \perp (\varepsilon_1 \perp \varepsilon_2)}) = \overline{\beta \perp \gamma \perp (\delta_1 \perp \delta_2) \perp \varepsilon_1\varepsilon_2}.$$

In other words,

$$\overline{\alpha \perp \gamma \perp \delta_1\delta_2 \perp (\varepsilon_1 \perp \varepsilon_2)} \text{ and } \overline{\beta \perp \gamma \perp (\delta_1 \perp \delta_2) \perp \varepsilon_1\varepsilon_2}$$

represent the same element of

$$G(\overline{A \perp C \perp D \perp E^2}) = G(\overline{B \perp C \perp D^2 \perp E}).$$

Passing to H, we therefore have

$$h(\overline{A \perp C \perp D \perp E^2})(\overline{\alpha \perp \gamma \perp \delta_1\delta_2 \perp (\varepsilon_1 \perp \varepsilon_2)})$$

$$= h(\overline{B \perp C \perp D^2 \perp E})(\overline{\beta \perp \gamma \perp (\delta_1 \perp \delta_2) \perp \varepsilon_1\varepsilon_2}) \qquad \ldots (1).$$

Now if (L, ρ), $(M, \sigma) \in \text{ob } \Omega\underline{C}$,

$$\begin{aligned} h(\overline{L \perp M})(\overline{\rho \perp \sigma}) &= h(\overline{L \perp M})(\overline{(\rho \perp 1)(1 \perp \sigma)}) \\ &= h(\overline{L \perp M})((\overline{\rho \perp 1})(\overline{1 \perp \sigma})) \\ &= h(\overline{L \perp M})(\overline{\rho \perp 1}) + h(\overline{L \perp M})(\overline{1 \perp \sigma}) \\ &= h(\overline{L \perp M})(\overline{\rho \perp 1}) + h(\overline{M \perp L})(\overline{\sigma \perp 1}) \end{aligned}$$

since there is an isomorphism $\phi : (L \perp M, 1 \perp \sigma) \to (M \perp L, \sigma \perp 1)$ in $\Omega\underline{C}$, by condition (ii), p.84. Thus

$$\begin{aligned} h(\overline{L \perp M})(\overline{\rho \perp \sigma}) &= h(\overline{L \perp M})g(\bar{L}, \bar{M})(\bar{\rho}) + h(\overline{M \perp L})g(\bar{M}, \bar{L})(\bar{\sigma}) \\ &= h(\bar{L})(\bar{\rho}) + h(\bar{M})(\bar{\sigma}) \qquad \ldots (2). \end{aligned}$$

Applying this to (1), we have

$$h(\bar{A})(\bar{\alpha}) + h(\bar{C})(\bar{\gamma}) + h(\bar{D})(\overline{\delta_1\delta_2}) + h(\bar{E})(\bar{\varepsilon}_1) + h(\bar{E})(\bar{\varepsilon}_2)$$

$$= h(\bar{B})(\bar{\beta}) + h(\bar{C})(\bar{\gamma}) + h(\bar{D})(\bar{\delta}_1) + h(\bar{D})(\bar{\delta}_2) + h(\bar{E})(\overline{\varepsilon_1\varepsilon_2})$$

but

$$h(\bar{D})(\overline{\delta_1\delta_2}) = h(\bar{D})(\bar{\delta}_1\bar{\delta}_2) = h(\bar{D})(\bar{\delta}_1) + h(\bar{D})(\bar{\delta}_2)$$

and similarly

$$h(\bar{E})(\overline{\varepsilon_1\varepsilon_2}) = h(\bar{E})(\bar{\varepsilon}_1) + h(\bar{E})(\bar{\varepsilon}_2)$$

so we have $h(\bar{A})(\bar{\alpha}) = h(\bar{B})(\bar{\beta})$, and thus θ is well defined. It is a homomorphism, since, for any (A, α), $(B, \beta) \in \text{ob } \Omega\underline{C}$,

$$\begin{aligned} \theta([A, \alpha] + [B, \beta]) &= \theta([A \perp B, \alpha \perp \beta]) = h(\overline{A \perp B})(\overline{\alpha \perp \beta}) \\ &= h(\bar{A})(\bar{\alpha}) + h(\bar{B})(\bar{\beta}) \text{ (by (2) above)} \\ &= \theta([A, \alpha]) + \theta([B, \beta]) \end{aligned}$$

and the proof is complete.

Exercise: Write down sufficient conditions on a subcategory $\underline{\underline{G}}'$ of $\underline{\underline{G}}$ to ensure $\varinjlim \underline{\underline{G}}' = \varinjlim \underline{\underline{G}}$, and hence give an alternative proof of Proposition 32.

Corollary 34. Let R be a commutative ring. Then $K_1\underline{\underline{\text{Pic}}}(R) \simeq R^{\bullet}$.
Proof. $\underline{\underline{\text{Pic}}}(R)$ has a full cofinal subcategory with just one object, R itself. Constructing $\underline{\underline{G}}$ for this subcategory, we see that $\underline{\underline{G}}$ has only one object, $G(\bar{R}) = \text{aut}(R)^{\text{ab}} \simeq (R^{\bullet})^{\text{ab}} = R^{\bullet}$, and only one morphism, $1 : G(\bar{R}) \to G(\bar{R})$. So $\varinjlim \underline{\underline{G}} \simeq R^{\bullet}$, and the result follows by Propositions 32 and 33.

6.2 *K_1 of a ring*

Let R be a ring. As we remarked in the example on p.101, $\underline{\underline{P}}(R)$ has a full cofinal subcategory $\underline{\underline{P}}'(R)$ whose objects are 0 and R^n, $n = 1,2,\ldots$, and so $K_1R = K_1\underline{\underline{P}}'(R)$. If we construct the category $\underline{\underline{G}}$ for $\underline{\underline{P}}'(R)$, as above, then the objects of $\underline{\underline{G}}$ are $G(\bar{0}) = 0$ (which we may ignore) and the groups $\text{aut}(R^n)^{\text{ab}} = \text{GL}_n(R)^{\text{ab}}$, where n runs through a set of positive integers such that each isomorphism class of finitely generated free R-modules occurs precisely once. So if R has IBN (for the definition see p.11) then n runs through *all* positive integers, and $\underline{\underline{G}}$ has infinitely many objects; but if R does not have IBN, then $\underline{\underline{G}}$ has only finitely many objects.

Write $G_n = \text{GL}_n(R)^{\text{ab}}$. Then if R has IBN, $\underline{\underline{G}}$ consists of the following objects and morphisms:

$$G_1 \to G_2 \to G_3 \to \ldots \to G_n \to \ldots$$

the morphisms being those shown, together with composites and identities (since $R^n \oplus R^m \simeq R^r$ if and only if $n + m = r$). There is exactly one morphism from G_n to G_r in $\underline{\underline{G}}$ if $n \leq r$, and none if $n > r$. On the other hand, if R does not have IBN, then there exist integers n, m with $1 \leq n < m$ such that $R^n \simeq R^m$ (n, m least such), and $\underline{\underline{G}}$ consists of the following objects and morphisms:

$$G_1 \to G_2 \to \ldots \to G_{n-1} \to G_n \to G_{n+1} \to \ldots \to G_{m-1}$$

the morphisms being those shown together with composites and identities (the morphisms shown being $g(\bar{R}^r, \bar{R})$, $1 \le r < m$).

Exercise: Show that there may be more than one morphism from G_r to G_s in $\underline{\underline{G}}$, but not more than $1 + (m - 1)/(m - n)$ such morphisms.

We now construct a new category, $\hat{\underline{\underline{G}}}$, whose objects are the general linear groups $\mathrm{GL}_n(R)$, $n = 1,2,\ldots$ (infinitely many objects, whether or not R has IBN). For each n, we choose the special homomorphism $i_n : \mathrm{GL}_n(R) \to \mathrm{GL}_{n+1}(R)$ induced by the embedding of R^n in R^{n+1} given by

$$(x_1, x_2, \ldots, x_n) \mapsto (x_1, x_2, \ldots, x_n, 0)$$

so that

$$i_n(M) = \begin{pmatrix} & & & 0 \\ & M & & \vdots \\ & & & 0 \\ 0 & \ldots & 0 & 1 \end{pmatrix} \quad (M \in \mathrm{GL}_n(R)).$$

We take as the morphisms of $\hat{\underline{\underline{G}}}$ all the i_n together with composites and identities. It is easy to see that $\hat{\underline{\underline{G}}}$ is a directed category of (non-abelian) groups. We leave it to the reader to fill in the details, and also to show that for such a category, just as in the abelian case, one can construct a direct limit $\varinjlim \hat{\underline{\underline{G}}}$. In this particular case, we can construct the limit in a very explicit manner, for each i_n is a monomorphism, so if we identify $\mathrm{GL}_n(R)$ with its image under i_n in $\mathrm{GL}_{n+1}(R)$, we have

$$\mathrm{GL}_1(R) \subset \mathrm{GL}_2(R) \subset \mathrm{GL}_3(R) \subset \ldots$$

and we can form the tower group

$$\mathrm{GL}(R) = \bigcup_{n=1}^{\infty} \mathrm{GL}_n(R)$$

the *stable general linear group* of R. One can think of $\mathrm{GL}(R)$ as consisting of all infinite invertible matrices (a_{ij}) with $a_{ij} \in R$, $1 \le i < \infty$, $1 \le j < \infty$, and $a_{ij} = \delta_{ij}$, the Kronecker delta, for all but a finite number of values of i and j. Then $\mathrm{GL}_n(R)$ is embedded in $\mathrm{GL}(R)$ as the subgroup of all $(a_{ij}) \in \mathrm{GL}(R)$ with $a_{ij} = \delta_{ij}$ for all $i, j > n$. It is easy to see that $\mathrm{GL}(R) = \varinjlim \underline{\underline{\hat{G}}}$; details are left to the reader.

Now there is a map $\mathrm{GL}_n(R) \to K_1R$ given by $M \mapsto [R^n, M]$, identifying $\mathrm{GL}_n(R)$ with $\mathrm{aut}(R^n)$, and every diagram

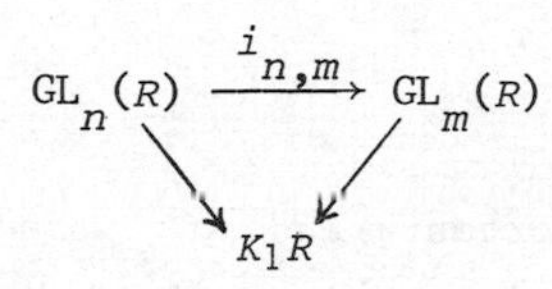

commutes, where $i_{n,n} = 1$ and $i_{n,m} = i_{m-1}i_{m-2}\ldots i_n$ $(n < m)$, since $[R^n, M] = [R^m, M \oplus I_{m-n}]$. So by the universal property of the direct limit there is a map $\mathrm{GL}(R) \to K_1R$, and since K_1R is abelian, this induces a map $\mathrm{GL}(R)^{\mathrm{ab}} \to K_1R$.

Proposition 35. For any ring R, $\mathrm{GL}(R)^{\mathrm{ab}} \cong K_1R$.

Proof. We have to show that the map constructed above is an isomorphism. Now $K_1R = K_1\underline{\underline{P}}'(R)$, so any element of K_1R is of the form $[R^n, M]$ for some n and $M \in \mathrm{GL}_n(R)$. Regarding $\mathrm{GL}_n(R)$ as a subgroup of $\mathrm{GL}(R)$, we see that the map $\mathrm{GL}(R)^{\mathrm{ab}} \to K_1R$ sends the image $\bar{M}$ of M in $\mathrm{GL}(R)^{\mathrm{ab}}$ to the given element $[R^n, M]$ of K_1R. So we have a surjection.

Before we can prove that the map is also injective, we need some more notation. As in the example at the beginning of this chapter (p.99) we let $B_{ij}(x) \in \mathrm{GL}_n(R)$ be the elementary matrix which differs from the $n \times n$ identity matrix I_n only in the i, j position $(i \ne j)$, where its entry is $x \in R$. The *elementary linear group* $\mathrm{E}_n(R)$ is the subgroup of the general linear group $\mathrm{GL}_n(R)$ generated by all $B_{ij}(x)$ $(1 \le i \le n,\ 1 \le j \le n,\ i \ne j,\ x \in R)$. Strictly, we should write $B_{ij}^{(n)}(x)$ for $B_{ij}(x)$, but if we identify $\mathrm{GL}_n(R)$ with its image in $\mathrm{GL}(R)$, it is not necessary to specify the value of n, since in $\mathrm{GL}(R)$, $B_{ij}^{(n)}(x) = B_{ij}^{(m)}(x)$, all $m > n$. Next, if i, j, k are distinct,

$$B_{ij}(x) = B_{ik}(x)B_{kj}(1)B_{ik}(x)^{-1}B_{kj}(1)^{-1}$$

so for $n \geq 3$, every element of $\mathrm{E}_n(R)$ can be written as a product of commutators: in other words, $\mathrm{E}_n(R)' = \mathrm{E}_n(R)$. Now let $A = (a_{ij})$ be any $n \times n$ matrix over R, not necessarily invertible. Working in $\mathrm{GL}_{2n}(R)$, we have, writing $I = I_n$,

$$\begin{pmatrix} I & A \\ 0 & I \end{pmatrix} = \prod_{i=1}^{n} \prod_{j=1}^{n} B_{i\ j+n}(a_{ij})$$

where the order of the terms is unimportant, and so

$$\begin{pmatrix} I & A \\ 0 & I \end{pmatrix} \in \mathrm{E}_{2n}(R), \text{ and similarly } \begin{pmatrix} I & 0 \\ A & I \end{pmatrix} \in \mathrm{E}_{2n}(R).$$

Then let $M \in \mathrm{GL}_n(R)$. So $M \oplus M^{-1} \in \mathrm{GL}_{2n}(R)$, but further,

$$M \oplus M^{-1} = \begin{pmatrix} M & 0 \\ 0 & M^{-1} \end{pmatrix} = \begin{pmatrix} I & 0 \\ M^{-1}-I & I \end{pmatrix} \begin{pmatrix} I & I \\ 0 & I \end{pmatrix} \begin{pmatrix} I & 0 \\ M-I & I \end{pmatrix} \begin{pmatrix} I & -M^{-1} \\ 0 & I \end{pmatrix}$$

and thus $M \oplus M^{-1} \in \mathrm{E}_{2n}(R)$. Passing to $\mathrm{GL}(R)$, $M \oplus M^{-1}$ can be written as a product of commutators, so it has trivial image in $\mathrm{GL}(R)^{\mathrm{ab}}$.

We can now complete the proof of Proposition 35. Let $A \in \mathrm{GL}(R)$, say $A \in \mathrm{GL}_n(R)$ for some n, and suppose $[R^n, A] = 0$ in $K_1 R$. We must show that A has trivial image in $\mathrm{GL}(R)^{\mathrm{ab}}$. By Proposition 31, working in $\Omega\underline{\underline{P}}'(R)$, we can find $B \in \mathrm{GL}_s(R)$, $C_1, C_2 \in \mathrm{GL}_t(R)$, and $D_1, D_2 \in \mathrm{GL}_u(R)$ such that

$$(R^m, A \oplus B \oplus C_1C_2 \oplus D_1 \oplus D_2) \simeq (R^r, B \oplus C_1 \oplus C_2 \oplus D_1D_2)$$

where $m = n + s + t + 2u$ and $r = s + 2t + u$. In particular, $R^m \simeq R^r$, though unless R has IBN we cannot deduce that $m = r$. But $(R^r, I_r) \simeq (R^m, I_m)$ in $\Omega\underline{\underline{P}}'(R)$, so we obtain

$$(R^{m+r}, A \oplus B \oplus C_1C_2 \oplus D_1 \oplus D_2 \oplus I_r)$$
$$\simeq (R^{m+r}, B \oplus C_1 \oplus C_2 \oplus D_1D_2 \oplus I_m)$$

and also

$$(R^{m+r}, I_n \oplus B^{-1} \oplus (C_1C_2)^{-1} \oplus D_1^{-1} \oplus D_2^{-1} \oplus I_r)$$
$$\simeq (R^{m+r}, B^{-1} \oplus I_t \oplus (C_1C_2)^{-1} \oplus D_2^{-1} \oplus D_1^{-1} \oplus I_{m-u})$$

the right-hand side being obtained by permuting the terms on the left-hand side. Combining, we have

$$(R^{m+r}, A \oplus I_{m+r-n})$$
$$\simeq (R^{m+r}, I_s \oplus C_1 \oplus C_1^{-1} \oplus D_1 \oplus D_1^{-1} \oplus I_{m-u})$$

so the matrices

$$A \oplus I_{m+r-n} \text{ and } I_s \oplus C_1 \oplus C_1^{-1} \oplus D_1 \oplus D_1^{-1} \oplus I_{m-u}$$

are conjugate in $\mathrm{GL}_{m+r}(R)$, and so have the same image in $\mathrm{GL}(R)^{\mathrm{ab}}$; but we have already seen that $C_1 \oplus C_1^{-1}$ and $D_1 \oplus D_1^{-1}$ have trivial image in $\mathrm{GL}(R)^{\mathrm{ab}}$, and hence so does A, and the proof is complete.

Note that the embedding $i_n : \mathrm{GL}_n(R) \hookrightarrow \mathrm{GL}_{n+1}(R)$ restricts to give an embedding $\mathrm{E}_n(R) \hookrightarrow \mathrm{E}_{n+1}(R)$; working in $\mathrm{GL}(R)$, the subgroup $\mathrm{E}(R) = \cup_{n=1}^{\infty} \mathrm{E}_n(R)$ is called the *stable elementary linear group* of R. Since $\mathrm{E}_n(R)' = \mathrm{E}_n(R)$ for $n \geq 3$, we have $\mathrm{E}(R)' = \mathrm{E}(R)$, and so $\mathrm{E}(R) \subset \mathrm{GL}(R)'$. But now if $A, B \in \mathrm{GL}_n(R)$, then in $\mathrm{GL}_{2n}(R)$,

$$\begin{pmatrix} ABA^{-1}B^{-1} & 0 \\ 0 & I \end{pmatrix} = \begin{pmatrix} A & 0 \\ 0 & A^{-1} \end{pmatrix} \begin{pmatrix} B & 0 \\ 0 & B^{-1} \end{pmatrix} \begin{pmatrix} (BA)^{-1} & 0 \\ 0 & BA \end{pmatrix}$$

where $I = I_n$, and as in the proof of Proposition 35, this lies in $\mathrm{E}_{2n}(R)$. Passing to the stable group, $ABA^{-1}B^{-1} \in \mathrm{E}(R)$, so $\mathrm{GL}(R)' \subset \mathrm{E}(R)$, and thus $\mathrm{GL}(R)' = \mathrm{E}(R)$. This result is known as the *Whitehead lemma*. It follows that $\mathrm{E}(R)$ is a normal subgroup of $\mathrm{GL}(R)$, and Proposition 35 can be restated as: for any ring R, $K_1R \simeq \mathrm{GL}(R)/\mathrm{E}(R)$. This is the construction of K_1R promised in the example on p.101; indeed the more usual definition of K_1R is just $K_1R = \mathrm{GL}(R)/\mathrm{E}(R)$. (It is then more natural to write the abelian group K_1R in multiplicative notation.)

Exercise: Let $\underline{\underline{G}}$ be a directed category of groups and group homomorphisms, and let $\underline{\underline{G}}^{\text{ab}}$ be the corresponding directed category of abelian groups and group homomorphisms given by taking ob $\underline{\underline{G}}^{\text{ab}} = \{G^{\text{ab}} : G \in \text{ob}\ \underline{\underline{G}}\}$, with morphisms induced by the morphisms of $\underline{\underline{G}}$. Show that $\varinjlim (\underline{\underline{G}}^{\text{ab}}) \simeq (\varinjlim \underline{\underline{G}})^{\text{ab}}$.

Remark: If R is a ring with IBN, and $\hat{\underline{\underline{G}}}$ is the category constructed on p.108, then $\hat{\underline{\underline{G}}}^{\text{ab}}$ is precisely the category $\underline{\underline{G}}$ of Proposition 33. Since $\varinjlim \hat{\underline{\underline{G}}} = \text{GL}(R)$, by definition, and $\varinjlim \underline{\underline{G}} = K_1R$, by Proposition 33, the result $K_1R \simeq \text{GL}(R)^{\text{ab}}$ follows from the above exercise. Thus in the special case when R has IBN we obtain a simpler proof of Proposition 35.

6.3 *Determinants and* SK_1

Let R be a commutative ring. The functor det : $\underline{\underline{P}}(R) \to \underline{\underline{\text{Pic}}}(R)$ is product-preserving: the quickest way to see this is to use the exercise on p.60. For if M, $N \in \text{ob}\ \underline{\underline{P}}(R)$ have rank maps f, g respectively, then $f + g$ is the rank map of $M \oplus N$, and so, in the notation of p.60,

$$\begin{aligned}\det(M \oplus N) = \wedge_R^{f+g} (M \oplus N) &\simeq (\wedge_R^f M) \otimes_R (\wedge_R^g N) \\ &= (\det M) \otimes_R (\det N).\end{aligned}$$

This depends on the generalization of Proposition 23 suggested on p.60; alternatively one can use idempotents, and we leave the details to the reader.

Since det : $\underline{\underline{P}}(R) \to \underline{\underline{\text{Pic}}}(R)$ is product-preserving, it induces $K_1(\det) : K_1R \to R^{\bullet}$ (using Corollary 34). Let us describe this map explicitly. A typical element of K_1R is of the form $[R^n, \alpha]$, with $\alpha \in \text{aut}(R^n)$. Now $\det(R^n) = \wedge_R^n R^n \simeq R$, and $\det(\alpha) = \wedge_R^n \alpha$, so $(\Omega\det)(R^n, \alpha) = (R, \wedge_R^n \alpha)$, whence $(K_1\det)[R^n, \alpha] = \wedge_R^n \alpha$, where we are identifying $\wedge_R^n R^n$ with R, and $K_1\underline{\underline{\text{Pic}}}(R)$ with $R^{\bullet} = \text{aut}(R)$. We thus have $\wedge_R^n \alpha \in R^{\bullet}$. Let

$\{v_1, v_2, \ldots, v_n\}$ be a basis for R^n: so $\{v_1 \wedge v_2 \wedge \ldots \wedge v_n\}$ is a basis for $\wedge_R^n R^n = R$. Let α have matrix $A = (a_{ij})$ with respect to this basis, that is, $\alpha(v_i) = \Sigma_j a_{ij}v_j$, $1 \le i \le n$. So

$$(\wedge_R^n \alpha)(v_1 \wedge v_2 \wedge \ldots \wedge v_n)$$
$$= \alpha(v_1) \wedge \alpha(v_2) \wedge \ldots \wedge \alpha(v_n)$$
$$= (\Sigma_j a_{1j}v_j) \wedge (\Sigma_j a_{2j}v_j) \wedge \ldots \wedge (\Sigma_j a_{nj}v_j)$$
$$= \Sigma_{\rho \in S_n} a_{1\rho_1}a_{2\rho_2}\ldots a_{n\rho_n}(v_{\rho_1} \wedge v_{\rho_2} \wedge \ldots \wedge v_{\rho_n}) \quad \text{(other terms vanishing)}$$
$$= \Sigma_{\rho \in S_n} \varepsilon(\rho)a_{1\rho_1}a_{2\rho_2}\ldots a_{n\rho_n}(v_1 \wedge v_2 \wedge \ldots \wedge v_n)$$
$$= (\det A)(v_1 \wedge v_2 \wedge \ldots \wedge v_n)$$

where S_n is the symmetric group, $\varepsilon(\rho)$ is the signature of $\rho \in S_n$, and det A is the usual matrix determinant. Thus det $\alpha = \wedge_R^n \alpha = \det A$, justifying the use of the word 'determinant' to describe the functor det : $\underline{\underline{P}}(R) \to \underline{\underline{Pic}}(R)$. Obviously (writing det for $K_1(\det)$) the map det : $K_1R \to R^{\cdot}$ is surjective; indeed, it is split by

$$R^{\cdot} \xrightarrow{\cong} \mathrm{GL}_1(R) \hookrightarrow \mathrm{GL}(R) \to K_1R$$
$$\alpha \longmapsto [R, \alpha].$$

Thus, if we define the *special Whitehead group*, or *special* K_1, of R to be $SK_1R = \ker(\det : K_1R \to R^{\cdot})$, and identify $R^{\cdot}$ with its image in K_1R, using the above map, we have:

Proposition 36. For any commutative ring R,

$$K_1R = R^{\cdot} \oplus SK_1R.$$

The *special linear groups* of R are the groups $\mathrm{SL}_n(R) = \ker(\det : \mathrm{GL}_n(R) \to R^{\cdot})$, and the *stable special linear group* of R is $\mathrm{SL}(R) = \cup_{n=1}^{\infty} \mathrm{SL}_n(R) = \ker(\det : \mathrm{GL}(R) \to R^{\cdot})$. Now elementary matrices always have trivial determinant, so $\mathrm{E}_n(R) \subset \mathrm{SL}_n(R)$, all n, and $\mathrm{E}(R) \subset \mathrm{SL}(R)$. Thus

$$SL(R)/E(R) = \ker(\det : GL(R)/E(R) \to R^{\bullet})$$

or in other words, since $GL(R)/E(R) = K_1R$, we have the alternative definition $SK_1R = SL(R)/E(R)$.

For any ring R, not necessarily commutative, we define $GE_n(R)$ to be the subgroup of $GL_n(R)$ generated by $E_n(R)$ and all $n \times n$ invertible diagonal matrices over R. Since the conjugate of an elementary matrix by a diagonal matrix is again elementary, any element of $GE_n(R)$ can be written as the product of a number of elementary matrices and one diagonal matrix. In other words, $GE_n(R)$ consists of those elements in $GL_n(R)$ which can be reduced to diagonal form by elementary row (or column) operations. We say R is a *GE_n-ring* if $GE_n(R) = GL_n(R)$, that is, if every $n \times n$ invertible matrix over R can be reduced to diagonal form by elementary row (or column) operations. Clearly, every field is a GE_n-ring, for all n; so is every Euclidean ring, local ring, and skew field. Every ring is a GE_1-ring; not every ring is a GE_n-ring for all n, as we shall see shortly. We say R is a *GE-ring* if it is a GE_n-ring for all n, and a *stable GE-ring* if it is a GE_n-ring for infinitely many values of n.

Exercise: Prove that a product of GE_n-rings is a GE_n-ring, and that if $R/\mathrm{rad}(R)$ is a GE_n-ring, so is R. Deduce that every semi-local ring is a GE-ring.

Now let R be a commutative ring. Every element of $GL_n(R)$ can be written as the product of a diagonal matrix and an element of $SL_n(R)$, from which it follows easily that R is a GE_n-ring if and only if $E_n(R) = SL_n(R)$. (The reason that this is not taken as the definition of GE_n-ring is that it does not make sense when R is not commutative.) So if R is a (stable) GE-ring, we have $E(R) = SL(R)$, or $SK_1R = 0$, and $K_1R = R^{\bullet}$.

We now exhibit a (commutative) ring R with $SK_1R \neq 0$. We take $R = \mathbb{R}[x, y]/(x^2 + y^2 - 1)$, where $\mathbb{R}$ is the field of real numbers. In R, $x^2 + y^2 = 1$, so any element of R can be regarded as a function $S^1 \to \mathbb{R}$, where $S^1 = \{(x, y) \in \mathbb{R}^2 : x^2 + y^2 = 1\}$. Explicitly, any polynomial in $\mathbb{R}[x, y]$ can be regarded as a function $\mathbb{R}^2 \to \mathbb{R}$, and if two such polynomials differ by a multiple of $x^2 + y^2 - 1$, then they agree on S^1. In a similar way, any element of $SL_n(R)$ can be regarded as a function $S^1 \to SL_n(\mathbb{R})$, that is, as a loop (or closed path) in $SL_n(\mathbb{R})$. This gives rise to a homomorphism $SL_n(R) \to \pi_1(SL_n(\mathbb{R}))$ (the fundamental group of $SL_n(\mathbb{R})$); further, loops coming from products of elementary matrices are easily seen to be contractible, so $E_n(R)$ lies in the kernel of this homomorphism. (For the definition of the fundamental group, see for example [10], Chapter 2.)

Now there is a retraction $SL_n(\mathbb{R}) \to SO(n)$ (the *special orthogonal group*, that is, all the orthogonal matrices in $SL_n(\mathbb{R})$) given by the Gram-Schmidt process, which therefore induces an isomorphism $\pi_1(SL_n(\mathbb{R})) \xrightarrow{\simeq} \pi_1(SO(n))$, the inverse isomorphism being induced by the inclusion $SO(n) \hookrightarrow SL_n(\mathbb{R})$. Then $\pi_1(SO(2)) \simeq \mathbb{Z}$ and $\pi_1(SO(n)) \simeq \mathbb{Z}/2\mathbb{Z}$ for $n \geq 3$, and the homomorphism $\pi_1(SO(n)) \to \pi_1(SO(n + 1))$ induced by $SO(n) \hookrightarrow SO(n + 1)$ is a surjection for $n = 2$ and an isomorphism for $n \geq 3$. (See [6], Chapter 7, sections 4 and 12.) We thus obtain a commutative diagram

$$\begin{array}{ccccccccc} SL_2(R) & \hookrightarrow & SL_3(R) & \hookrightarrow & \cdots & \hookrightarrow & SL_n(R) & \hookrightarrow & \cdots \\ \downarrow & & \downarrow & & & & \downarrow & & \\ \mathbb{Z} & \twoheadrightarrow & \mathbb{Z}/2\mathbb{Z} & \simeq & \cdots & \simeq & \mathbb{Z}/2\mathbb{Z} & \simeq & \cdots \end{array}$$

and on passing to the limit, we obtain a homomorphism $SL(R) \to \mathbb{Z}/2\mathbb{Z}$, with $E(R)$ contained in the kernel; hence we obtain a homomorphism $SK_1R = SL(R)/E(R) \to \mathbb{Z}/2\mathbb{Z}$. To see that SK_1R is nontrivial, it thus suffices to find a matrix in $SL_2(R)$ mapping to $1 \in \mathbb{Z}$ in the above diagram. But $S^1 \simeq SO(2)$ by

$$f : e^{i\theta} \mapsto \begin{pmatrix} \cos\theta & \sin\theta \\ -\sin\theta & \cos\theta \end{pmatrix}$$

and $\pi_1(S^1) \simeq \mathbb{Z}$ is generated by the homotopy class of the identity map $1 : S^1 \to S^1$, so $\pi_1(SO(2))$ is generated by the homotopy class of f, which may be written

$$f : (x, y) \mapsto \begin{pmatrix} x & y \\ -y & x \end{pmatrix}$$

where $x^2 + y^2 = 1$, and so is the image of

$$M = \begin{pmatrix} x & y \\ -y & x \end{pmatrix} \in SL_2(R).$$

So this matrix has non-trivial image in SK_1R, and $SK_1R \neq 0$.

Note that, for every $n \geq 2$, the above ring R is not a GE_n-ring. Indeed, we have found a matrix $M \in SL_2(R)$ whose image in the embedding $SL_2(R) \hookrightarrow SL_n(R)$ lies outside $E_n(R)$ for all $n \geq 2$.

6.4 *The non-stable K_1 of a commutative ring*

We now give another way of constructing K_1R when R is commutative.

Proposition 37. Let R be a commutative ring, and let $n \geq 3$. Then $E_n(R)$ is a normal subgroup of $GL_n(R)$.

Proof. Suppose first that U is an $n \times 1$ matrix over R, and that V is a $1 \times n$ matrix over R such that $VU = 0$. Suppose further that U is *unimodular*, that is, if $U = (u_1, u_2, \ldots, u_n)^T$ (where T denotes transposition) then $u_1R + u_2R + \ldots + u_nR = R$. Then we claim $1 + UV \in E_n(R)$. For suppose first that V has a zero entry, say $V = (x, 0)$, where x is a $1 \times m$ matrix over R, and $m < n$. Let $U = \begin{pmatrix} y \\ y_1 \end{pmatrix}$ where y is an $m \times 1$ matrix over R. Since $VU = 0$, we have $xy = 0$, and then (partitioning the matrices appropriately),

$$1 + UV = 1 + \begin{pmatrix} y \\ y_1 \end{pmatrix} (x \quad 0) = \begin{pmatrix} 1+yx & 0 \\ y_1x & 1 \end{pmatrix}$$

$$= \begin{pmatrix} 1 & y \\ 0 & 1 \end{pmatrix} \begin{pmatrix} 1 & 0 \\ x & 1 \end{pmatrix} \begin{pmatrix} 1 & -y \\ 0 & 1 \end{pmatrix} \begin{pmatrix} 1 & 0 \\ -x & 1 \end{pmatrix} \begin{pmatrix} 1 & 0 \\ y_1x & 1 \end{pmatrix}$$

and this is in $E_n(R)$, since $E_n(R)$ contains all uni-triangular matrices. A similar argument holds if V contains any zero entry, by permuting rows and columns appropriately. In the general case, since U is unimodular, there is a $1 \times n$ matrix W such that $WU = 1$. Put $S = W^T V - V^T W$, and note that $U^T S = U^T W^T V - U^T V^T W = (WU)^T V - (VU)^T W$ (since R is commutative), and so $U^T S = V$. Let e_{ij} be the usual 'matrix units', that is, e_{ij} is the $n \times n$ matrix with 1 in the i, j position and 0 elsewhere, and then put

$$V(i, j) = U^T(e_{ii} S e_{jj} + e_{jj} S e_{ii}).$$

Now since R is commutative, $(W^T V)^T = V^T W$, and so the diagonal entries of S are all zeros, so $e_{ii} S e_{ii} = 0$, all i. Also $1 = e_{11} + e_{22} + \ldots + e_{nn}$, and so

$$V = U^T S = \Sigma_i \Sigma_j U^T e_{ii} S e_{jj} = \Sigma_{i<j} V(i, j).$$

Further,

$$(U^T e_{ii} S e_{jj} U)^T = U^T e_{jj} S^T e_{ii} U = -U^T e_{jj} S e_{ii} U$$

so $V(i, j)U = 0$, all i, j, $i < j$. Thus

$$1 + UV = 1 + U(\Sigma_{i<j} V(i, j)) = \Pi_{i<j} (1 + UV(i, j))$$

where the order of the terms in the product is unimportant, and now the result follows from the first part, since $n \geq 3$, and for each i, j, $i < j$, $V(i, j)$ has at most two non-zero entries. So we have shown that $1 + UV \in \mathrm{E}_n(R)$.

We can now complete the proof of the proposition. Let $A \in \mathrm{GL}_n(R)$, $r \in R$, and $i \neq j$. We must show $A^{-1} B_{ij}(r) A \in \mathrm{E}_n(R)$. Let e_i be the $n \times 1$ matrix with 1 in the ith position and 0 elsewhere. Then

$$A^{-1} B_{ij}(r) A = A^{-1}(1 + re_{ij})A = 1 + A^{-1}(re_{ij})A = 1 + UV$$

where $U = A^{-1} e_i$ and $V = re_j^T A$. Now e_i is unimodular, and $A^{-1} \in \mathrm{GL}_n(R)$, so U is unimodular, and $VU = re_j^T A A^{-1} e_i = re_j^T e_i = 0$, since $i \neq j$, and thus $1 + UV \in \mathrm{E}_n(R)$, as above, and the proof is complete.

From the above proposition, if R is commutative and $n \geq 3$, it makes sense to define the *non-stable Whitehead group* $K_1(n, R) = \mathrm{GL}_n(R)/\mathrm{E}_n(R)$, and then $\mathrm{GL}_n(R) \hookrightarrow \mathrm{GL}_{n+1}(R)$ induces $K_1(n, R) \to K_1(n + 1, R)$. Further, $\mathrm{GL}_n(R) \hookrightarrow \mathrm{GL}(R)$ induces $K_1(n, R) \to K_1 R$, and every diagram

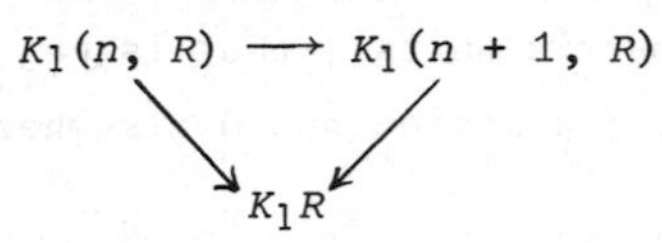

commutes, so there is a map $\varinjlim K_1(n, R) \to K_1 R$. We leave it to the reader to show that this is an isomorphism. The determinant map induces $\det : K_1(n, R) \to R^{\cdot}$, with kernel the *non-stable special Whitehead group* $SK_1(n, R) = \mathrm{SL}_n(R)/\mathrm{E}_n(R)$, and this is split by the composite $R^{\cdot} \simeq \mathrm{GL}_1(R) \hookrightarrow \mathrm{GL}_n(R) \to K_1(n, R)$, giving $K_1(n, R) \simeq R^{\cdot} \oplus SK_1(n, R)$. Further, $K_1(n, R) \to K_1(n + 1, R)$ is the direct sum of $1 : R^{\cdot} \to R^{\cdot}$ and the map $SK_1(n, R) \to SK_1(n + 1, R)$ induced by $\mathrm{SL}_n(R) \hookrightarrow \mathrm{SL}_{n+1}(R)$, and it is clear that $SK_1 R \simeq \varinjlim SK_1(n, R)$.

Problems. When is the map $K_1(n, R) \to K_1(n + 1, R)$: (a) injective, (b) surjective ?

When is the map $K_1(n, R) \to K_1 R$: (a) injective, (b) surjective ? (These questions are equivalent to the corresponding questions for SK_1.)

Is $K_1(n, R)$ (equivalently, $SK_1(n, R)$) abelian, either for all $n \geq 3$, or perhaps for sufficiently large n (maybe depending on R) ?

If R is a non-commutative ring, is $\mathrm{E}_n(R)$ a normal subgroup of $\mathrm{GL}_n(R)$ for $n \geq 3$, or perhaps for sufficiently large n ?

Note that $\mathrm{E}_2(R)$ need not be a normal subgroup of $\mathrm{GL}_2(R)$, even when R is commutative. For example, let k be a field, and let R be a ring containing k as a subring such that $R^{\cdot} = k^{\cdot}$. Suppose R has a degree function $d : R \to \mathbb{Z} \cup \{-\infty\}$ such that:

(i) $d(a) = -\infty$ if $a = 0$ and $d(a) \geq 0$ if $a \neq 0$;

(ii) $d(a) = 0$ if and only if $a \in R^{\bullet}$;

(iii) $d(a + b) \le \max(d(a), d(b))$, all $a, b \in R$;

(iv) $d(ab) = d(a) + d(b)$, all $a, b \in R$.

It follows that, if $d(b) < d(a)$, then $d(a + b) = d(a)$, and so if $d(a + b) < d(a)$, then $d(a) = d(b)$.

For $a \in R$, put $|a| = 2^{d(a)}$. Let $M \in GL_2(R)$, and suppose M has first row (a, b). Define $\sigma(M) = |a| + |b|$. Now let $M \in GE_2(R)$, and suppose

$$M = A_1 A_2 \dots A_r P \qquad \dots (*)$$

where $A_i = B_{12}(a_i)$ or $B_{21}(a_i)$, some $a_i \in R$, each i, and P is of the form

$$\begin{pmatrix} \alpha & 0 \\ 0 & \beta \end{pmatrix} \text{ or } \begin{pmatrix} 0 & \alpha \\ \beta & 0 \end{pmatrix}$$

for some $\alpha, \beta \in R^{\bullet}$. Put $\sigma_0 = 1$ and $\sigma_i = \sigma(A_1 A_2 \dots A_i)$, $1 \le i \le r$, and call

$$(\sigma_0, \sigma_1, \dots, \sigma_r)$$

the *diagram* of the expression (*) for M. Now either

$$\sigma_0 \le \sigma_1 \le \dots \le \sigma_r$$

or else $\sigma_\alpha > \sigma_{\alpha+1}$ for some α. In the former case, put $\lambda = \mu = 1$, and in the latter case, put

$$\lambda = \max\{\sigma_\alpha : \sigma_\alpha > \sigma_{\alpha+1}\} \text{ and } \mu = \max\{\alpha : \sigma_\alpha = \lambda > \sigma_{\alpha+1}\}.$$

We show that there is an expression of the form (*) for M whose diagram is obtained from the above diagram either by omitting or by reducing the term σ_μ. In either case, the new diagram will be monotonic increasing, or the new value of λ will be less than the old value, or else it will be unchanged and the new value of μ will be less than the old value. After a finite number of repetitions of the argument, we must obtain an expression of the form (*) for M with monotonic increasing diagram.

Suppose that $A_1 A_2 \dots A_\mu$ has first row (a, b), and suppose $A_\mu = B_{12}(p)$. Then $A_1 A_2 \dots A_{\mu-1}$ has first row $(a, b - ap)$, and since $\sigma_{\mu-1} \le \sigma_\mu$, we have $|b - ap| \le |b|$. Let $A_{\mu+1} = B_{12}(q)$. Then

$B_{12}(p)B_{12}(q) = B_{12}(p + q)$, so

$$M = A_1A_2\ldots A_{\mu-1}A'_\mu A_{\mu+2}\ldots A_rP$$

where $A'_\mu = B_{12}(p + q)$, and this expression has diagram

$$(\sigma_0, \sigma_1, \ldots, \sigma_{\mu-1}, \sigma_{\mu+1}, \ldots, \sigma_r).$$

Next suppose $A_{\mu+1} = B_{21}(q)$. So $A_1A_2\ldots A_{\mu+1}$ has first row $(a + bq, b)$, and since $\sigma_{\mu+1} < \sigma_\mu$, we have $|a + bq| < |a|$, whence $|a| = |bq|$. Next, if $|ap| > |b|$, then $|b - ap| = |ap| > |b|$, a contradiction; so $|ap| \le |b|$. Thus $|apq| \le |bq| = |a|$; but $a \ne 0$ since $|a + bq| < |a|$, and so $|pq| \le 1$, whence $pq \in k$, and so $1 + pq \in k$.

Suppose that $1 + pq = 0$, so $pq = -1$ and $|p||q| = 1$, so $p, q \in R^\bullet$. Then

$$B_{12}(p)B_{21}(q) = B_{21}(-q)Q, \text{ where } Q = \begin{pmatrix} 0 & p \\ q & 0 \end{pmatrix}.$$

Thus $M = A_1A_2\ldots A_{\mu-1}A'_\mu A'_{\mu+2}\ldots A'_rP'$, where $A'_\mu = B_{21}(-q)$, $A'_i = QA_iQ^{-1}$, $\mu + 2 \le i \le r$, and $P' = QP$. Each A'_i is an elementary matrix, and $A_1A_2\ldots A'_\mu A'_{\mu+2}\ldots A'_iQ = A_1A_2\ldots A_i$, whence

$$\sigma(A_1A_2\ldots A'_\mu A'_{\mu+2}\ldots A'_i) = \sigma(A_1A_2\ldots A_i) = \sigma_i, \ \mu + 2 \le i \le r.$$

Also $A_1A_2\ldots A_{\mu-1}A'_\mu$ has first row $(a - (b - ap)q, b - ap) = (-bq, b - ap)$, and $|-bq| = |b| = |a|$, $|b - ap| = |(b - ap)q| = |a + bq|$, whence $\sigma(A_1A_2\ldots A_{\mu-1}A'_\mu) = \sigma_{\mu+1}$, and the diagram for the new expression for M is $(\sigma_0, \sigma_1, \ldots, \sigma_{\mu-1}, \sigma_{\mu+1}, \ldots, \sigma_r)$.

Now suppose $1 + pq \ne 0$, so $1 + pq = \alpha \in k^\bullet = R^\bullet$. Then

$$B_{12}(p)B_{21}(q) = B_{21}(q\alpha^{-1})B_{12}(\alpha p)Q, \text{ where } Q = \begin{pmatrix} \alpha & 0 \\ 0 & \alpha^{-1} \end{pmatrix}.$$

So $M = A_1A_2\ldots A_{\mu-1}A'_\mu A'_{\mu+1}A'_{\mu+2}\ldots A'_rP'$, where $A'_\mu = B_{21}(q\alpha^{-1})$, $A'_{\mu+1} = B_{12}(\alpha p)$, $A'_i = QA_iQ^{-1}$, $\mu + 2 \le i \le r$, and $P' = QP$. As before, each A'_i is elementary, and $\sigma(A_1A_2\ldots A_{\mu-1}A'_\mu\ldots A'_i) = \sigma_i$ for $\mu + 1 \le i \le r$. Then $A_1A_2\ldots A_{\mu-1}A'_\mu$ has first row

$$(a + (b - ap)q\alpha^{-1}, b - ap) = ((a + bq)\alpha^{-1}, b - ap)$$

and $|(a + bq)\alpha^{-1}| = |a + bq| < |a|$, and $|b - ap| \le |b|$, so if $\sigma'_\mu = \sigma(A_1A_2\ldots A_{\mu-1}A'_\mu)$, then $\sigma'_\mu < \sigma_\mu$, and the new expression for M has diagram $(\sigma_0, \sigma_1, \ldots, \sigma_{\mu-1}, \sigma'_\mu, \sigma_{\mu+1}, \ldots, \sigma_r)$.

The above argument was based on the assumption that $A_\mu = B_{12}(p)$; a similar argument holds if $A_\mu = B_{21}(p)$. So now we may assume that in the expression (*) for M, $\sigma_0 \le \sigma_1 \le \ldots \le \sigma_r$. If, for some i, $\sigma_i = \sigma_{i+1}$, and $A_1A_2\ldots A_{i+1}$ has first row (a, b), then $A_1A_2\ldots A_i$ has first row $(a, b - ap)$ or $(a - bp, b)$, with $|b - ap| = |b|$ or $|a - bp| = |a|$ respectively, and if $|a| = |b|$, it follows that $p \in k$. Thus, if M has first row (a, b) with $|a| = |b|$, and if $\sigma_i = \sigma_{i+1} = \ldots = \sigma_r$, then $A_1A_2\ldots A_i$ has first row (c, d), where $|c| = |d| = |a| = |b|$, and $c = ap + bq$, $d = ar + bs$ for some $p, q, r, s \in k$ with $ps - rq \ne 0$. If we now choose $i \le r$ so that $\sigma_{i-1} < \sigma_i = \sigma_{i+1} = \ldots = \sigma_r$, then $A_1A_2\ldots A_{i-1}$ has first row $(c - dt, d)$ or $(c, d - ct)$ for some $t \in R$ with $|c - dt| < |c|$ or $|d - ct| < |d|$, respectively. In either case, $t \in k$, and on substituting for c, d, we see that there exist $u, v \in k$, not both zero, such that $|au + bv| < |a|$ $(= |b|)$.

Now let us take a particular example. Let $R = k[x, y]$, with the usual degree function, and let

$$M = \begin{pmatrix} 1+xy & x^2 \\ -y^2 & 1-xy \end{pmatrix}.$$

Then $M \in \mathrm{GL}_2(R)$, and $1 + xy$ and x^2 both have degree 2, but if $u, v \in k$, not both zero, then $(1 + xy)u + x^2v$ has degree 2 also, and so $M \notin \mathrm{GE}_2(R)$. Thus R is not a GE_2-ring. Further,

$$MB_{12}(1)M^{-1} = \begin{pmatrix} 1+xy & x^2 \\ -y^2 & 1-xy \end{pmatrix}\begin{pmatrix} 1 & 1 \\ 0 & 1 \end{pmatrix}\begin{pmatrix} 1-xy & -x^2 \\ y^2 & 1+xy \end{pmatrix}$$

$$= \begin{pmatrix} 1+y^2+xy^3 & 1+2xy+x^2y^2 \\ -y^4 & 1-y^2-xy^3 \end{pmatrix}.$$

Then $1 + y^2 + xy^3$ and $1 + 2xy + x^2y^2$ both have degree 4, and if

$u, v \in k$, not both zero, then

$$(1 + y^2 + xy^3)u + (1 + 2xy + x^2y^2)v$$

has degree 4 also, whence $MB_{12}(1)M^{-1} \notin GE_2(R)$, and thus $MB_{12}(1)M^{-1} \notin E_2(R)$, so that $E_2(R)$ is not a normal subgroup of $GL_2(R)$.

Exercise: Let R, M be as above. Show that the image of M in the embedding $SL_2(R) \hookrightarrow SL_3(R)$ lies inside $E_3(R)$ (cf. the situation on p.116).

6.5 *Dieudonné determinants, and K_1 of semi-local rings*

Let R be a commutative ring. Recall that the homomorphisms $\det : GL_n(R) \to R^{\cdot}$ have the following properties:

(a) $\ker(\det) \supset E_n(R)$;

(b)
$$\det\begin{pmatrix} \alpha_1 & & & 0 \\ & \alpha_2 & & \\ & & \ddots & \\ 0 & & & \alpha_n \end{pmatrix} = \alpha_1\alpha_2\cdots\alpha_n \quad (\alpha_i \in R^{\cdot}, \text{ all } i);$$

(c) The diagram

$$\begin{array}{ccc} GL_n(R) & \longrightarrow & GL_{n+1}(R) \\ \det \searrow & & \swarrow \det \\ & R^{\cdot} & \end{array}$$

commutes.

Indeed, if R is a GE-ring, properties (a) and (b) characterize the determinant map. We now examine the situation when R is not (necessarily) commutative.

Proposition 38. (i) Let R be a GE-ring. For $n \geq 2$, $GL_n(R)' \subset E_n(R)$.

(ii) For any ring R, and for $n \geq 3$, $GL_n(R)' \supset E_n(R)$.

(iii) Let R be a ring such that for any $x \in R$ there exist $\alpha, \beta \in R^{\cdot}$ with $x = \alpha + \beta$. Then $GL_2(R)' \supset E_2(R)$.

Proof. (i) By assumption, any matrix in $GL_n(R)$ can be written as a product of elementary and diagonal matrices. Now

$$\begin{pmatrix} \alpha_1 & & & 0 \\ & \alpha_2 & & \\ & & \ddots & \\ 0 & & & \alpha_n \end{pmatrix} B_{ij}(x) = B_{ij}(\alpha_i x \alpha_j^{-1}) \begin{pmatrix} \alpha_1 & & & 0 \\ & \alpha_2 & & \\ & & \ddots & \\ 0 & & & \alpha_n \end{pmatrix}$$

where $\alpha_i \in R^{\bullet}$, all i, and $x \in R$, and a product of diagonal matrices is again diagonal, so any matrix in $GL_n(R)$ can be written as a product of a diagonal matrix and a matrix in $E_n(R)$. So let $M_1 = D_1E_1$ and $M_2 = D_2E_2$, where D_1, D_2 are diagonal (and invertible) and $E_1, E_2 \in E_n(R)$. Then

$$M_1M_2M_1^{-1}M_2^{-1} = D_1E_1D_2E_2E_1^{-1}D_1^{-1}E_2^{-1}D_2^{-1} = D_1D_2D_1^{-1}D_2^{-1}E$$

where $E \in E_n(R)$. The diagonal entries of the diagonal matrix $D = D_1D_2D_1^{-1}D_2^{-1}$ are commutators of units; but, taking $n = 2$ to simplify the notation, if $\alpha \in R^{\bullet}$,

$$\begin{pmatrix} \alpha & 0 \\ 0 & \alpha^{-1} \end{pmatrix} = \begin{pmatrix} 1 & \alpha-1 \\ 0 & 1 \end{pmatrix}\begin{pmatrix} 1 & 0 \\ 1 & 1 \end{pmatrix}\begin{pmatrix} 1 & \alpha^{-1}-1 \\ 0 & 1 \end{pmatrix}\begin{pmatrix} 1 & 0 \\ -\alpha & 1 \end{pmatrix} \in E_2(R)$$

and so, if $\alpha, \beta \in R^{\bullet}$,

$$\begin{pmatrix} \alpha\beta\alpha^{-1}\beta^{-1} & 0 \\ 0 & 1 \end{pmatrix} = \begin{pmatrix} \alpha & 0 \\ 0 & \alpha^{-1} \end{pmatrix}\begin{pmatrix} \beta & 0 \\ 0 & \beta^{-1} \end{pmatrix}\begin{pmatrix} (\beta\alpha)^{-1} & 0 \\ 0 & \beta\alpha \end{pmatrix} \in E_2(R).$$

Similarly, if $\gamma, \delta \in R^{\bullet}$, then

$$\begin{pmatrix} 1 & 0 \\ 0 & \gamma\delta\gamma^{-1}\delta^{-1} \end{pmatrix} \in E_2(R), \text{ so } \begin{pmatrix} \alpha\beta\alpha^{-1}\beta^{-1} & 0 \\ 0 & \gamma\delta\gamma^{-1}\delta^{-1} \end{pmatrix} \in E_2(R).$$

So $D \in E_2(R)$ when $n = 2$, and essentially the same argument shows that in general $D \in E_n(R)$. Thus $M_1M_2M_1^{-1}M_2^{-1} \in E_n(R)$.

(ii) This is immediate from the equation

$$B_{ij}(x) = B_{ik}(x)B_{kj}(1)B_{ik}(x)^{-1}B_{kj}(1)^{-1}$$

where i, j, k are distinct subscripts and $x \in R$.

(iii) Let $x \in R$, $x = \alpha + \beta$, where $\alpha, \beta \in R^{\bullet}$. Then

$$\begin{pmatrix}1 & x\\ 0 & 1\end{pmatrix} = \begin{pmatrix}-\alpha\beta^{-1} & 0\\ 0 & 1\end{pmatrix}\begin{pmatrix}1 & -\beta\\ 0 & 1\end{pmatrix}\begin{pmatrix}-\beta\alpha^{-1} & 0\\ 0 & 1\end{pmatrix}\begin{pmatrix}1 & \beta\\ 0 & 1\end{pmatrix}$$

so $B_{12}(x) \in GL_2(R)'$, and similarly $B_{21}(x) \in GL_2(R)'$, and the result follows.

Note that, by the above proposition, we cannot hope to have homomorphisms det : $GL_n(R) \to R^\bullet$ satisfying (a) - (c) (p.122) when R is not commutative, or rather when $R^\bullet$ is not abelian. For by (b) such homomorphisms must be surjective, and by (a) and Proposition 38(i), if R is a GE-ring, then $GL_n(R)' \subset \ker(\det)$, so $R^\bullet$ must be abelian. However, for suitable non-commutative rings R we can construct a homomorphism det : $GL_n(R) \to (R^\bullet)^{ab}$, for each n, such that

(a) $\ker(\det) \supset E_n(R)$;

(b) $$\det\begin{pmatrix}\alpha_1 & & & 0\\ & \alpha_2 & & \\ & & \ddots & \\ 0 & & & \alpha_n\end{pmatrix} = \overline{\alpha_1\alpha_2\ldots\alpha_n},$$ where $\alpha_i \in R^\bullet$, all i, and

$\alpha \mapsto \bar{\alpha}$ is the natural map $R^\bullet \to (R^\bullet)^{ab}$;

(c) The diagram

$$\begin{array}{ccc} GL_n(R) & \longrightarrow & GL_{n+1}(R) \\ \det \searrow & & \swarrow \det \\ & (R^\bullet)^{ab} & \end{array}$$

commutes.

Such a map is called a *Dieudonné determinant*. Of course, every commutative ring has a Dieudonné determinant, namely the ordinary matrix determinant map. Note also that if R is a GE-ring and has a Dieudonné determinant, the determinant is characterized by properties (a) and (b).

Proposition 39. Let R be a skew field. Then R has a Dieudonné determinant.

Proof. We use induction on n. For $n = 1$, $GL_1(R) = R^{\bullet}$, and $\det : R^{\bullet} \to (R^{\bullet})^{ab}$ is the natural map $\alpha \mapsto \bar{\alpha}$. Clearly (a) and (b) are satisfied (remember $E_1(R) = 1$), and (c) does not apply. So now assume $n \geq 1$, and assume $\det : GL_m(R) \to (R^{\bullet})^{ab}$ has been defined for $m \leq n$, satisfying (a) - (c) where appropriate. Let $M \in GL_{n+1}(R)$. The last row of M must have at least one non-zero entry, so we can write $M = M_1M_2$, where M_1 has a non-zero entry in the $(n + 1, n + 1)$ position, and $M_2 = B_{i\ n+1}(z)$, some $z \in R$ and $i \leq n$. Then $M_1 = N_1N_2$, where N_1 has last row $(0, 0, \ldots, 0, \alpha)$, some $\alpha \in R^{\bullet}$, and $N_2 = \Pi_{i=1}^{n} B_{n+1\ i}(y_i)$, some $y_i \in R$. Thus we can write (partitioning as appropriate)

$$M = \begin{pmatrix} A & \underline{x} \\ 0 & \alpha \end{pmatrix} \begin{pmatrix} I & \underline{0} \\ \underline{y} & 1 \end{pmatrix} \begin{pmatrix} I & \underline{z} \\ 0 & 1 \end{pmatrix}$$

where $A \in GL_n(R)$, $\alpha \in R^{\bullet}$, and $\underline{x}, \underline{y}, \underline{z} \in R^n$ (rows or columns, as appropriate). This will be called a *normal expression* for M; it is not unique, in general. The *determinant* of this expression is defined to be $\bar{\alpha} \det A$. We must show first that this depends only on M, and not on the choice of normal expression for M. We need a lemma:

Lemma. Let

$$M = \begin{pmatrix} I & \underline{0} \\ \underline{a} & 1 \end{pmatrix} \begin{pmatrix} I & \underline{b} \\ \underline{0} & 1 \end{pmatrix} \begin{pmatrix} I & \underline{0} \\ \underline{c} & 1 \end{pmatrix}$$

where $\underline{a}, \underline{b}, \underline{c} \in R^n$ (rows or columns, as appropriate). Then M has a normal expression with determinant 1.

Proof. Multiplying up, we see

$$M = \begin{pmatrix} I+\underline{bc} & \underline{b} \\ \underline{a}+\underline{c}+\underline{abc} & 1+\underline{ab} \end{pmatrix}.$$

First suppose $1 + \underline{ab} = \beta \in R^{\bullet}$. Thus

$$M = \begin{pmatrix} I+\underline{bc} & \underline{b} \\ \underline{a}+\beta\underline{c} & \beta \end{pmatrix} = \begin{pmatrix} I-\underline{b}\beta^{-1}\underline{a} & \underline{b} \\ \underline{0} & \beta \end{pmatrix} \begin{pmatrix} I & \underline{0} \\ \beta^{-1}\underline{a}+\underline{c} & 1 \end{pmatrix}.$$

This is a normal expression for M, with determinant $\bar{\beta}\det(I - \underline{b}\beta^{-1}\underline{a})$. Now since R is a skew field, we can find $N \in GL_n(R)$ such that $N\underline{b} = \begin{pmatrix} r \\ \underline{0} \end{pmatrix}$, some $r \in R$, and then $\underline{a}N^{-1} = (s, \underline{t})$, some $s \in R$, $\underline{t} \in R^{n-1}$. So $\beta = 1 + (\underline{a}N^{-1})(N\underline{b}) = 1 + sr$. Then

$$\begin{aligned}
\det(I - \underline{b}\beta^{-1}\underline{a}) &= \det(N(I - \underline{b}\beta^{-1}\underline{a})N^{-1}) \\
&= \det(I - (N\underline{b})\beta^{-1}(\underline{a}N^{-1})) \\
&= \det\begin{pmatrix} 1 - r\beta^{-1}s & r\beta^{-1}\underline{t} \\ \underline{0} & I \end{pmatrix} \\
&= \det\begin{pmatrix} 1 - r\beta^{-1}s & \underline{0} \\ \underline{0} & I \end{pmatrix}, \text{ by (a)}, \\
&= \overline{1 - r\beta^{-1}s}, \text{ by (b)}.
\end{aligned}$$

If $r = 0$, then $\beta = 1$ and $\overline{1 - r\beta^{-1}s} = 1$. If $r \neq 0$, then

$$\begin{aligned}
\overline{1 - r\beta^{-1}s} &= \bar{r}\,\overline{(r^{-1} - \beta^{-1}s)} = \overline{(r^{-1} - \beta^{-1}s)}\,\bar{r} \\
&= \overline{1 - \beta^{-1}sr} = \bar{\beta}^{-1}\,\overline{(\beta - sr)} = \bar{\beta}^{-1}
\end{aligned}$$

and in either case we have $\bar{\beta}\det(I - \underline{b}\beta^{-1}\underline{a}) = 1$, as required. Secondly, suppose $1 + \underline{\underline{ab}} \notin R^{\bullet}$, that is, $1 + \underline{\underline{ab}} = 0$, since R is a skew field. Thus

$$M = \begin{pmatrix} I + \underline{\underline{bc}} & \underline{b} \\ \underline{a} & 0 \end{pmatrix} = \begin{pmatrix} I + \underline{\underline{bc}} + \underline{\underline{bcba}} & -\underline{\underline{bcb}} \\ \underline{0} & 1 \end{pmatrix}\begin{pmatrix} I & \underline{0} \\ \underline{a} & 1 \end{pmatrix}\begin{pmatrix} I & \underline{b} \\ \underline{0} & 1 \end{pmatrix}.$$

This is a normal expression for M; we must show $\det(I + \underline{\underline{bc}} + \underline{\underline{bcba}}) = 1$. As before, choose $N \in GL_n(R)$ so that $N\underline{b} = \begin{pmatrix} r \\ \underline{0} \end{pmatrix}$, some $r \in R$, and let $\underline{a}N^{-1} = (s, \underline{t})$ and $\underline{c}N^{-1} = (u, \underline{v})$, where $s, u \in R$ and $\underline{t}, \underline{v} \in R^{n-1}$. Then

$$0 = 1 + \underline{\underline{ab}} = 1 + (\underline{a}N^{-1})(N\underline{b}) = 1 + sr$$

so $sr = -1$, and thus $rs = -1$ also. Thus

$$\det(I + \underline{\underline{bc}} + \underline{\underline{bcba}}) = \det(N(I + \underline{\underline{bc}} + \underline{\underline{bcba}})N^{-1})$$

$$= \det(I + (N\underline{b})(\underline{c}N^{-1}) + (N\underline{b})(\underline{c}N^{-1})(N\underline{b})(\underline{a}N^{-1}))$$

$$= \det\begin{pmatrix} 1+ru+rurs & r\underline{v}+rur\underline{t} \\ \underline{0} & I \end{pmatrix}$$

$$= \det\begin{pmatrix} 1 & r\underline{v}+rur\underline{t} \\ \underline{0} & I \end{pmatrix}, \text{ since } rs = -1,$$

$$= 1, \text{ by (a).}$$

This completes the proof of the lemma.

To complete the proof of Proposition 39, suppose $M \in \mathrm{GL}_{n+1}(R)$, and suppose

$$M = \begin{pmatrix} A_1 & \underline{x}_1 \\ \underline{0} & \alpha_1 \end{pmatrix}\begin{pmatrix} I & \underline{0} \\ \underline{y}_1 & 1 \end{pmatrix}\begin{pmatrix} I & \underline{z}_1 \\ \underline{0} & 1 \end{pmatrix} = \begin{pmatrix} A_2 & \underline{x}_2 \\ \underline{0} & \alpha_2 \end{pmatrix}\begin{pmatrix} I & \underline{0} \\ \underline{y}_2 & 1 \end{pmatrix}\begin{pmatrix} I & \underline{z}_2 \\ \underline{0} & 1 \end{pmatrix}$$

are two normal expressions for M. We must show $\bar{\alpha}_1 \det A_1 = \bar{\alpha}_2 \det A_2$. We have

$$\begin{pmatrix} A_2 & \underline{x}_2 \\ \underline{0} & \alpha_2 \end{pmatrix}^{-1}\begin{pmatrix} A_1 & \underline{x}_1 \\ \underline{0} & \alpha_1 \end{pmatrix} = \begin{pmatrix} I & \underline{0} \\ \underline{y}_2 & 1 \end{pmatrix}\begin{pmatrix} I & \underline{z}_2 \\ \underline{0} & 1 \end{pmatrix}\begin{pmatrix} I & \underline{z}_1 \\ \underline{0} & 1 \end{pmatrix}^{-1}\begin{pmatrix} I & \underline{0} \\ \underline{y}_1 & 1 \end{pmatrix}^{-1}$$

or

$$\begin{pmatrix} A_2^{-1}A_1 & A_2^{-1}\underline{x}_1 - A_2^{-1}\underline{x}_2\alpha_2^{-1}\alpha_1 \\ \underline{0} & \alpha_2^{-1}\alpha_1 \end{pmatrix}$$

$$= \begin{pmatrix} I & \underline{0} \\ \underline{y}_2 & 1 \end{pmatrix}\begin{pmatrix} I & \underline{z}_2-\underline{z}_1 \\ \underline{0} & 1 \end{pmatrix}\begin{pmatrix} I & \underline{0} \\ -\underline{y}_1 & 1 \end{pmatrix}$$

$$= \begin{pmatrix} A & \underline{x} \\ \underline{0} & \alpha \end{pmatrix}\begin{pmatrix} I & \underline{0} \\ \underline{y} & 1 \end{pmatrix}\begin{pmatrix} I & \underline{z} \\ \underline{0} & 1 \end{pmatrix}$$

for some $A \in \mathrm{GL}_n(R)$, $\alpha \in R^{\bullet}$, and $\underline{x}$, $\underline{y}$, $\underline{z} \in R^n$, with $\bar{\alpha} \det A = 1$, by the lemma. So

$$\begin{pmatrix} A_2^{-1}A_1 & A_2^{-1}\underline{x}_1 - A_2^{-1}\underline{x}_2\alpha_2^{-1}\alpha_1 - A_2^{-1}A_1\underline{z} \\ \underline{0} & \alpha_2^{-1}\alpha_1 \end{pmatrix}$$

$$= \begin{pmatrix} A & \underline{x} \\ \underline{0} & \alpha \end{pmatrix} \begin{pmatrix} I & \underline{0} \\ \underline{y} & 1 \end{pmatrix} = \begin{pmatrix} A+\underline{x}\underline{y} & \underline{x} \\ \alpha\underline{y} & \alpha \end{pmatrix}$$

and thus $\alpha\underline{y} = 0$, $\underline{y} = 0$, $A = A_2^{-1}A_1$, and $\alpha = \alpha_2^{-1}\alpha_1$. Therefore

$$1 = \bar{\alpha} \det A = \overline{\alpha_2^{-1}\alpha_1} \det(A_2^{-1}A_1)$$

$$= (\bar{\alpha}_2 \det A_2)^{-1}(\bar{\alpha}_1 \det A_1)$$

and so $\bar{\alpha}_1 \det A_1 = \bar{\alpha}_2 \det A_2$ and we have a well-defined map det : $GL_{n+1}(R) \to (R^{\cdot})^{ab}$. We must show that it is a homomorphism.

Let $M_i \in GL_{n+1}(R)$, and suppose M_i has a normal expression

$$M_i = \begin{pmatrix} A_i & \underline{x}_i \\ \underline{0} & \alpha_i \end{pmatrix} \begin{pmatrix} I & \underline{0} \\ \underline{y}_i & 1 \end{pmatrix} \begin{pmatrix} I & \underline{z}_i \\ \underline{0} & 1 \end{pmatrix}, \ i = 1, 2.$$

Then

$$M_1M_2 = \begin{pmatrix} A_1 & \underline{x}_1 \\ \underline{0} & \alpha_1 \end{pmatrix} \begin{pmatrix} I & \underline{0} \\ \underline{y}_1 & 1 \end{pmatrix} \begin{pmatrix} I & \underline{z}_1 \\ \underline{0} & 1 \end{pmatrix} \begin{pmatrix} A_2 & \underline{x}_2 \\ \underline{0} & \alpha_2 \end{pmatrix} \begin{pmatrix} I & \underline{0} \\ \underline{y}_2 & 1 \end{pmatrix} \begin{pmatrix} I & \underline{z}_2 \\ \underline{0} & 1 \end{pmatrix}$$

$$= \begin{pmatrix} A_1 & \underline{x}_1 \\ \underline{0} & \alpha_1 \end{pmatrix} \begin{pmatrix} I & \underline{0} \\ \underline{y}_1 & 1 \end{pmatrix} \begin{pmatrix} A_2 & \underline{x}_2+\underline{z}_1\alpha_2 \\ \underline{0} & \alpha_2 \end{pmatrix} \begin{pmatrix} I & \underline{0} \\ \underline{y}_2 & 1 \end{pmatrix} \begin{pmatrix} I & \underline{z}_2 \\ \underline{0} & 1 \end{pmatrix}$$

$$= \begin{pmatrix} A_1 & \underline{x}_1 \\ \underline{0} & \alpha_1 \end{pmatrix} \begin{pmatrix} I & \underline{0} \\ \underline{y}_1 & 1 \end{pmatrix} \begin{pmatrix} A_2 & \underline{0} \\ \underline{0} & \alpha_2 \end{pmatrix} \begin{pmatrix} I & \underline{x}_2' \\ \underline{0} & 1 \end{pmatrix} \begin{pmatrix} I & \underline{0} \\ \underline{y}_2 & 1 \end{pmatrix} \begin{pmatrix} I & \underline{z}_2 \\ \underline{0} & 1 \end{pmatrix}$$

(where $\underline{x}_2' = A_2^{-1}(\underline{x}_2 + \underline{z}_1\alpha_2)$)

$$= \begin{pmatrix} A_1A_2 & \underline{x}_1' \\ \underline{0} & \alpha_1\alpha_2 \end{pmatrix} \begin{pmatrix} I & \underline{0} \\ \underline{y}_1' & 1 \end{pmatrix} \begin{pmatrix} I & \underline{x}_2' \\ \underline{0} & 1 \end{pmatrix} \begin{pmatrix} I & \underline{0} \\ \underline{y}_2 & 1 \end{pmatrix} \begin{pmatrix} I & \underline{z}_2 \\ \underline{0} & 1 \end{pmatrix}$$

(where $\underline{x}_1' = \underline{x}_1\alpha_2$ and $\underline{y}_1' = \alpha_2^{-1}\underline{y}_1A_2$)

$$= \begin{pmatrix} A_1A_2 & \underline{x}_1' \\ \underline{0} & \alpha_1\alpha_2 \end{pmatrix} \begin{pmatrix} A & \underline{x} \\ \underline{0} & \alpha \end{pmatrix} \begin{pmatrix} I & \underline{0} \\ \underline{y} & 1 \end{pmatrix} \begin{pmatrix} I & \underline{z} \\ \underline{0} & 1 \end{pmatrix} \begin{pmatrix} I & \underline{z}_2 \\ \underline{0} & 1 \end{pmatrix}$$

(say, by the lemma, where $\bar{\alpha} \det A = 1$)

$$= \begin{pmatrix} A_1A_2A & \underline{x}' \\ \underline{0} & \alpha_1\alpha_2\alpha \end{pmatrix} \begin{pmatrix} I & \underline{0} \\ \underline{y} & 1 \end{pmatrix} \begin{pmatrix} I & \underline{z}' \\ \underline{0} & 1 \end{pmatrix}$$

(where $\underline{x}' = A_1A_2\underline{x} + \underline{x}_1'\alpha$ and $\underline{z}' = \underline{z} + \underline{z}_2$). This is a normal expression for M_1M_2, and so

$$\begin{aligned} \det(M_1M_2) &= \overline{\alpha_1\alpha_2\alpha}\ \det(A_1A_2A) \\ &= (\bar{\alpha}_1 \det A_1)(\bar{\alpha}_2 \det A_2)(\bar{\alpha} \det A) \\ &= (\det M_1)(\det M_2). \end{aligned}$$

So we have a homomorphism. It is clear that $\det(B_{ij}(x)) = 1$, all $x \in R$ and all i, j, and so $E_{n+1}(R) \subset \ker(\det)$. Also

$$\begin{aligned} \det\begin{pmatrix} \alpha_1 & & & 0 \\ & \alpha_2 & & \\ & & \ddots & \\ 0 & & & \alpha_{n+1} \end{pmatrix} &= \bar{\alpha}_{n+1} \det\begin{pmatrix} \alpha_1 & & & 0 \\ & \alpha_2 & & \\ & & \ddots & \\ 0 & & & \alpha_n \end{pmatrix} \\ &= \bar{\alpha}_{n+1}(\overline{\alpha_1\alpha_2\cdots\alpha_n}) \\ &= \overline{\alpha_1\alpha_2\cdots\alpha_n\alpha_{n+1}}. \end{aligned}$$

Finally, if $A \in GL_n(R)$, then

$$\det\begin{pmatrix} A & \underline{0} \\ \underline{0} & 1 \end{pmatrix} = \det A$$

so the diagram

$$\begin{array}{ccc} GL_n(R) & \longrightarrow & GL_{n+1}(R) \\ \det \searrow & & \swarrow \det \\ & (R^\cdot)^{ab} & \end{array}$$

commutes. This completes the proof of Proposition 39.

Corollary 40. Let R be a skew field with Dieudonné determinant $\det : GL_n(R) \to (R^\cdot)^{ab}$. For $n \geq 2$, $\ker(\det) = E_n(R)$.

Proof. We use induction on n. Let $M \in GL_2(R)$, so M has a normal expression

$$M = \begin{pmatrix} \alpha & x \\ 0 & \beta \end{pmatrix} \begin{pmatrix} 1 & 0 \\ y & 1 \end{pmatrix} \begin{pmatrix} 1 & z \\ 0 & 1 \end{pmatrix}$$

with det $M = \overline{\alpha\beta}$. Thus

$$M = \begin{pmatrix} \alpha & 0 \\ 0 & \beta \end{pmatrix} E$$

where $E \in \mathrm{E}_2(R)$, and so

$$M = \begin{pmatrix} \alpha\beta & 0 \\ 0 & 1 \end{pmatrix} \begin{pmatrix} \beta^{-1} & 0 \\ 0 & \beta \end{pmatrix} E$$

and furthermore

$$\begin{pmatrix} \beta^{-1} & 0 \\ 0 & \beta \end{pmatrix} \in \mathrm{E}_2(R)$$

as on p.123. If det $M = 1$, then $\alpha\beta \in (R^{\bullet})'$, so

$$\begin{pmatrix} \alpha\beta & 0 \\ 0 & 1 \end{pmatrix}$$

is a product of matrices of the form

$$\begin{pmatrix} \gamma\delta\gamma^{-1}\delta^{-1} & 0 \\ 0 & 1 \end{pmatrix}$$

each of which is in $\mathrm{E}_2(R)$, as on p.123. So $M \in \mathrm{E}_2(R)$, and thus ker(det) $= \mathrm{E}_2(R)$.

Now suppose $M \in \mathrm{GL}_{n+1}(R)$, with normal expression

$$M = \begin{pmatrix} A & \underline{x} \\ \underline{0} & \alpha \end{pmatrix} \begin{pmatrix} I & \underline{0} \\ \underline{y} & 1 \end{pmatrix} \begin{pmatrix} I & \underline{z} \\ \underline{0} & 1 \end{pmatrix}.$$

Thus

$$M = \begin{pmatrix} A & \underline{0} \\ \underline{0} & \alpha \end{pmatrix} E$$

where $E \in \mathrm{E}_{n+1}(R)$, and then

$$M = \begin{pmatrix} A_1 & \underline{0} \\ \underline{0} & 1 \end{pmatrix} E_1$$

where

$$A_1 = A\begin{pmatrix} I & \underline{0} \\ \underline{0} & \alpha \end{pmatrix} \in \mathrm{GL}_n(R)$$

and

$$E_1 = \begin{pmatrix} I & \underline{0} & \underline{0} \\ \underline{0} & \alpha^{-1} & 0 \\ \underline{0} & 0 & \alpha \end{pmatrix} E.$$

Since

$$\begin{pmatrix} \alpha^{-1} & 0 \\ 0 & \alpha \end{pmatrix} \in \mathrm{E}_2(R)$$

we have $E_1 \in \mathrm{E}_{n+1}(R)$. Then

$$\det A_1 = (\det A)(\det\begin{pmatrix} I & \underline{0} \\ \underline{0} & \alpha \end{pmatrix}) = (\det A)\bar{\alpha} = \det M$$

so if $M \in \ker(\det)$, we have $\det A_1 = 1$, and so $A_1 \in \mathrm{E}_n(R)$ by the inductive hypothesis. Thus

$$\begin{pmatrix} A_1 & \underline{0} \\ \underline{0} & 1 \end{pmatrix} \in \mathrm{E}_{n+1}(R)$$

and therefore $M \in \mathrm{E}_{n+1}(R)$, and the proof is complete.

Exercise: Let R be a skew field, and let

$$M = \begin{pmatrix} a & b \\ c & d \end{pmatrix} \in \mathrm{GL}_2(R).$$

Show that, if $d = 0$, then $\det M = -\overline{bc}$, and if $d \neq 0$, then

$$\det M = \overline{ad - bd^{-1}cd}.$$

Give an example to show that this last expression need not be equal to any of $\overline{ad - bc}$, $\overline{ad - cb}$, $\overline{da - bc}$, or $\overline{da - cb}$.

The above results show that, if R is a skew field and $n \geq 2$, there is an exact sequence

$$1 \to \mathrm{E}_n(R) \to \mathrm{GL}_n(R) \xrightarrow{\det} (R^{\bullet})^{\mathrm{ab}} \to 1.$$

Now if $n \geq 3$, or if $n = 2$ and $|R| > 2$, then $\mathrm{E}_n(R) = \mathrm{GL}_n(R)'$, by

Proposition 38. So except in the case $n = 2 = |R|$, the sequence

$$1 \to GL_n(R)' \to GL_n(R) \xrightarrow{\det} (R^{\bullet})^{ab} \to 1$$

is exact, or in other words, $\det : GL_n(R) \to (R^{\bullet})^{ab}$ induces an isomorphism $\overline{\det} : GL_n(R)^{ab} \to (R^{\bullet})^{ab}$. Note that $\overline{\det}^{-1} : (R^{\bullet})^{ab} \to GL_n(R)^{ab}$ is the map induced by the natural map $R^{\bullet} \to GL_n(R)$, regarding $R^{\bullet}$ as $GL_1(R)$.

Let R be any ring. We shall call R a *D_n-ring* if the natural map $R^{\bullet} \to GL_n(R)$ induces an isomorphism $(R^{\bullet})^{ab} \to GL_n(R)^{ab}$. R is a *D-ring* (*Dieudonné ring*) if it is a D_n-ring for all $n \geq 1$.

Proposition 41. (i) Every ring is a D_1-ring.

(ii) Every skew field R with $|R| > 2$ is a D-ring.

(iii) If $|R| = 2$, R is a D_n-ring if and only if $n \neq 2$.

Proof. (i) is obvious, and (ii) and (iii) are done already, except to note that the case $n = 2 = |R|$ is a genuine exception, since then $GL_2(R)$ is non-abelian of order 6, so $|GL_2(R)^{ab}| = 2$, whereas $R^{\bullet} = 1$, so $(R^{\bullet})^{ab} = 1$.

Exercise: Show that every D-ring has a Dieudonné determinant.

Proposition 42. If R is a D_n-ring for all $n \geq m$, some m, then $K_1 R \simeq (R^{\bullet})^{ab}$.

Proof. The diagram

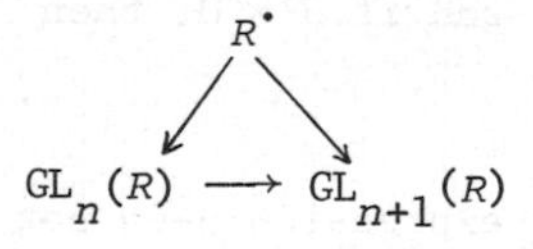

commutes, and for $n \geq m$ it gives, on abelianizing,

$$\begin{array}{ccc} & (R^{\bullet})^{ab} & \\ \simeq \swarrow & & \searrow \simeq \\ GL_n(R)^{ab} & \longrightarrow & GL_{n+1}(R)^{ab} \end{array}$$

so $GL_n(R)^{ab} \to GL_{n+1}(R)^{ab}$ is an isomorphism for $n \geq m$. Then

$$K_1R \simeq \mathrm{GL}(R)^{\mathrm{ab}} = (\varinjlim \mathrm{GL}_n(R))^{\mathrm{ab}} \simeq \varinjlim (\mathrm{GL}_n(R)^{\mathrm{ab}}) \simeq (R^{\bullet})^{\mathrm{ab}}.$$

Corollary 43. If R is a skew field, $K_1R \simeq (R^{\bullet})^{\mathrm{ab}}$.

Proof. Immediate from Propositions 41 and 42.

Now if R is any ring, write $R^{(m)}$ for the ring of all $m \times m$ matrices with coefficients in R. Note that $(R^{(m)})^{\bullet} = \mathrm{GL}_m(R)$, $(R^{(m)})^{(n)} \simeq R^{(mn)}$, $\mathrm{GL}_n(R^{(m)}) \simeq \mathrm{GL}_{nm}(R)$, and $\mathrm{GL}(R^{(m)}) \simeq \mathrm{GL}(R)$.

Proposition 44. Let S be a (left) Artinian ring (defined in Lemma 12) with $\mathrm{rad}(S) = 0$, say

$$S \simeq R_1^{(n_1)} \times R_2^{(n_2)} \times \ldots \times R_m^{(n_m)}$$

where each R_i is a skew field. Then

$$K_1S \simeq (R_1^{\bullet} \times R_2^{\bullet} \times \ldots \times R_m^{\bullet})^{\mathrm{ab}}.$$

Proof. Note first that the existence of the given decomposition of S is the Wedderburn-Artin theorem (the corollary to Proposition 13).

Now if S_1, S_2 are rings, $\mathrm{GL}_n(S_1 \times S_2) \simeq \mathrm{GL}_n(S_1) \times \mathrm{GL}_n(S_2)$, so $\mathrm{GL}(S_1 \times S_2) \simeq \mathrm{GL}(S_1) \times \mathrm{GL}(S_2)$, and on abelianizing, we have $K_1(S_1 \times S_2) \simeq K_1S_1 \oplus K_1S_2$. Then if R is any ring, $K_1(R^{(m)}) \simeq K_1R$, since $\mathrm{GL}(R^{(m)}) \simeq \mathrm{GL}(R)$. Putting this together,

$$\begin{aligned} K_1S &\simeq K_1R_1 \oplus K_1R_2 \oplus \ldots \oplus K_1R_m \\ &\simeq (R_1^{\bullet})^{\mathrm{ab}} \oplus (R_2^{\bullet})^{\mathrm{ab}} \oplus \ldots \oplus (R_m^{\bullet})^{\mathrm{ab}} \end{aligned}$$

by Corollary 43, and the result follows.

We now show that, with a few exceptions, the conclusion of the above proposition can be amended to read $K_1S \simeq (S^{\bullet})^{\mathrm{ab}}$.

Proposition 45. (i) If R is a D_m-ring and a D_{mn}-ring, and $S = R^{(m)}$, then S is a D_n-ring.

(ii) If R, S are D_n-rings, so is $R \times S$.

Proof. (i) The diagram

$$\begin{array}{ccc} & R^{\cdot} & \\ \swarrow & & \searrow \\ \mathrm{GL}_m(R) & \longrightarrow & \mathrm{GL}_{mn}(R) \end{array}$$

commutes, and on abelianizing it gives

$$\begin{array}{ccc} & (R^{\cdot})^{\mathrm{ab}} & \\ \cong\swarrow & & \searrow\cong \\ \mathrm{GL}_m(R)^{\mathrm{ab}} & \longrightarrow & \mathrm{GL}_{mn}(R)^{\mathrm{ab}}. \end{array}$$

So the natural map $\mathrm{GL}_m(R) \to \mathrm{GL}_{mn}(R)$ induces an isomorphism $\mathrm{GL}_m(R)^{\mathrm{ab}} \to \mathrm{GL}_{mn}(R)^{\mathrm{ab}}$, that is, the natural map $S^{\cdot} \to \mathrm{GL}_n(S)$ induces an isomorphism $(S^{\cdot})^{\mathrm{ab}} \to \mathrm{GL}_n(S)^{\mathrm{ab}}$.

(ii) The diagram

$$\begin{array}{ccc} R^{\cdot} \times S^{\cdot} & \longrightarrow & \mathrm{GL}_n(R) \times \mathrm{GL}_n(S) \\ \cong\downarrow & & \downarrow\cong \\ (R \times S)^{\cdot} & \longrightarrow & \mathrm{GL}_n(R \times S) \end{array}$$

commutes, and on abelianizing gives

$$\begin{array}{ccc} (R^{\cdot})^{\mathrm{ab}} \times (S^{\cdot})^{\mathrm{ab}} & \xrightarrow{\cong} & \mathrm{GL}_n(R)^{\mathrm{ab}} \times \mathrm{GL}_n(S)^{\mathrm{ab}} \\ \cong\downarrow & & \downarrow\cong \\ (R^{\cdot} \times S^{\cdot})^{\mathrm{ab}} & \longrightarrow & (\mathrm{GL}_n(R) \times \mathrm{GL}_n(S))^{\mathrm{ab}} \\ \cong\downarrow & & \downarrow\cong \\ ((R \times S)^{\cdot})^{\mathrm{ab}} & \longrightarrow & \mathrm{GL}_n(R \times S)^{\mathrm{ab}} \end{array}$$

whence the result.

Proposition 46. Let R be a skew field, and let $S = R^{(m)}$.

(i) If $|R| > 2$, S is a D-ring.

(ii) If $|R| = 2$ and $m \geq 3$, S is a D-ring.

(iii) If $|R| = 2 = m$, S is not a D_n-ring for any $n \geq 2$.

(iv) If $|R| = 2$ and $m = 1$, S is a D_n-ring if and only if $n \neq 2$.

Proof. Immediate from Propositions 41 and 45.

Corollary 47. Let S be a (left) Artinian ring with $\mathrm{rad}(S) = 0$, and suppose S has no direct factor isomorphic to $R^{(2)}$, where $|R| = 2$. Then S is a D_n-ring for all $n \neq 2$, and $K_1S \simeq (S^{\bullet})^{ab}$.

Proof. Immediate from Propositions 42, 45, and 46.

The condition on the direct factors of S in Corollary 47 is a little unsatisfactory, and we now describe a way of overcoming this. Let R be a ring, and suppose $x, y \in R$ with $1 + xy \in R^{\bullet}$. Then it is easy to check that

$$(1 + yx)(1 - y\alpha^{-1}x) = 1 = (1 - y\alpha^{-1}x)(1 + yx)$$

where $\alpha = 1 + xy$, and so $1 + yx \in R^{\bullet}$. An alternative way of seeing this is to note that, for any $x, y \in R$,

$$\begin{pmatrix}1 & x\\ 0 & 1\end{pmatrix}\begin{pmatrix}1 & 0\\ y & 1\end{pmatrix}\begin{pmatrix}1 & 0\\ 0 & 1+yx\end{pmatrix} = \begin{pmatrix}1+xy & 0\\ 0 & 1\end{pmatrix}\begin{pmatrix}1 & 0\\ y & 1\end{pmatrix}\begin{pmatrix}1 & x\\ 0 & 1\end{pmatrix}$$

from which

$$\begin{pmatrix}1 & 0\\ 0 & 1+yx\end{pmatrix} \in \mathrm{GL}_2(R) \text{ if and only if } \begin{pmatrix}1+xy & 0\\ 0 & 1\end{pmatrix} \in \mathrm{GL}_2(R).$$

An immediate consequence of the above relation is that, if $1 + xy \in R^{\bullet}$, then

$$\begin{pmatrix}1+xy & 0\\ 0 & (1+yx)^{-1}\end{pmatrix} \in \mathrm{E}_2(R)$$

and hence, since

$$\begin{pmatrix}\alpha & 0\\ 0 & \alpha^{-1}\end{pmatrix} \in \mathrm{E}_2(R), \text{ all } \alpha \in R^{\bullet}$$

we have

$$\begin{pmatrix}(1+xy)(1+yx)^{-1} & 0\\ 0 & 1\end{pmatrix} \in \mathrm{E}_2(R)$$

and indeed, if $n \geq 2$, the $n \times n$ diagonal matrix with diagonal entries $(1 + xy)(1 + yx)^{-1}, 1, 1, \ldots, 1$ belongs to $\mathrm{E}_n(R)$. Let us write $V(R)$ for the subgroup of $R^{\bullet}$ generated by all expressions $(1 + xy)(1 + yx)^{-1}$ $(x, y \in R,\ 1 + xy \in R^{\bullet})$. If $\alpha, \beta \in R^{\bullet}$, then putting $x = \alpha - \beta^{-1}$, $y = \beta$, we see $1 + xy = \alpha\beta \in R^{\bullet}$, and

$1 + yx = \beta\alpha$, so $(1 + xy)(1 + yx)^{-1} = \alpha\beta\alpha^{-1}\beta^{-1}$, and thus $(R^{\bullet})' \subset V(R)$, and also $V(R)$ is a normal subgroup of $R^{\bullet}$.

Proposition 48. If R is a D_n-ring for some $n \geq 3$, then $V(R) = (R^{\bullet})'$.

Proof. We saw above that the natural map $R^{\bullet} \to GL_n(R)$ maps $V(R)$ into $E_n(R)$. But $E_n(R) \subset GL_n(R)'$, by Proposition 38, and the result follows.

Let us write $\mathcal{V}(R) = R^{\bullet}/V(R)$. We shall say that R is a *D_n'-ring* if the natural map $R^{\bullet} \to GL_n(R)$ induces an isomorphism $\mathcal{V}(R) \to GL_n(R)^{ab}$. By the last proposition, if R is a D_n-ring, and $n \geq 3$, then R is a D_n'-ring.

Proposition 49. If R is a D_n'-ring for all $n \geq m$, some m, then $K_1R \simeq \mathcal{V}(R)$.

Proof. The proof is similar to that of Proposition 42, and details are left to the reader.

Proposition 50. Let R be a skew field, and let $S = R^{(m)}$. Then S is a D_n'-ring for all $n \geq 3$.

Proof. If $|R| > 2$, or if $|R| = 2$ and $m \neq 2$, then R is a D_n-ring for all $n \geq 3$, by Proposition 46, and the result follows. Now suppose $|R| = 2 = m$. Then $R^{\bullet} = 1$, and $E_2(R) = GL_2(R)$. Working in S, put

$$x = \begin{pmatrix} 1 & 0 \\ 0 & 0 \end{pmatrix} \text{ and } y = \begin{pmatrix} 0 & 1 \\ 0 & 0 \end{pmatrix}$$

so $xy = y$ and $yx = 0$, so $(1 + xy)(1 + yx)^{-1} = B_{12}(1) \in V(S)$. Similarly $B_{21}(1) \in V(S)$, and so $E_2(R) \subset V(S)$, and hence $V(S) = GL_2(R) = S^{\bullet}$, that is, $\mathcal{V}(S) = 1$. But then $GL_n(S) = GL_{2n}(R)$, and since $n \geq 3$, $GL_{2n}(R)^{ab} \simeq (R^{\bullet})^{ab} = 1$, by Proposition 41. Thus $GL_n(S)^{ab} = 1$, and so S is a D_n'-ring.

Proposition 51. If R, S are D_n'-rings, so is $R \times S$.

Proof. The natural isomorphism $R^{\bullet} \times S^{\bullet} \to (R \times S)^{\bullet}$ restricts to give an isomorphism $V(R) \times V(S) \to V(R \times S)$, and so induces an isomorphism $\mathcal{V}(R) \times \mathcal{V}(S) \to \mathcal{V}(R \times S)$. The result now follows from the commutativity of the diagram

$$\begin{array}{ccc} \mathcal{V}(R) \times \mathcal{V}(S) & \xrightarrow{\simeq} & \mathrm{GL}_n(R)^{\mathrm{ab}} \times \mathrm{GL}_n(S)^{\mathrm{ab}} \\ \simeq\downarrow & & \downarrow\simeq \\ \mathcal{V}(R \times S) & \longrightarrow & \mathrm{GL}_n(R \times S)^{\mathrm{ab}}. \end{array}$$

Corollary. Let S be a (left) Artinian ring with $\mathrm{rad}(S) = 0$. Then S is a D_n'-ring for all $n \geq 3$.

Proof. Immediate from the Wedderburn-Artin theorem (p.27, corollary), and Propositions 50 and 51.

It follows immediately from this corollary and Proposition 49 that $K_1S \simeq \mathcal{V}(S)$; in fact, we shall show that this isomorphism holds whenever S is semi-local (Proposition 53, below).

Proposition 52. Let R be any ring, with $J = \mathrm{rad}(R)$, and suppose R/J is a D_m'-ring, where $m \geq 3$. Then R is a D_m'-ring.

Proof. There is a subgroup $C \subset R^{\bullet}$ such that the diagram

$$\begin{array}{ccccccccc} 1 & \longrightarrow & C & \longrightarrow & R^{\bullet} & \longrightarrow & \mathrm{GL}_m(R)^{\mathrm{ab}} & \longrightarrow & 1 \\ & & \downarrow & & \downarrow & & \downarrow & & \\ 1 & \to & V(R/J) & \to & (R/J)^{\bullet} & \to & \mathrm{GL}_m(R/J)^{\mathrm{ab}} & \to & 1 \end{array}$$

commutes and has exact rows. The hypothesis $m \geq 3$ is needed to ensure that $R^{\bullet} \to \mathrm{GL}_m(R)^{\mathrm{ab}}$ is surjective, which follows from $E_m(R) \subset \mathrm{GL}_m(R)'$ (Proposition 38); details are left to the reader. We must show $C = V(R)$; now clearly $V(R) \subset C$, and $V(R) \to V(R/J)$ is surjective, so it is sufficient to prove that $C \cap (1 + J) \subset V(R)$, that is, if $\alpha \in 1 + J$ and the diagonal matrix with diagonal entries $\alpha, 1, 1, \ldots, 1$ belongs to $E_m(R)$, then $\alpha \in V(R)$.

Write $\mathrm{GL}_n(R, J) = \ker(\mathrm{GL}_n(R) \to \mathrm{GL}_n(R/J))$ and $\mathrm{E}_n(R, J) = \mathrm{E}_n(R) \cap \mathrm{GL}_n(R, J)$, all n.

Lemma. There is a homomorphism $\det : \mathrm{GL}_n(R, J) \to V(R)$ with

(a) $\ker(\det) \supset \mathrm{E}_n(R, J)$;

(b)
$$\det\begin{pmatrix} \alpha_1 & & & 0 \\ & \alpha_2 & & \\ & & \ddots & \\ 0 & & & \alpha_n \end{pmatrix} = \overline{\alpha_1\alpha_2\cdots\alpha_n}, \text{ where } \alpha_i \in 1 + J, \text{ all } i,$$

and $\alpha \mapsto \bar{\alpha}$ is the natural map $R^{\bullet} \to V(R)$;

(c) The diagram

$$\begin{array}{ccc} \mathrm{GL}_n(R, J) & \longrightarrow & \mathrm{GL}_{n+1}(R, J) \\ \det \searrow & & \swarrow \det \\ & V(R) & \end{array}$$

commutes;

(d) If $\underline{a}, \underline{b} \in J^n$ (row and column, respectively), then $\det(I + \underline{ba}) = \overline{1 + \underline{ab}}$.

Proof. This is on similar lines to the proof of Proposition 39. We use induction on n. For $n = 1$, $\det : 1 + J \to V(R)$ is the natural map, which satisfies (a), (b), (d); and (c) does not apply.

So now let $n \geq 1$ and let $M \in \mathrm{GL}_{n+1}(R, J)$. Since $M \equiv I \pmod{J}$ and $1 + J \subset R^{\bullet}$, we can write *uniquely*

$$M = \begin{pmatrix} A & \underline{x} \\ \underline{0} & \alpha \end{pmatrix}\begin{pmatrix} I & \underline{0} \\ \underline{y} & 1 \end{pmatrix}$$

where $A \in \mathrm{GL}_n(R, J)$, $\alpha \in 1 + J$, and $\underline{x}, \underline{y} \in J^n$. We then define $\det M = \bar{\alpha} \det A$, and this is well defined.

Next, let $\underline{a}, \underline{b}, \underline{c} \in J^n$ (rows or columns, as appropriate), and let

$$M = \begin{pmatrix} I & \underline{0} \\ \underline{a} & 1 \end{pmatrix}\begin{pmatrix} I & \underline{b} \\ \underline{0} & 1 \end{pmatrix}\begin{pmatrix} I & \underline{0} \\ \underline{c} & 1 \end{pmatrix}.$$

Put $\beta = 1 + \underline{a}\underline{b} \in 1 + J$; then, as in the proof of the lemma in the proof of Proposition 39,

$$\det M = \overline{\beta \det(I - \underline{b}\beta^{-1}\underline{a})}$$
$$= \overline{\beta(1 - \beta^{-1}\underline{a}\underline{b})} \text{ (by (d), in } GL_n)$$
$$= \overline{\beta - \underline{a}\underline{b}} = 1.$$

The proof that det is a homomorphism now proceeds exactly as in the proof of Proposition 39 (except that the $\underline{z}$-terms are missing). Conditions (a), (b), (c) are clearly satisfied; for (d), let $\underline{a}$, $\underline{b} \in J^n$ (row and column, respectively) and $p, q \in J$. We must show

$$\det\begin{pmatrix} I+\underline{b}\underline{a} & \underline{b}p \\ q\underline{a} & 1+qp \end{pmatrix} = \overline{1 + \underline{a}\underline{b} + pq}.$$

Put

$$M = \begin{pmatrix} I+\underline{b}\underline{a} & \underline{b}p \\ q\underline{a} & 1+qp \end{pmatrix} = \begin{pmatrix} A & \underline{b}p \\ \underline{0} & 1+qp \end{pmatrix}\begin{pmatrix} I & \underline{0} \\ (1+qp)^{-1}q\underline{a} & 1 \end{pmatrix}$$

say, where

$$I + \underline{b}\underline{a} = A + \underline{b}p(1 + qp)^{-1}q\underline{a}$$

or

$$A = I + \underline{b}(1 - p(1 + qp)^{-1}q)\underline{a} = I + \underline{b}(1 + pq)^{-1}\underline{a}.$$

So

$$\det M = \overline{(1 + qp)}\ \overline{\det(I + \underline{b}(1 + pq)^{-1}\underline{a})}$$
$$= \overline{(1 + qp)}\,\overline{(1 + (1 + pq)^{-1}\underline{a}\underline{b})} \text{ (by (d), in } GL_n)$$
$$= \overline{(1 + qp)}\ \overline{(1 + pq)}^{-1}\ \overline{(1 + pq + \underline{a}\underline{b})}$$
$$= \overline{1 + pq + \underline{a}\underline{b}}$$

as required, and the lemma is proved.

To complete the proof of Proposition 52, note that if $\alpha \in 1 + J$ and the diagonal matrix with diagonal entries $\alpha, 1, 1, \ldots, 1$ belongs to $E_m(R)$, then applying det gives $\bar{\alpha} = 1$, by (a) and (b), whence $\alpha \in V(R)$, as required.

Corollary. Let R be a semi-local ring. Then R is a D'_n-ring for all $n \geq 3$.

Proof. Immediate from Proposition 52 and the corollary to Proposition 51.

Proposition 53. Let R be a semi-local ring. Then $K_1R \simeq V(R)$.

Proof. Immediate from Proposition 49 and the corollary to Proposition 52.

Exercise: Let R be a D'_n-ring for all $n \geq m$. Prove that there is a homomorphism det : $GL_n(R) \to V(R)$, all n, satisfying the three conditions (a), (b), (c) for a Dieudonné determinant (p.124), but with $V(R)$ in place of $(R^{\bullet})^{ab}$ throughout.

CHAPTER SEVEN

Exact Sequences

Let $(\underline{\underline{C}}, \perp)$, $(\underline{\underline{D}}, \perp)$ be categories with product, and let $f : \underline{\underline{C}} \to \underline{\underline{D}}$ be a product-preserving functor. The *fibre category* Φf has as objects all triples (M, N, α) with $M, N \in \text{ob } \underline{\underline{C}}$ and $\alpha : \overline{M} \to \overline{N}$ an isomorphism in $\underline{\underline{D}}$, where we are writing $\overline{M} = f(M)$, $\overline{N} = f(N)$. A morphism in Φf is a pair $(\beta, \gamma) : (M, N, \alpha) \to (M', N', \alpha')$, where $\beta : M \to M'$ and $\gamma : N \to N'$ are isomorphisms in $\underline{\underline{C}}$ such that the diagram

$$\begin{array}{ccc} \overline{M} & \xrightarrow{\alpha} & \overline{N} \\ \overline{\beta}\downarrow & & \downarrow\overline{\gamma} \\ \overline{M'} & \xrightarrow{\alpha'} & \overline{N'} \end{array}$$

commutes, where we are writing $\overline{\beta} = f(\beta)$, $\overline{\gamma} = f(\gamma)$. Morphisms are composed componentwise. If $(M, N, \alpha), (M', N', \alpha') \in \text{ob } \Phi f$, then $\alpha \perp \alpha' : \overline{M} \perp \overline{M'} \to \overline{N} \perp \overline{N'}$ is an isomorphism in $\underline{\underline{D}}$, and since f is product-preserving, there are isomorphisms $\psi : \overline{M \perp M'} \to \overline{M} \perp \overline{M'}$, $\psi : \overline{N \perp N'} \to \overline{N} \perp \overline{N'}$ (see p.86). Thus we have an isomorphism $\psi^{-1}(\alpha \perp \alpha')\psi : \overline{M \perp M'} \to \overline{N \perp N'}$. and we may thus define

$$(M, N, \alpha) \perp (M', N', \alpha') = (M \perp M', N \perp N', \psi^{-1}(\alpha \perp \alpha')\psi).$$

For $(M, N, \alpha), (N, P, \beta) \in \text{ob } \Phi f$ we define

$$(M, N, \alpha) \circ (N, P, \beta) = (M, P, \beta\alpha)$$

and we leave it to the reader to check that this makes Φf into a category with product and composition.

Proposition 54. Let $f : \underline{\underline{C}} \to \underline{\underline{D}}$ be a product-preserving functor of categories with product. Then there is a homomorphism d : $K_0\Phi f \to K_0\underline{\underline{C}}$ given by $d : [M, N, \alpha] \mapsto [M] - [N]$, where $[M, N, \alpha]$ is the generator of $K_0\Phi f$ corresponding to $(M, N, \alpha) \in$ ob Φf.

Proof. Note first that, by Lemma 1, any element of $K_0\Phi f$ can be put in the form $[M, N, \alpha] - [P, Q, \beta]$; but

$$\begin{aligned}[P, Q, \beta] + [Q, P, \beta^{-1}] &= [(P, Q, \beta) \circ (Q, P, \beta^{-1})] \\ &= [P, P, 1]\end{aligned}$$

and

$$\begin{aligned}[P, P, 1] + [P, P, 1] &= [(P, P, 1) \circ (P, P, 1)] \\ &= [P, P, 1]\end{aligned}$$

so $[P, P, 1] = 0$, and thus

$$\begin{aligned}[M, N, \alpha] - [P, Q, \beta] &= [M, N, \alpha] + [Q, P, \beta^{-1}] \\ &= [M \perp Q, N \perp P, \psi^{-1}(\alpha \perp \beta^{-1})\psi].\end{aligned}$$

So any element of $K_0\Phi f$ can be put in the form $[M, N, \alpha]$, for some $(M, N, \alpha) \in$ ob Φf.

We must show that d is well defined. Suppose $[M, N, \alpha] = [P, Q, \beta]$; by Proposition 31 there exist (A, B, γ), (C, D, δ_1), (D, E, δ_2), (F, G, ε_1), $(G, H, \varepsilon_2) \in$ ob Φf such that, writing the triples as columns,

$$\begin{pmatrix} M \\ N \\ \alpha \end{pmatrix} \perp \begin{pmatrix} A \\ B \\ \gamma \end{pmatrix} \perp \left(\begin{pmatrix} C \\ D \\ \delta_1 \end{pmatrix} \circ \begin{pmatrix} D \\ E \\ \delta_2 \end{pmatrix}\right) \perp \begin{pmatrix} F \\ G \\ \varepsilon_1 \end{pmatrix} \perp \begin{pmatrix} G \\ H \\ \varepsilon_2 \end{pmatrix}$$

$$\cong \begin{pmatrix} P \\ Q \\ \beta \end{pmatrix} \perp \begin{pmatrix} A \\ B \\ \gamma \end{pmatrix} \perp \begin{pmatrix} C \\ D \\ \delta_1 \end{pmatrix} \perp \begin{pmatrix} D \\ E \\ \delta_2 \end{pmatrix} \perp \left(\begin{pmatrix} F \\ G \\ \varepsilon_1 \end{pmatrix} \circ \begin{pmatrix} G \\ H \\ \varepsilon_2 \end{pmatrix}\right)$$

or

$$\begin{pmatrix} M \perp A \perp C \perp F \perp G \\ N \perp B \perp E \perp G \perp H \\ \psi^{-1}(\alpha \perp \gamma \perp \delta_2\delta_1 \perp \varepsilon_1 \perp \varepsilon_2)\psi \end{pmatrix}$$

$$\simeq \begin{pmatrix} P \perp A \perp C \perp D \perp F \\ Q \perp B \perp D \perp E \perp H \\ \psi^{-1}(\beta \perp \gamma \perp \delta_1 \perp \delta_2 \perp \varepsilon_2\varepsilon_1)\psi \end{pmatrix}$$

where ψ stands for the various isomorphisms $\overline{M \perp A \perp C \perp F \perp G} \to \bar{M} \perp \bar{A} \perp \bar{C} \perp \bar{F} \perp \bar{G}$, $\overline{N \perp B \perp E \perp G \perp H} \to \bar{N} \perp \bar{B} \perp \bar{E} \perp \bar{G} \perp \bar{H}$, etc., as appropriate. It follows that, in $\underline{\underline{C}}$,

$$M \perp A \perp C \perp F \perp G \simeq P \perp A \perp C \perp D \perp F$$

and

$$N \perp B \perp E \perp G \perp H \simeq Q \perp B \perp D \perp E \perp H.$$

Passing to $K_0\underline{\underline{C}}$, we have

$$[M] + [A] + [C] + [F] + [G] = [P] + [A] + [C] + [D] + [F]$$

and

$$[N] + [B] + [E] + [G] + [H] = [Q] + [B] + [D] + [E] + [H].$$

Subtracting and cancelling, we see $[M] - [N] = [P] - [Q]$, and so d is well defined.

Finally, let (M, N, α), $(P, Q, \beta) \in$ ob Φf. Then

$$\begin{aligned} d([M, N, \alpha] + [P, Q, \beta]) &= d[M \perp P, N \perp Q, \psi^{-1}(\alpha \perp \beta)\psi] \\ &= [M \perp P] \quad [N \perp Q] \\ &= [M] + [P] - [N] - [Q] \\ &= [M] - [N] + [P] - [Q] \\ &= d[M, N, \alpha] + d[P, Q, \beta] \end{aligned}$$

and so d is a homomorphism.

We say that the product-preserving functor $f : \underline{\underline{C}} \to \underline{\underline{D}}$ is *cofinal* if $f(\underline{\underline{C}})$ is a cofinal subcategory of $\underline{\underline{D}}$, that is, if for each $A \in$ ob $\underline{\underline{D}}$ there exists $A' \in$ ob $\underline{\underline{D}}$ and $B \in$ ob $\underline{\underline{C}}$ such that $A \perp A' \simeq \bar{B}$.

Proposition 55. Let $f : \underline{\underline{C}} \to \underline{\underline{D}}$ be a cofinal product-preserving functor of categories with product. Then there is a homomorphism $d' : K_1\underline{\underline{D}} \to K_0\Phi f$ given by $d' : [\bar{M}, \alpha] \mapsto [M, M, \alpha]$.

Proof. First recall that, by the remark on p.101, any element of $K_1\underline{\underline{D}}$ can be put in the form $[M, \alpha]$, for some $(M, \alpha) \in$ ob $\Omega\underline{\underline{D}}$. Since f is cofinal, there exists $M' \in$ ob $\underline{\underline{D}}$ and $N \in$ ob $\underline{\underline{C}}$ and an isomorphism $\beta : M \perp M' \to \bar{N}$. Then

$$[M, \alpha] = [M, \alpha] + [M', 1] = [M \perp M', \alpha \perp 1]$$
$$= [\bar{N}, \beta(\alpha \perp 1)\beta^{-1}]$$

and so any element of $K_1\underline{D}$ can be put in the form $[\bar{M}, \alpha]$, for some $(\bar{M}, \alpha) \in \text{ob } \Omega\underline{D}$, where $M \in \text{ob } \underline{C}$.

We must show that d' is well defined. Suppose that $[\bar{M}, \alpha] = [\bar{N}, \beta]$: we must show that $[M, M, \alpha] = [N, N, \beta]$. By Proposition 31, there exist (A, γ), (B, δ_1), (B, δ_2), (C, ε_1), (C, ε_2) ob $\Omega\underline{D}$ such that, writing the pairs as columns,

$$\begin{pmatrix}\bar{M}\\ \alpha\end{pmatrix} \perp \begin{pmatrix}A\\ \gamma\end{pmatrix} \perp \left(\begin{pmatrix}B\\ \delta_1\end{pmatrix} \circ \begin{pmatrix}B\\ \delta_2\end{pmatrix}\right) \perp \begin{pmatrix}C\\ \varepsilon_1\end{pmatrix} \perp \begin{pmatrix}C\\ \varepsilon_2\end{pmatrix}$$
$$\simeq \begin{pmatrix}\bar{N}\\ \beta\end{pmatrix} \perp \begin{pmatrix}A\\ \gamma\end{pmatrix} \perp \begin{pmatrix}B\\ \delta_1\end{pmatrix} \perp \begin{pmatrix}B\\ \delta_2\end{pmatrix} \perp \left(\begin{pmatrix}C\\ \varepsilon_1\end{pmatrix} \circ \begin{pmatrix}C\\ \varepsilon_2\end{pmatrix}\right)$$

or

$$\begin{pmatrix}\bar{M} \perp A \perp B \perp C \perp C\\ \alpha \perp \gamma \perp \delta_1\delta_2 \perp \varepsilon_1 \perp \varepsilon_2\end{pmatrix} \simeq \begin{pmatrix}\bar{N} \perp A \perp B \perp B \perp C\\ \beta \perp \gamma \perp \delta_1 \perp \delta_2 \perp \varepsilon_1\varepsilon_2\end{pmatrix} \quad \ldots(1).$$

In particular, $\bar{M} \perp A \perp B \perp C \perp C \simeq \bar{N} \perp A \perp B \perp B \perp C$, in $\underline{D}$, and so in $\Omega\underline{D}$,

$$\begin{pmatrix}\bar{M} \perp A \perp B \perp C \perp C\\ 1 \perp 1 \perp 1 \perp 1 \perp 1\end{pmatrix} \simeq \begin{pmatrix}\bar{N} \perp A \perp B \perp B \perp C\\ 1 \perp 1 \perp 1 \perp 1 \perp 1\end{pmatrix} \quad \ldots(2).$$

Forming the product of the left-hand side of (1) with the right-hand side of (2), and vice versa, we obtain

$$\begin{pmatrix}\bar{M} \perp A \perp B \perp C \perp C \perp \bar{N} \perp A \perp B \perp B \perp C\\ \alpha \perp \gamma \perp \delta_1\delta_2 \perp \varepsilon_1 \perp \varepsilon_2 \perp 1 \perp 1 \perp 1 \perp 1 \perp 1\end{pmatrix}$$
$$\simeq \begin{pmatrix}\bar{N} \perp A \perp B \perp B \perp C \perp \bar{M} \perp A \perp B \perp C \perp C\\ \beta \perp \gamma \perp \delta_1 \perp \delta_2 \perp \varepsilon_1\varepsilon_2 \perp 1 \perp 1 \perp 1 \perp 1 \perp 1\end{pmatrix}$$

and so, using the coherence relations to permute the terms,

$$\begin{pmatrix}\bar{M} \perp \bar{N} \perp A \perp A \perp B \perp B \perp B \perp C \perp C \perp C\\ \alpha \perp 1 \perp \gamma \perp 1 \perp \delta_1\delta_2 \perp 1 \perp 1 \perp \varepsilon_1 \perp \varepsilon_2 \perp 1\end{pmatrix}$$
$$\simeq \begin{pmatrix}\bar{M} \perp \bar{N} \perp A \perp A \perp B \perp B \perp B \perp C \perp C \perp C\\ 1 \perp \beta \perp \gamma \perp 1 \perp \delta_1 \perp \delta_2 \perp 1 \perp \varepsilon_1\varepsilon_2 \perp 1 \perp 1\end{pmatrix} \quad \ldots(3).$$

Since f is cofinal, we can find $A', B', C' \in \text{ob } \underline{D}$, $D, E, F \in \text{ob } \underline{C}$ and isomorphisms $\xi : A \perp A' \to \bar{D}$, $\eta : B \perp B' \to \bar{E}$, and ζ :

$C \perp C' \to \bar{F}$, in $\underline{\underline{D}}$. Then, in $\Omega\underline{\underline{D}}$,

$$(A, \gamma) \perp (A', 1) = (A \perp A', \gamma \perp 1) \simeq (\bar{D}, \gamma')$$

where $\gamma' = \xi(\gamma \perp 1)\xi^{-1}$, and similarly

$$(B, \delta_i) \perp (B', 1) \simeq (\bar{E}, \delta_i')$$

and

$$(C, \varepsilon_i) \perp (C', 1) \simeq (\bar{F}, \varepsilon_i')$$

where $\delta_i' = \eta(\delta_i \perp 1)\eta^{-1}$ and $\varepsilon_i' = \zeta(\varepsilon_i \perp 1)\zeta^{-1}$, $i = 1, 2$. Note that

$$\delta_1'\delta_2' = \xi(\delta_1 \perp 1)\xi^{-1}\xi(\delta_2 \perp 1)\xi^{-1} = \xi(\delta_1\delta_2 \perp 1)\xi^{-1}$$

so

$$(B, \delta_1\delta_2) \perp (B', 1) \simeq (\bar{E}, \delta_1'\delta_2')$$

and similarly

$$(C, \varepsilon_1\varepsilon_2) \perp (C', 1) \simeq (\bar{F}, \varepsilon_1'\varepsilon_2').$$

If we now form the product of each side of (3) with

$$\begin{pmatrix} A' \perp A' \perp B' \perp B' \perp B' \perp C' \perp C' \perp C' \\ 1 \perp 1 \perp 1 \perp 1 \perp 1 \perp 1 \perp 1 \perp 1 \end{pmatrix}$$

and permute the terms, we obtain, on replacing $A \perp A'$ by $\bar{D}$, and so on,

$$\begin{pmatrix} \bar{M} \perp \bar{N} \perp \bar{D} \perp \bar{D} \perp \bar{E} \perp \bar{E} \perp \bar{E} \perp \bar{F} \perp \bar{F} \perp \bar{F} \\ \alpha \perp 1 \perp \gamma' \perp 1 \perp \delta_1'\delta_2' \perp 1 \perp 1 \perp \varepsilon_1' \perp \varepsilon_2' \perp 1 \end{pmatrix}$$

$$\simeq \begin{pmatrix} \bar{M} \perp \bar{N} \perp \bar{D} \perp \bar{D} \perp \bar{E} \perp \bar{E} \perp \bar{E} \perp \bar{F} \perp \bar{F} \perp \bar{F} \\ 1 \perp \beta \perp \gamma' \perp 1 \perp \delta_1' \perp \delta_2' \perp 1 \perp \varepsilon_1'\varepsilon_2' \perp 1 \perp 1 \end{pmatrix} \quad \ldots(4).$$

Put

$$P = \bar{M} \perp \bar{N} \perp \bar{D} \perp \bar{D} \perp \bar{E} \perp \bar{E} \perp \bar{E} \perp \bar{F} \perp \bar{F} \perp \bar{F}$$

$$\pi = \alpha \perp 1 \perp \gamma' \perp 1 \perp \delta_1'\delta_2' \perp 1 \perp 1 \perp \varepsilon_1' \perp \varepsilon_2' \perp 1$$

and

$$\sigma = 1 \perp \beta \perp \gamma' \perp 1 \perp \delta_1' \perp \delta_2' \perp 1 \perp \varepsilon_1'\varepsilon_2' \perp 1 \perp 1.$$

Then (4) says there is an isomorphism $\theta : P \to P$ such that the diagram

$$\begin{array}{ccc} P & \xrightarrow{\theta} & P \\ \pi \downarrow & & \downarrow \sigma \\ P & \xrightarrow{\theta} & P \end{array}$$

commutes, that is, $\theta\pi = \sigma\theta$. Put

$$Q = M \perp N \perp D \perp D \perp E \perp E \perp E \perp F \perp F \perp F$$

and then there is an isomorphism $\psi : \bar{Q} \to P$. (Strictly, we should choose a bracketing of Q, and then ψ will be a rather complicated expression built up from isomorphisms $\psi : \overline{M \perp N} \to \bar{M} \perp \bar{N}$, $\psi : \overline{(M \perp N) \perp D} \to \overline{M \perp N} \perp \bar{D}$, and so on. We have used this simplification once before, without comment, on p.143.) Since $\theta\pi = \sigma\theta$, we have $(\psi^{-1}\theta\psi)(\psi^{-1}\pi\psi) = (\psi^{-1}\sigma\psi)(\psi^{-1}\theta\psi)$, all four bracketed expressions being automorphisms of $\bar{Q}$. It follows that, in Φf,

$$(Q, Q, (\psi^{-1}\theta\psi)(\psi^{-1}\pi\psi)) = (Q, Q, (\psi^{-1}\sigma\psi)(\psi^{-1}\theta\psi))$$

and so

$$(Q, Q, \psi^{-1}\pi\psi) \circ (Q, Q, \psi^{-1}\theta\psi)$$
$$= (Q, Q, \psi^{-1}\theta\psi) \circ (Q, Q, \psi^{-1}\sigma\psi).$$

Passing to $K_0\Phi f$, we have

$$[Q, Q, \psi^{-1}\pi\psi] + [Q, Q, \psi^{-1}\theta\psi]$$
$$= [Q, Q, \psi^{-1}\theta\psi] + [Q, Q, \psi^{-1}\sigma\psi]$$

so

$$[Q, Q, \psi^{-1}\pi\psi] = [Q, Q, \psi^{-1}\sigma\psi].$$

Substituting back for Q, π, and σ, and splitting up the products, we obtain

$$\begin{aligned}
&[M, M, \alpha] + [N, N, 1] + [D, D, \gamma'] + [D, D, 1] \\
&+ [E, E, \delta_1'\delta_2'] + [E, E, 1] + [E, E, 1] \\
&+ [F, F, \varepsilon_1'] + [F, F, \varepsilon_2'] + [F, F, 1] \\
&= [M, M, 1] + [N, N, \beta] + [D, D, \gamma'] + [D, D, 1] \\
&+ [E, E, \delta_1'] + [E, E, \delta_2'] + [E, E, 1] \\
&+ [F, F, \varepsilon_1'\varepsilon_2'] + [F, F, 1] + [F, F, 1].
\end{aligned}$$

On cancelling and removing zero terms, this reads

$$[M, M, \alpha] + [E, E, \delta_1'\delta_2'] + [F, F, \varepsilon_1'] + [F, F, \varepsilon_2']$$
$$= [N, N, \beta] + [E, E, \delta_1'] + [E, E, \delta_2'] + [F, F, \varepsilon_1'\varepsilon_2'].$$

Then

$$\begin{aligned}
[E, E, \delta_1'\delta_2'] &= [(E, E, \delta_2') \circ (E, E, \delta_1')] \\
&= [E, E, \delta_2'] + [E, E, \delta_1']
\end{aligned}$$

and similarly

$$[F, F, \varepsilon_1'\varepsilon_2'] = [F, F, \varepsilon_2'] + [F, F, \varepsilon_1']$$

and so finally we obtain $[M, M, \alpha] = [N, N, \beta]$, and we have proved that d' is well defined.

Let $(\bar{M}, \alpha)$, $(\bar{N}, \beta) \in \text{ob } \Omega\underline{\underline{D}}$, where $M, N \in \text{ob } \underline{\underline{C}}$. Then

$$\begin{aligned} d'([\bar{M}, \alpha] + [\bar{N}, \beta]) &= d'[\bar{M} \perp \bar{N}, \alpha \perp \beta] \\ &= d'[\overline{M \perp N}, \psi^{-1}(\alpha \perp \beta)\psi] \\ &= [M \perp N, M \perp N, \psi^{-1}(\alpha \perp \beta)\psi] \\ &= [M, M, \alpha] + [N, N, \beta] \\ &= d'[\bar{M}, \alpha] + d'[\bar{N}, \beta]. \end{aligned}$$

Thus d' is a homomorphism, and the proof is complete.

Recall that a sequence

$$\cdots \xrightarrow{f} G \xrightarrow{g} \cdots$$

of group homomorphisms is said to be *exact* (at G) if the image of f coincides with the kernel of g.

Proposition 56. Let $f : \underline{\underline{C}} \to \underline{\underline{D}}$ be a cofinal product-preserving functor of categories with product. Then the sequence

$$K_1\underline{\underline{C}} \xrightarrow{K_1 f} K_1\underline{\underline{D}} \xrightarrow{d'} K_0\Phi f \xrightarrow{d} K_0\underline{\underline{C}} \xrightarrow{K_0 f} K_0\underline{\underline{D}}$$

is exact.

Proof. 1. Exactness at $K_0\underline{\underline{C}}$.

(i) $(K_0 f)d[M, N, \alpha] = (K_0 f)([M] - [N]) = [\bar{M}] - [\bar{N}]$. But $\alpha : \bar{M} \to \bar{N}$ is an isomorphism, so $[\bar{M}] = [\bar{N}]$, and therefore $(K_0 f)d = 0$.

(ii) A typical element of $K_0\underline{\underline{C}}$ is of the form $[M] - [N]$, by Lemma 1, and if $[M] - [N] \in \ker K_0 f$, then $[\bar{M}] = [\bar{N}]$ in $K_0\underline{\underline{D}}$. By Proposition 2, there exists $P \in \text{ob } \underline{\underline{D}}$ such that $\bar{M} \perp P \simeq \bar{N} \perp P$, and since f is cofinal there exist $P' \in \text{ob } \underline{\underline{D}}$ and $Q \in \text{ob } \underline{\underline{C}}$ such that $P \perp P' \simeq \bar{Q}$. Then

$$\overline{M \perp Q} \simeq \bar{M} \perp \bar{Q} \simeq \bar{M} \perp P \perp P' \simeq \bar{N} \perp P \perp P' \simeq \bar{N} \perp \bar{Q} \simeq \overline{N \perp Q}.$$

So let $\alpha : \overline{M \perp Q} \to \overline{N \perp Q}$ be an isomorphism; then

$$(M \perp Q, N \perp Q, \alpha) \in \text{ob } \Phi f$$

and

$$d[M \perp Q, N \perp Q, \alpha] = [M \perp Q] - [N \perp Q] = [M] - [N]$$

and so $[M] - [N] \in d(K_0\Phi f)$.

2. Exactness at $K_0\Phi f$.

(i) $dd'[\bar{M}, \alpha] = d[M, M, \alpha] = [M] - [M] = 0$, so $dd' = 0$.

(ii) Let $[M, N, \alpha] \in \ker d$, that is, $[M] = [N]$ in $K_0\underline{\underline{C}}$. By Proposition 2 there exists $P \in \text{ob } \underline{\underline{C}}$ and an isomorphism $\beta : M \perp P \to N \perp P$ in $\underline{\underline{C}}$. Define $\gamma = \psi^{-1}(\alpha \perp 1)\psi\bar{\beta}^{-1}$, so that the diagram

$$\begin{array}{ccccccc} \overline{M \perp P} & \xrightarrow{\psi} & \bar{M} \perp \bar{P} & \xrightarrow{\alpha \perp 1} & \bar{N} \perp \bar{P} & \xrightarrow{\psi^{-1}} & \overline{N \perp P} \\ \bar{\beta}\downarrow & & & & & & \downarrow \bar{1} = 1 \\ \overline{N \perp P} & & & \xrightarrow{\gamma} & & & \overline{N \perp P} \end{array}$$

commutes, or in other words,

$$(\beta, 1) : (M \perp P, N \perp P, \psi^{-1}(\alpha \perp 1)\psi) \to (N \perp P, N \perp P, \gamma)$$

is an isomorphism in Φf. Thus

$$\begin{aligned} [M, N, \alpha] &= [M, N, \alpha] + [P, P, 1] \\ &= [M \perp P, N \perp P, \psi^{-1}(\alpha \perp 1)\psi] \\ &= [N \perp P, N \perp P, \gamma] \\ &= d'[\overline{N \perp P}, \gamma] \end{aligned}$$

and so $[M, N, \alpha] \in d'(K_1\underline{\underline{D}})$.

3. Exactness at $K_1\underline{\underline{D}}$.

(i) $d'(K_1 f)[M, \alpha] = d'[\bar{M}, \bar{\alpha}] = [M, M, \bar{\alpha}]$. The diagram

$$\begin{array}{ccc} \bar{M} & \xrightarrow{\bar{\alpha}} & \bar{M} \\ \bar{\alpha}\downarrow & & \downarrow \bar{1} = 1 \\ \bar{M} & \xrightarrow{1} & \bar{M} \end{array}$$

commutes, so $(\alpha, 1) : (M, M, \bar{\alpha}) \to (M, M, 1)$ is an isomorphism in Φf, so in $K_0\Phi f$, $[M, M, \bar{\alpha}] = [M, M, 1] = 0$, and therefore $d'(K_1 f) = 0$.

(ii) Let $[\bar{M}, \alpha] \in \ker d'$; so $[M, M, \alpha] = 0$ in $K_0\Phi f$. By Proposition 31, there exist (A, B, β), (C, D, γ), (D, E, γ_1), (F, G, δ), $(G, H, \delta_1) \in \text{ob } \Phi f$ such that, writing the triples as columns,

$$\begin{pmatrix} M \\ M \\ \alpha \end{pmatrix} \perp \begin{pmatrix} A \\ B \\ \beta \end{pmatrix} \perp \begin{pmatrix} C \\ D \\ \gamma \end{pmatrix} \perp \begin{pmatrix} D \\ E \\ \gamma_1 \end{pmatrix} \perp \left(\begin{pmatrix} F \\ G \\ \delta \end{pmatrix} \circ \begin{pmatrix} G \\ H \\ \delta_1 \end{pmatrix}\right)$$

$$\simeq \begin{pmatrix} A \\ B \\ \beta \end{pmatrix} \perp \left(\begin{pmatrix} C \\ D \\ \gamma \end{pmatrix} \circ \begin{pmatrix} D \\ E \\ \gamma_1 \end{pmatrix}\right) \perp \begin{pmatrix} F \\ G \\ \delta \end{pmatrix} \perp \begin{pmatrix} G \\ H \\ \delta_1 \end{pmatrix}$$

or

$$\begin{pmatrix} M \perp A \perp C \perp D \perp F \\ M \perp B \perp D \perp E \perp H \\ \psi^{-1}(\alpha \perp \beta \perp \gamma \perp \gamma_1 \perp \delta_1\delta)\psi \end{pmatrix} \simeq \begin{pmatrix} A \perp C \perp F \perp G \\ B \perp E \perp G \perp H \\ \psi^{-1}(\beta \perp \gamma_1\gamma \perp \delta \perp \delta_1)\psi \end{pmatrix} \quad \ldots(1).$$

In particular, from the second row,

$$M \perp B \perp D \perp E \perp H \simeq B \perp E \perp G \perp H$$

and so

$$\begin{pmatrix} M \perp B \perp D \perp E \perp H \\ M \perp B \perp D \perp E \perp H \\ \psi^{-1}(1 \perp 1 \perp 1 \perp 1 \perp 1)\psi \end{pmatrix} \simeq \begin{pmatrix} B \perp E \perp G \perp H \\ B \perp E \perp G \perp H \\ \psi^{-1}(1 \perp 1 \perp 1 \perp 1)\psi \end{pmatrix} \quad \ldots(2).$$

Forming the product of the left-hand side of (1) with the right-hand side of (2), and vice versa, gives

$$\begin{pmatrix} M \perp A \perp C \perp D \perp F \perp B \perp E \perp G \perp H \\ M \perp B \perp D \perp E \perp H \perp B \perp E \perp G \perp H \\ \psi^{-1}(\alpha \perp \beta \perp \gamma \perp \gamma_1 \perp \delta_1\delta \perp 1 \perp 1 \perp 1 \perp 1)\psi \end{pmatrix}$$

$$\simeq \begin{pmatrix} A \perp C \perp F \perp G \perp M \perp B \perp D \perp E \perp H \\ B \perp E \perp G \perp H \perp M \perp B \perp D \perp E \perp H \\ \psi^{-1}(\beta \perp \gamma_1\gamma \perp \delta \perp \delta_1 \perp 1 \perp 1 \perp 1 \perp 1 \perp 1)\psi \end{pmatrix} \quad \ldots(3).$$

From the coherence relations, we have

$$\begin{pmatrix} M \perp B \perp D \perp E \perp H \perp B \perp E \perp G \perp H \\ M \perp A \perp C \perp D \perp F \perp B \perp E \perp G \perp H \\ \psi^{-1}(1 \perp \beta^{-1} \perp \gamma^{-1} \perp \gamma_1^{-1} \perp (\delta_1\delta)^{-1} \perp 1 \perp 1 \perp 1 \perp 1)\psi \end{pmatrix}$$

$$\simeq \begin{pmatrix} B \perp E \perp G \perp H \perp M \perp B \perp D \perp E \perp H \\ A \perp D \perp G \perp F \perp M \perp B \perp C \perp E \perp H \\ \psi^{-1}(\beta^{-1} \perp \gamma_1^{-1} \perp 1 \perp (\delta_1\delta)^{-1} \perp 1 \perp 1 \perp \gamma^{-1} \perp 1 \perp 1)\psi \end{pmatrix} \quad \ldots(4)$$

where the right-hand side is obtained by permuting the objects on the left-hand side. Next, forming the composite of the left-hand side of (3) with the left-hand side of (4), and of the corresponding right-hand sides, gives

$$\begin{pmatrix} M \perp A \perp C \perp D \perp F \perp B \perp E \perp G \perp H \\ M \perp A \perp C \perp D \perp F \perp B \perp E \perp G \perp H \\ \psi^{-1}(\alpha \perp 1 \perp 1 \perp 1 \perp 1 \perp 1 \perp 1 \perp 1 \perp 1)\psi \end{pmatrix}$$

$$\simeq \begin{pmatrix} A \perp C \perp F \perp G \perp M \perp B \perp D \perp E \perp H \\ A \perp D \perp G \perp F \perp M \perp B \perp C \perp E \perp H \\ \psi^{-1}(1 \perp \gamma \perp \delta \perp \delta^{-1} \perp 1 \perp 1 \perp \gamma^{-1} \perp 1 \perp 1)\psi \end{pmatrix} \quad \ldots(5).$$

Put

$$N = A \perp C \perp D \perp F \perp B \perp E \perp G \perp H$$

$$P = A \perp M \perp B \perp E \perp H,\ Q = C \perp F,\ R = D \perp G,\ \xi = \gamma \perp \delta$$

and note that $\xi^{-1} = \gamma^{-1} \perp \delta^{-1}$. Then from (5), permuting the terms,

$$\begin{pmatrix} M \perp N \\ M \perp N \\ \psi^{-1}(\alpha \perp 1)\psi \end{pmatrix} \simeq \begin{pmatrix} P \perp Q \perp R \\ P \perp R \perp Q \\ \psi^{-1}(1 \perp \xi \perp \xi^{-1})\psi \end{pmatrix} \quad \ldots(6).$$

Now the diagram

$$\begin{array}{ccc} \bar{Q} \perp \bar{R} & \xrightarrow{\ \xi \perp \xi^{-1}\ } & \bar{R} \perp \bar{Q} \\ 1 \downarrow & & \downarrow \phi \\ \bar{Q} \perp \bar{R} & \xrightarrow{\ \phi(\xi \perp \xi^{-1})\ } & \bar{Q} \perp \bar{R} \end{array}$$

commutes, where ϕ is the transposition isomorphism (defined on p.84), and so the diagram

$$\begin{array}{ccc} \overline{Q \perp R} & \xrightarrow{\ \psi^{-1}(\xi \perp \xi^{-1})\psi\ } & \overline{R \perp Q} \\ 1 \downarrow & & \downarrow \psi^{-1}\phi\psi \\ \overline{Q \perp R} & \xrightarrow{\ \psi^{-1}\phi(\xi \perp \xi^{-1})\psi\ } & \overline{Q \perp R} \end{array}$$

commutes. By abuse of notation, we use ϕ to denote the transposition isomorphism in both $\underline{C}$ and $\underline{D}$, and then $\psi^{-1}\phi\psi = \bar{\phi}$, and of course $\bar{1} = 1$, so $(1, \phi) : (Q \perp R, R \perp Q, \psi^{-1}(\xi \perp \xi^{-1})\psi) \to (Q \perp R, Q \perp R, \psi^{-1}\phi(\xi \perp \xi^{-1})\psi)$ is an isomorphism in Φf. Put $S = Q \perp R$ and $\eta = \psi^{-1}\phi(\xi \perp \xi^{-1})\psi$. Then in (6) we have

$$\begin{pmatrix} M \perp N \\ M \perp N \\ \psi^{-1}(\alpha \perp 1)\psi \end{pmatrix} \simeq \begin{pmatrix} P \perp S \\ P \perp S \\ \psi^{-1}(1 \perp \eta)\psi \end{pmatrix} \qquad \ldots(7).$$

From (7), there are isomorphisms $\pi, \sigma : M \perp N \to P \perp S$ in $\underline{C}$ such that the diagram

$$\begin{array}{ccc} \overline{M \perp N} & \xrightarrow{\psi^{-1}(\alpha \perp 1)\psi} & \overline{M \perp N} \\ \bar{\pi}\downarrow & & \downarrow\bar{\sigma} \\ \overline{P \perp S} & \xrightarrow{\psi^{-1}(1 \perp \eta)\psi} & \overline{P \perp S} \end{array}$$

commutes, in $\underline{D}$, and so the diagram

$$\begin{array}{ccc} \overline{M \perp N} & \xrightarrow{\psi^{1}(\alpha \perp 1)\psi} & \overline{M \perp N} \\ \bar{\pi}\downarrow & & \downarrow\bar{\pi} \\ \overline{P \perp S} & \xrightarrow{(\bar{\pi}^{-1}\bar{\sigma})\psi^{-1}(1 \perp \eta)\psi} & \overline{P \perp S} \end{array}$$

commutes. It follows that, in $\Omega\underline{D}$,

$$(\overline{M \perp N}, \psi^{-1}(\alpha \perp 1)\psi) \simeq (\overline{P \perp S}, (\bar{\pi}^{-1}\bar{\sigma})\psi^{-1}(1 \perp \eta)\psi) \qquad \ldots(8).$$

Now

$$(\overline{M \perp N}, \psi^{-1}(\alpha \perp 1)\psi) \simeq (\bar{M} \perp \bar{N}, \alpha \perp 1) = (\bar{M}, \alpha) \perp (\bar{N}, 1)$$

and

$$(\overline{P \perp S}, (\bar{\pi}^{-1}\bar{\sigma})\psi^{-1}(1 \perp \eta)\psi)$$

$$= (\overline{P \perp S}, \bar{\pi}^{-1}\bar{\sigma}) \circ (\overline{P \perp S}, \psi^{-1}(1 \perp \eta)\psi).$$

Then

$$(\overline{P \perp S}, \psi^{-1}(1 \perp \eta)\psi) \simeq (\bar{P} \perp \bar{S}, 1 \perp \eta) = (\bar{P}, 1) \perp (\bar{S}, \eta).$$

Putting all this back into (8), and passing to $K_1\underline{D}$, gives

$$[\bar{M}, \alpha] + [\bar{N}, 1] = [\overline{P \perp S}, \bar{\pi}^{-1}\bar{\sigma}] + [\bar{P}, 1] + [\bar{S}, \eta] \qquad \ldots(9).$$

Now $[\bar{N}, 1] = 0$ and $[\bar{P}, 1] = 0$, and

$$[\overline{P \perp S}, \bar{\pi}^{-1}\bar{\sigma}] = (K_1 f)[P \perp S, \pi^{-1}\sigma]$$

which is in the image of $K_1 f$. We want to show that $[\bar{M}, \alpha]$ is in the image of $K_1 f$, so from (9) it remains only to show that $[\bar{S}, \eta]$ is in the image of $K_1 f$. Now

$$(\bar{S}, \eta) = (\overline{Q \perp R}, \psi^{-1}\phi(\xi \perp \xi^{-1})\psi) \simeq (\bar{Q} \perp \bar{R}, \phi(\xi \perp \xi^{-1}))$$

and the diagram

$$\begin{array}{ccc} \bar{R} \perp \bar{Q} & \xrightarrow{\phi} & \bar{Q} \perp \bar{R} \\ {\scriptstyle 1 \perp \xi} \downarrow & & \downarrow {\scriptstyle \xi \perp 1} \\ \bar{R} \perp \bar{R} & \xrightarrow{\phi} & \bar{R} \perp \bar{R} \end{array}$$

commutes, and so the diagram

$$\begin{array}{ccc} \bar{Q} \perp \bar{R} & \xrightarrow{\phi(\xi \perp \xi^{-1})} & \bar{Q} \perp \bar{R} \\ {\scriptstyle \xi \perp 1} \downarrow & & \downarrow {\scriptstyle \xi \perp 1} \\ \bar{R} \perp \bar{R} & \xrightarrow{\phi} & \bar{R} \perp \bar{R} \end{array}$$

commutes, whence $(\bar{Q} \perp \bar{R}, \phi(\xi \perp \xi^{-1})) \simeq (\bar{R} \perp \bar{R}, \phi)$. Then

$$(\bar{R} \perp \bar{R}, \phi) \simeq (\overline{R \perp R}, \psi^{-1}\phi\psi) = (\overline{R \perp R}, \bar{\phi})$$

and so

$$[\bar{S}, \eta] = [\overline{R \perp R}, \bar{\phi}] = (K_1 f)[R \perp R, \phi]$$

which is in the image of $K_1 f$, and the proof of Proposition 56 is complete.

Exercise: Extend the above exact sequence by one extra term on the left by constructing a suitable homomorphism $K_1 \Phi f \to K_1 \underline{\underline{C}}$. (Identify $\Omega\Phi f$ with $\Phi\Omega f$ in the obvious way, and imitate the proof of Proposition 54 and Section 1 of the proof of Proposition 56, with $\Omega f : \Omega\underline{\underline{C}} \to \Omega\underline{\underline{D}}$ in place of $f : \underline{\underline{C}} \to \underline{\underline{D}}$.) (Note that this is *not* the way we shall in fact extend our sequence; see Proposition 67, the exercise on p.177, and Proposition 74.)

Corollary 57. Let R, S be rings, and let $f : R \to S$ be a ring homomorphism. Then there is an exact sequence

$$K_1 R \xrightarrow{K_1 f} K_1 S \xrightarrow{d'} K_0 \Phi f \xrightarrow{d} K_0 R \xrightarrow{K_0 f} K_0 S$$

where, by abuse of notation, we are also using f to denote the functor $S \otimes_f \cdot : \underline{\underline{P}}(R) \to \underline{\underline{P}}(S)$.

Proof. This is immediate once we note that f is cofinal, which follows from the fact that the image of f in $\underline{\underline{P}}(S)$ contains all the free S-modules of finite rank.

Corollary 58. Let R, S be commutative rings, and let $f : R \to S$ be a ring homomorphism. Then the diagram

$$\begin{array}{ccccccccc} K_1R & \xrightarrow{K_1f} & K_1S & \xrightarrow{d'} & K_0\Phi f & \xrightarrow{d} & K_0R & \xrightarrow{K_0f} & K_0S \\ \downarrow{\scriptstyle \det} & & \downarrow{\scriptstyle \det} & & \downarrow{\scriptstyle \det} & & \downarrow{\scriptstyle \det} & & \downarrow{\scriptstyle \det} \\ R^{\bullet} & \xrightarrow[f^*]{} & S^{\bullet} & \xrightarrow[d']{} & K_0\Phi f' & \xrightarrow[d]{} & \operatorname{Pic} R & \xrightarrow[\operatorname{Pic} f]{} & \operatorname{Pic} S \end{array}$$

commutes and has exact rows, where $f^* : R^{\bullet} \to S^{\bullet}$ is the restriction of $f : R \to S$, and $f' : \underline{\underline{\mathrm{Pic}}}(R) \to \underline{\underline{\mathrm{Pic}}}(S)$ is the restriction of $f : \underline{\underline{P}}(R) \to \underline{\underline{P}}(S)$.

Proof. The exactness of the rows follows from Proposition 56, noting that f, f' are both cofinal. The map $\det : K_0\Phi f \to K_0\Phi f'$ is defined by

$$\det[M, N, \alpha] = [\det M, \det N, \det \alpha]$$

where $M, N \in \mathrm{ob}\,\underline{\underline{P}}(R)$ and $\alpha : \bar{M} \to \bar{N}$ is an isomorphism in $\underline{\underline{P}}(S)$. It is straightforward to check that this gives a well defined homomorphism of groups, making the diagram commutative, and we leave the details to the reader.

Examples. 1. Let R be a commutative ring. Then $\det : \underline{\underline{P}}(R) \to \underline{\underline{\mathrm{Pic}}}(R)$ is a cofinal product-preserving functor, so by Proposition 56 there is an exact sequence

$$K_1R \xrightarrow{\det} R^{\bullet} \xrightarrow{d'} K_0\Phi(\det) \xrightarrow{d} K_0R \xrightarrow{\det} \operatorname{Pic} R.$$

But $\det : K_0R \to \operatorname{Pic} R$ and $\det : K_1R \to R^{\bullet}$ are surjective, so we obtain an exact sequence

$$0 \to K_0\Phi(\det) \xrightarrow{d} K_0R \xrightarrow{\det} \operatorname{Pic} R \to 0.$$

2. Let R, S be rings, and suppose there are ring homomorphisms $f : R \to S$ and $g : S \to R$ with $gf = 1 : R \to R$. Then $(K_ig)(K_if) = 1$, $i = 0, 1$, whence K_if is injective, K_ig is surjective, and $K_iS \simeq K_iR \oplus \ker(K_ig)$, $i = 0, 1$. By Corollary 57, the sequence

$$K_1S \xrightarrow{K_1g} K_1R \xrightarrow{d'} K_0\Phi g \xrightarrow{d} K_0S \xrightarrow{K_0g} K_0R$$

is exact, so $d' = 0$ and therefore d is injective, whence

$$K_0S \simeq K_0R \oplus K_0\Phi g.$$

Again, by Corollary 57, the sequence

$$K_1R \xhookrightarrow{K_1f} K_1S \xrightarrow{d'} K_0\Phi f \xrightarrow{d} K_0R \xhookrightarrow{K_0f} K_0S$$

is exact, so $d = 0$ and therefore d' is surjective, whence

$$K_1S \simeq K_1R \oplus K_0\Phi f.$$

These isomorphisms would enable us to calculate K_0S and K_1S from K_0R and K_1R provided we could first calculate the relative terms $K_0\Phi g$ and $K_0\Phi f$. Unfortunately, this may not be very easy; in the next section we describe another exact sequence with no such relative terms: however, it involves four rings, not two, though a special case of it yields the above sequence again.

7.1 *Cartesian squares, and the Mayer-Vietoris sequence*

Let

$$\begin{array}{ccc} R & \xrightarrow{i_2} & R_2 \\ {\scriptstyle i_1}\downarrow & & \downarrow{\scriptstyle j_2} \\ R_1 & \xrightarrow{j_1} & S \end{array}$$

be a commutative square of rings and ring homomorphisms. We say that the square is *cartesian* if, for all $(r_1, r_2) \in R_1 \times R_2$ with $j_1r_1 = j_2r_2$, there is a unique $r \in R$ with $i_1r = r_1$ and $i_2r = r_2$.

Given a cartesian square, as above, let us suppose further that j_2 is surjective. Let P_1, P_2 be finitely generated projective modules over R_1, R_2 respectively, and suppose there is an S-module isomorphism $h : S \otimes_{j_1} P_1 \to S \otimes_{j_2} P_2$. We write

$$(P_1, P_2, h) = \{(p_1, p_2) \in P_1 \times P_2 : h(1 \otimes p_1) = 1 \otimes p_2\}.$$

This is an additive subgroup of $P_1 \times P_2$, and becomes an R-module if we define

$$r(p_1, p_2) = ((i_1r)p_1, (i_2r)p_2).$$

This makes sense, since $h(1 \otimes (i_1r)p_1) = h(j_1i_1r \otimes p_1) = h((j_1i_1r)(1 \otimes p_1)) = (j_1i_1r)h(1 \otimes p_1) = (j_2i_2r)(1 \otimes p_2) = j_2i_2r \otimes p_2 = 1 \otimes (i_2r)p_2$, and it is easy to see that the axioms for an R-module are satisfied. With all the above data, we then have:

Proposition 59. (P_1, P_2, h) is a finitely generated projective R-module.

Proof. Let $P_1 \oplus N_1 \simeq R_1{}^r$ and $P_2 \oplus N_2 \simeq R_2{}^s$. Put $Q_1 = N_1 \oplus R_1{}^s$ and $Q_2 = N_2 \oplus R_2{}^r$; so $P_1 \oplus Q_1 \simeq R_1{}^n$ and $P_2 \oplus Q_2 \simeq R_2{}^n$, where $n = r + s$. Further, we have

$$\begin{aligned}
S \otimes_{j_1} Q_1 &= S \otimes_{j_1} (N_1 \oplus R_1{}^s) \\
&\simeq (S \otimes_{j_1} N_1) \oplus (S \otimes_{j_1} R_1)^s \\
&\simeq (S \otimes_{j_1} N_1) \oplus S^s \\
&\simeq (S \otimes_{j_1} N_1) \oplus (S \otimes_{j_2} R_2)^s \\
&\simeq (S \otimes_{j_1} N_1) \oplus (S \otimes_{j_2} (P_2 \oplus N_2)) \\
&\simeq (S \otimes_{j_1} N_1) \oplus (S \otimes_{j_2} P_2) \oplus (S \otimes_{j_2} N_2) \\
&\simeq (S \otimes_{j_1} N_1) \oplus (S \otimes_{j_1} P_1) \oplus (S \otimes_{j_2} N_2) \\
&\simeq (S \otimes_{j_1} (N_1 \oplus P_1)) \oplus (S \otimes_{j_2} N_2) \\
&\simeq (S \otimes_{j_1} R_1)^r \oplus (S \otimes_{j_2} N_2) \\
&\simeq S^r \oplus (S \otimes_{j_2} N_2) \\
&\simeq (S \otimes_{j_2} R_2)^r \oplus (S \otimes_{j_2} N_2) \\
&\simeq S \otimes_{j_2} (N_2 \oplus R_2{}^r) \\
&= S \otimes_{j_2} Q_2.
\end{aligned}$$

So let $k : S \otimes_{j_1} Q_1 \to S \otimes_{j_2} Q_2$ be an S-module isomorphism, and we have R-module isomorphisms

$$\begin{aligned}
(P_1, P_2, h) \oplus (Q_1, Q_2, k) &\simeq (P_1 \oplus Q_1, P_2 \oplus Q_2, h \oplus k) \\
&\simeq (R_1{}^n, R_2{}^n, A)
\end{aligned}$$

for some $A \in \mathrm{GL}_n(S)$. Here we are using the obvious fact that, given isomorphisms $\beta : P_1 \to P_1'$ and $\gamma : P_2 \to P_2'$ such that the diagram

$$\begin{array}{ccc} S \otimes_{j_1} P_1 & \xrightarrow{h} & S \otimes_{j_2} P_2 \\ 1 \otimes \beta \downarrow & & \downarrow 1 \otimes \gamma \\ S \otimes_{j_1} P_1' & \xrightarrow{h'} & S \otimes_{j_2} P_2' \end{array}$$

commutes, there is an isomorphism $(\beta, \gamma) : (P_1, P_2, h) \to (P_1', P_2', h')$, given by $(\beta, \gamma)(p_1, p_2) = (\beta p_1, \gamma p_2)$. The reader is advised to compare the above with the definition of the fibre category of a product-preserving functor, on p.141. Note also that strictly we should not write $(R_1{}^n, R_2{}^n, A)$, but $(R_1{}^n, R_2{}^n, g)$, where $g : S \otimes_{j_1} (R_1{}^n) \to S \otimes_{j_2} (R_2{}^n)$ is the composite of the isomorphisms

$$S \otimes_{j_1} (R_1{}^n) \to (S \otimes_{j_1} R_1)^n \to S^n \xrightarrow{A} S^n$$
$$\to (S \otimes_{j_2} R_2)^n \to S \otimes_{j_2} (R_2{}^n).$$

If, however, we agree to identify $S \otimes_{j_1} (R_1{}^n)$ with S^n, and similarly for $S \otimes_{j_2} (R_2{}^n)$, then the notation makes sense.

We have shown that (P_1, P_2, h) is isomorphic to a direct summand of $(R_1{}^n, R_2{}^n, A)$. This in turn is a direct summand of

$$(R_1{}^n, R_2{}^n, A) \oplus (R_1{}^n, R_2{}^n, A^{-1}) \simeq (R_1{}^{2n}, R_2{}^{2n}, A \oplus A^{-1}).$$

To complete the proof, we shall show that the latter module is isomorphic to R^{2n}.

Now $A \oplus A^{-1} \in \mathrm{E}_{2n}(S)$, by the argument on p.110. Further, $j_2 : R_2 \to S$ is surjective, and so the induced map $j_2 : \mathrm{E}_{2n}(R_2) \to \mathrm{E}_{2n}(S)$ is also surjective, since $j_2(B_{ij}(x)) = B_{ij}(j_2 x)$, $x \in R_2$, and thus the image of j_2 contains all the generators of $\mathrm{E}_{2n}(S)$. So we can find $B \in \mathrm{E}_{2n}(R_2)$ with $j_2(B) = A \oplus A^{-1}$, and it follows that there is an isomorphism

$$(1, B) : (R_1{}^{2n}, R_2{}^{2n}, 1) \to (R_1{}^{2n}, R_2{}^{2n}, A \oplus A^{-1}).$$

Then $(R_1{}^{2n}, R_2{}^{2n}, 1) \simeq (R_1, R_2, 1)^{2n}$, and to complete the proof we must show $(R_1, R_2, 1) \simeq R$. Now the map $R \to R_1 \times R_2$ given by $r \mapsto (i_1 r, i_2 r)$ is clearly additive, and since, identifying $S \otimes_{j_1} R_1$ and $S \otimes_{j_2} R_2$ with S, we have $1 \otimes_{j_1} i_1 r = j_1 i_1 r \otimes_{j_1} 1 =$

$j_1 i_1 r = j_2 i_2 r = j_2 i_2 r \otimes_{j_2} 1 = 1 \otimes_{j_2} i_2 r$, it follows that $(i_1 r, i_2 r) \in (R_1, R_2, 1)$. The map is an R-module homomorphism, since $rr' \mapsto (i_1(rr'), i_2(rr')) = ((i_1 r)(i_1 r'), (i_2 r)(i_2 r')) = r(i_1 r', i_2 r')$. Then, given $(r_1, r_2) \in (R_1, R_2, 1)$, we have $j_1 r_1 = j_1 r_1 \otimes_{j_1} 1 = 1 \otimes_{j_1} r_1 = 1 \otimes_{j_2} r_2 = j_2 r_2 \otimes_{j_2} 1 = j_2 r_2$, and so there is a unique element $r \in R$ with $i_1 r = r_1$ and $i_2 r = r_2$, that is, such that $r \mapsto (r_1, r_2)$. Thus we have an isomorphism, as required.

We shall now show that every finitely generated projective R-module arises in the above manner. Let M be a finitely generated projective R-module, and write $M_1 = R_1 \otimes_{i_1} M$, $M_2 = R_2 \otimes_{i_2} M$. This notation is unambiguous when $M = R$ provided we agree to identify R_1 with $R_1 \otimes_{i_1} R$ by $r_1 = r_1 \otimes 1$, and similarly for R_2. Let $h_M : S \otimes_{j_1} M_1 \to S \otimes_{j_2} M_2$ be the composite of the obvious isomorphisms

$$S \otimes_{j_1} R_1 \otimes_{i_1} M \to S \otimes_{j_1 i_1} M = S \otimes_{j_2 i_2} M \to S \otimes_{j_2} R_2 \otimes_{i_2} M.$$

Then the map $f_M : M \to (M_1, M_2, h_M)$ given by $f_M(m) = (1 \otimes m, 1 \otimes m)$ is obviously well defined and additive, and then $f_M(rm) = (1 \otimes rm, 1 \otimes rm) = (i_1 r \otimes m, i_2 r \otimes m) = r(1 \otimes m, 1 \otimes m) = rf_M(m)$ $(r \in R, m \in M)$, so we have an R-module homomorphism. With all the above data, we then have:

Proposition 60. $f_M : M \to (M_1, M_2, h_M)$ is an isomorphism.

Proof. First note that if $\alpha : M \to N$ is an isomorphism of R-modules, the diagram

$$\begin{array}{ccc} M & \xrightarrow{f_M} & (M_1, M_2, h_M) \\ \alpha \downarrow & & \downarrow (1 \otimes \alpha, 1 \otimes \alpha) \\ N & \xrightarrow{f_N} & (N_1, N_2, h_N) \end{array}$$

commutes, and so f_M is an isomorphism if and only if f_N is an isomorphism. Next, given finitely generated projective R-modules M, N, the diagram

$$\begin{array}{ccc} M \oplus N & \xrightarrow{f_M \oplus f_N} & (M_1, M_2, h_M) \oplus (N_1, N_2, h_N) \\ f_{M\oplus N} \downarrow & & \downarrow \simeq \\ ((M \oplus N)_1, (M \oplus N)_2, h_{M\oplus N}) & \xrightarrow{\simeq} & (M_1 \oplus N_1, M_2 \oplus N_2, h_M \oplus h_N) \end{array}$$

commutes, where the unnamed arrows are the obvious isomorphisms, and so $f_{M\oplus N}$ is an isomorphism if and only if $f_M \oplus f_N$ is an isomorphism.

Now let M, N be R-modules such that $M \oplus N \simeq R^n$. The map $f_R : R \to (R_1, R_2, h_R)$ is $r \mapsto (i_1 r, i_2 r)$, and we saw in the last part of the proof of Proposition 59 that this is an isomorphism. (Note that h_R becomes 1 when we make the identifications $S \otimes_{j_1} R_1 = S = S \otimes_{j_2} R_2$.) So f_R is an isomorphism, and hence so is $f_R \oplus f_R \oplus \ldots \oplus f_R$ (n terms). Thus f_{R^n} is an isomorphism, and hence $f_{M\oplus N}$ is an isomorphism. It follows that $f_M \oplus f_N$ is an isomorphism, so f_M is an isomorphism, and the proof is complete.

Proposition 61. Let $M = (P_1, P_2, h)$. Then $P_1 \simeq R_1 \otimes_{i_1} M$ and $P_2 \simeq R_2 \otimes_{i_2} M$.

Proof. The map $R_1 \times M \to P_1$ given by $(r_1, (p_1, p_2)) \mapsto r_1 p_1$ is bilinear and balanced (regarding R_1 as an R_1-R-bimodule, via i_1), and so induces an R_1-module homomorphism $f_M : R_1 \otimes_{i_1} M \to P_1$, $f_M(\Sigma\, r_1 \otimes (p_1, p_2)) = \Sigma\, r_1 p_1$. The proof now proceeds on lines similar to the proof of Proposition 60:

(a) Let $(\alpha, \beta) : M \to N$ be an isomorphism, where $M = (P_1, P_2, h)$ and $N = (Q_1, Q_2, k)$. Then the diagram

$$\begin{array}{ccc} R_1 \otimes_{i_1} M & \xrightarrow{f_M} & P_1 \\ 1 \otimes (\alpha, \beta) \downarrow & & \downarrow \alpha \\ R_1 \otimes_{i_1} N & \xrightarrow{f_N} & Q_1 \end{array}$$

commutes, so f_M is an isomorphism if and only if f_N is an isomorphism.

(b) Let $M = (P_1, P_2, h)$, $M' = (P_1', P_2', h')$, and $N = (P_1 \oplus P_1', P_2 \oplus P_2', h \oplus h')$. Then the diagram

$$\begin{array}{ccc}
(R_1 \otimes_{i_1} M) \oplus (R_1 \otimes_{i_1} M') & \xrightarrow{f_M \oplus f_{M'}} & P_1 \oplus P_1' \\
\simeq \downarrow & & \uparrow f_N \\
R_1 \otimes_{i_1} (M \oplus M') & \xrightarrow{\simeq} & R_1 \otimes_{i_1} N
\end{array}$$

commutes, where the unnamed arrows are the obvious isomorphisms, and so f_N is an isomorphism if and only if $f_M \oplus f_{M'}$ is an isomorphism.

(c) Let $M = (R_1, R_2, 1)$. We have seen that the map $\gamma : R \to M$, $r \mapsto (i_1 r, i_2 r)$ is an isomorphism of R-modules. Then the diagram

$$\begin{array}{ccc}
R_1 & \xrightarrow{1} & R_1 \\
\simeq \downarrow & & \uparrow f_M \\
R_1 \otimes_{i_1} R & \xrightarrow{1 \otimes \gamma} & R_1 \otimes_{i_1} M
\end{array}$$

commutes, and so f_M is an isomorphism. By (b), if $M = (R_1{}^n, R_2{}^n, 1)$, then f_M is an isomorphism.

Now let $M = (P_1, P_2, h)$. As in the proof of Proposition 59, there exists $M' = (P_1', P_2', h')$ and an isomorphism $(\alpha, \beta) : N \to N'$, where $N = (P_1 \oplus P_1', P_2 \oplus P_2', h \oplus h')$ and $N' = (R_1{}^m, R_2{}^m, 1)$. By (c), $f_{N'}$ is an isomorphism, and so by (a), f_N is an isomorphism. Thus by (b), $f_M \oplus f_{M'}$ is an isomorphism, and finally we deduce f_M is an isomorphism.

Similarly, the map $R_2 \times M \to P_2$, $(r_2, (p_1, p_2)) \mapsto r_2 p_2$, induces an isomorphism $g_M : R_2 \otimes_{i_2} M \to P_2$.

Corollary 62. Let $\alpha : M \to N$ be an isomorphism of R-modules, where $M = (P_1, P_2, h)$ and $N = (Q_1, Q_2, k)$. Then there exist isomorphisms $\beta : P_1 \to Q_1$ and $\gamma : P_2 \to Q_2$ such that $\alpha = (\beta, \gamma)$.

Proof. We define β, γ by making the diagrams

$$\begin{array}{ccc}
R_1 \otimes_{i_1} M & \xrightarrow{f_M} & P_1 \\
1 \otimes \alpha \downarrow & & \downarrow \beta \\
R_1 \otimes_{i_1} N & \xrightarrow{f_N} & Q_1
\end{array}
\quad \text{and} \quad
\begin{array}{ccc}
R_2 \otimes_{i_2} M & \xrightarrow{g_M} & P_2 \\
1 \otimes \alpha \downarrow & & \downarrow \gamma \\
R_2 \otimes_{i_2} N & \xrightarrow{g_N} & Q_2
\end{array}$$

commute, where f_M, g_M, etc., are as in the proof of Proposition 61, and so β and γ are isomorphisms. The diagram

$$\begin{array}{ccc} S \otimes_{j_1} P_1 & \xrightarrow{h} & S \otimes_{j_2} P_2 \\ {\scriptstyle 1 \otimes f_M^{-1}} \downarrow & & \downarrow {\scriptstyle 1 \otimes g_M^{-1}} \\ S \otimes_{j_1} R_1 \otimes_{i_1} M & & S \otimes_{j_2} R_2 \otimes_{i_2} M \\ {\scriptstyle 1 \otimes 1 \otimes \alpha} \downarrow & & \downarrow {\scriptstyle 1 \otimes 1 \otimes \alpha} \\ S \otimes_{j_1} R_1 \otimes_{i_1} N & & S \otimes_{j_2} R_2 \otimes_{i_2} N \\ {\scriptstyle 1 \otimes f_N} \downarrow & & \downarrow {\scriptstyle 1 \otimes g_N} \\ S \otimes_{j_1} Q_1 & \xrightarrow{k} & S \otimes_{j_2} Q_2 \end{array}$$

commutes, and so the diagram

$$\begin{array}{ccc} S \otimes_{j_1} P_1 & \xrightarrow{h} & S \otimes_{j_2} P_2 \\ {\scriptstyle 1 \otimes \beta} \downarrow & & \downarrow {\scriptstyle 1 \otimes \gamma} \\ S \otimes_{j_1} Q_1 & \xrightarrow{k} & S \otimes_{j_2} Q_2 \end{array}$$

commutes. Thus $(\beta, \gamma) : M \to N$ is an isomorphism, and it is easy to check that $(\beta, \gamma) = \alpha$.

We are now ready to describe the *Mayer-Vietoris sequence*

$$K_1R \xrightarrow{f_1} K_1R_1 \oplus K_1R_2 \xrightarrow{g_1} K_1S \xrightarrow{\delta}$$

$$\xrightarrow{\delta} K_0R \xrightarrow{f_0} K_0R_1 \oplus K_0R_2 \xrightarrow{g_0} K_0S.$$

We must define the group homomorphisms in the above sequence. As always in this section, we are assuming that

$$\begin{array}{ccc} R & \xrightarrow{i_2} & R_2 \\ {\scriptstyle i_1} \downarrow & & \downarrow {\scriptstyle j_2} \\ R_1 & \xrightarrow{j_1} & S \end{array}$$

is a cartesian square, with j_2 surjective. Now the map

$$K_0i_1 \oplus K_0i_2 : K_0R \oplus K_0R \to K_0R_1 \oplus K_0R_2$$

is a homomorphism, and so is the diagonal map $K_0R \to K_0R \oplus K_0R$,

$x \mapsto (x, x)$. The composite yields the homomorphism $f_0 : K_0R \to K_0R_1 \oplus K_0R_2$. More explicitly, if $[M]$ is a generator of K_0R, then $M \simeq (P_1, P_2, h)$, say, by Proposition 60. Thus $[M] = [P_1, P_2, h]$, and

$$(K_0i_1)[P_1, P_2, h] = [R_1 \otimes_{i_1} (P_1, P_2, h)] = [P_1]$$

by Proposition 61, and similarly $(K_0i_2)[P_1, P_2, h] = [P_2]$. Thus $f_0[M] = f_0[P_1, P_2, h] = ([P_1], [P_2])$.

Somewhat similarly, the map

$$K_1i_1 \oplus K_1i_2 : K_1R \oplus K_1R \to K_1R_1 \oplus K_1R_2$$

is a homomorphism, and so is the diagonal map $K_1R \to K_1R \oplus K_1R$, $x \mapsto (x, x)$. The composite yields the homomorphism $f_1 : K_1R \to K_1R_1 \oplus K_1R_2$. More explicitly, a typical element of K_1R is of the form $[M, \alpha]$, and by Proposition 60 there is an isomorphism $\beta : M \to (P_1, P_2, h)$, say. Thus $[M, \alpha] = [(P_1, P_2, h), \beta\alpha\beta^{-1}]$. But by Corollary 62, there are isomorphisms $\gamma_1 : P_1 \to P_1$ and $\gamma_2 : P_2 \to P_2$ such that $\beta\alpha\beta^{-1} = (\gamma_1, \gamma_2)$. Then

$$(K_1i_1)[(P_1, P_2, h), (\gamma_1, \gamma_2)]$$
$$= [R_1 \otimes_{i_1} (P_1, P_2, h), 1 \otimes (\gamma_1, \gamma_2)] = [P_1, \gamma_1]$$

by Proposition 61, and similarly

$$(K_1i_2)[(P_1, P_2, h), (\gamma_1, \gamma_2)] = [P_2, \gamma_2].$$

So

$$f_1[M, \alpha] = f_1[(P_1, P_2, h), (\gamma_1, \gamma_2)]$$
$$= ([P_1, \gamma_1], [P_2, \gamma_2]).$$

Next, the map

$$K_0j_1 \oplus K_0j_2 : K_0R_1 \oplus K_0R_2 \to K_0S \oplus K_0S$$

is a homomorphism, and so is the difference map $K_0S \oplus K_0S \to K_0S$, $(x, y) \mapsto x - y$. The composite yields the homomorphism $g_0 : K_0R_1 \oplus K_0R_2 \to K_0S$, whose effect on generators is given by

$$g_0([P_1], [P_2]) = [S \otimes_{j_1} P_1] - [S \otimes_{j_2} P_2].$$

Similarly, the map

$$K_1j_1 \oplus K_1j_2 : K_1R_1 \oplus K_1R_2 \to K_1S \oplus K_1S$$

is a homomorphism, and so is the difference map $K_1S \oplus K_1S \to K_1S$, $(x, y) \mapsto x - y$. The composite yields the homomorphism g_1 :

$K_1R_1 \oplus K_1R_2 \to K_1S$, given by

$$g_1([P_1, \alpha_1], [P_2, \alpha_2])$$
$$= [S \otimes_{j_1} P_1, 1 \otimes \alpha_1] - [S \otimes_{j_2} P_2, 1 \otimes \alpha_2].$$

Finally, the homomorphism $\delta : K_1S \to K_0R$ is given by

$$\delta[S^n, A] = [R_1{}^n, R_2{}^n, A] - [R^n]$$

where $A \in \mathrm{GL}_n(S)$. To see that this is well defined, let $[S^n, A] = [S^m, B]$, where $B \in \mathrm{GL}_m(S)$. Passing to $\mathrm{GL}(S)$, we have $BA^{-1} \in \mathrm{E}(S)$, and as in the proof of Proposition 59, the surjectivity of j_2 implies that $j_2(C) = BA^{-1}$ for some $C \in \mathrm{E}(R_2)$. Now choose r large enough so that $C \in \mathrm{E}_r(R_2)$, and also $r > n$, $r > m$, and we have an isomorphism

$$(1, C) : (R_1{}^r, R_2{}^r, A \oplus 1) \to (R_1{}^r, R_2{}^r, B \oplus 1)$$

whence

$$\begin{aligned}[R_1{}^n, R_2{}^n, A] - [R^n] &= [R_1{}^r, R_2{}^r, A \oplus 1] - [R^r] \\ &= [R_1{}^r, R_2{}^r, B \oplus 1] - [R^r] \\ &= [R_1{}^m, R_2{}^m, B] - [R^m]\end{aligned}$$

and so δ is well defined. It is a homomorphism, since

$$\begin{aligned}&\delta([S^n, A] + [S^m, B]) \\ &= \delta[S^{n+m}, A \oplus B] \\ &= [R_1{}^{n+m}, R_2{}^{n+m}, A \oplus B] - [R^{n+m}] \\ &= [R_1{}^n, R_2{}^n, A] + [R_1{}^m, R_2{}^m, B] - [R^n] - [R^m] \\ &= \delta[S^n, A] + \delta[S^m, B].\end{aligned}$$

Proposition 63. The sequence

$$K_1R \xrightarrow{f_1} K_1R_1 \oplus K_1R_2 \xrightarrow{g_1} K_1S \xrightarrow{\delta}$$
$$\xrightarrow{\delta} K_0R \xrightarrow{f_0} K_0R_1 \oplus K_0R_2 \xrightarrow{g_0} K_0S$$

is exact.

Proof. 1. Exactness at $K_0R_1 \oplus K_0R_2$.

(i) $g_0f_0[P_1, P_2, h] = g_0([P_1], [P_2])$

$$= [S \otimes_{j_1} P_1] - [S \otimes_{j_2} P_2] = 0$$

since $h : S \otimes_{j_1} P_1 \to S \otimes_{j_2} P_2$ is an isomorphism. Since $g_0 f_0$ kills the generators of $K_0 R$, we deduce $g_0 f_0 = 0$.

(ii) A typical element of $K_0 R_1 \oplus K_0 R_2$ is of the form

$$([P_1] - [R_1{}^n], [P_2] - [R_2{}^m])$$

and then

$$g_0([P_1] - [R_1{}^n], [P_2] - [R_2{}^m])$$

$$= [S \otimes_{j_1} P_1] - [S^n] - [S \otimes_{j_2} P_2] + [S^m].$$

So if $([P_1] - [R_1{}^n], [P_2] - [R_2{}^m]) \in \ker g_0$, we have

$$[(S \otimes_{j_1} P_1) \oplus S^m] = [(S \otimes_{j_2} P_2) \oplus S^n]$$

whence

$$(S \otimes_{j_1} P_1) \oplus S^{m+r} \cong (S \otimes_{j_2} P_2) \oplus S^{n+r}$$

for some r, and so there is an isomorphism

$$h : S \otimes_{j_1} (P_1 \oplus R_1{}^{m+r}) \to S \otimes_{j_2} (P_2 \oplus R_2{}^{n+r}).$$

We then have $[P_1 \oplus R_1{}^{m+r}, P_2 \oplus R_2{}^{n+r}, h] \in K_0 R$, and

$$([P_1] - [R_1{}^n], [P_2] - [R_2{}^m])$$

$$= ([P_1 \oplus R_1{}^{m+r}], [P_2 \oplus R_2{}^{n+r}]) - ([R_1{}^{n+m+r}], [R_2{}^{n+m+r}])$$

$$= f_0[P_1 \oplus R_1{}^{m+r}, P_2 \oplus R_2{}^{n+r}, h] - f_0[R_1{}^{n+m+r}, R_2{}^{n+m+r}, 1]$$

which is in the image of f_0.

2. Exactness at $K_0 R$.

(i) $f_0 \delta[S^n, A] = f_0([R_1{}^n, R_2{}^n, A] - [R^n])$

$$= f_0[R_1{}^n, R_2{}^n, A] - f_0[R_1{}^n, R_2{}^n, 1]$$

$$= ([R_1{}^n], [R_2{}^n]) - ([R_1{}^n], [R_2{}^n]) = 0$$

so $f_0 \delta = 0$.

(ii) A typical element of $K_0 R$ is of the form $[M] - [R^n]$, and by Proposition 60 we may write $[M] = [P_1, P_2, h]$, say. Then

$$f_0([P_1, P_2, h] - [R^n]) = f_0[P_1, P_2, h] - f_0[R_1{}^n, R_2{}^n, 1]$$

$$= ([P_1], [P_2]) - ([R_1{}^n], [R_2{}^n])$$

so if $[P_1, P_2, h] - [R^n] \in \ker f_0$, we have $[P_1] = [R_1{}^n]$ and $[P_2] = [R_2{}^n]$. Thus there are isomorphisms $\alpha : P_1 \oplus R_1{}^m \to R_1{}^{n+m}$ and $\beta : P_2 \oplus R_2{}^r \to R_2{}^{n+r}$, and without loss of generality we may take $r = m$. We define the isomorphism $h' : S^{n+m} \to S^{n+m}$ to make the diagram

$$\begin{array}{ccc}
(S \otimes_{j_1} P_1) \oplus S^m & \xrightarrow{h \oplus 1} & (S \otimes_{j_2} P_2) \oplus S^m \\
\simeq \downarrow & & \downarrow \simeq \\
S \otimes_{j_1} (P_1 \oplus R_1{}^m) & & S \otimes_{j_2} (P_2 \oplus R_2{}^m) \\
1 \otimes \alpha \downarrow & & \downarrow 1 \otimes \beta \\
S \otimes_{j_1} R_1{}^{n+m} & & S \otimes_{j_2} R_2{}^{n+m} \\
\simeq \downarrow & & \downarrow \simeq \\
S^{n+m} & \xrightarrow{h'} & S^{n+m}
\end{array}$$

commute. This gives

$$(P_1 \oplus R_1{}^m, P_2 \oplus R_2{}^m, h \oplus 1) \simeq (R_1{}^{n+m}, R_2{}^{n+m}, h')$$

whence

$$\begin{aligned}
[P_1, P_2, h] - [R^n] &= [P_1 \oplus R_1{}^m, P_2 \oplus R_2{}^m, h \oplus 1] - [R^{n+m}] \\
&= [R_1{}^{n+m}, R_2{}^{n+m}, h'] - [R^{n+m}] \\
&= \delta[S^{n+m}, h'] \in \delta(K_1 S).
\end{aligned}$$

3. Exactness at $K_1 S$.

(i) $\delta g_1([R_1{}^n, A_1], [R_2{}^m, A_2])$

$$\begin{aligned}
&= \delta([S^n, j_1 A_1] - [S^m, j_2 A_2]) \\
&= \delta[S^n, j_1 A_1] - \delta[S^m, j_2 A_2] \\
&= [R_1{}^n, R_2{}^n, j_1 A_1] - [R^n] - [R_1{}^m, R_2{}^m, j_2 A_2] + [R^m]
\end{aligned}$$

where $A_1 \in \mathrm{GL}_n(R_1)$ and $A_2 \in \mathrm{GL}_m(R_2)$. Then

$$(A_1, 1) : (R_1{}^n, R_2{}^n, j_1 A_1) \to (R_1{}^n, R_2{}^n, 1)$$

and

$$(1, A_2) : (R_1{}^m, R_2{}^m, 1) \to (R_1{}^m, R_2{}^m, j_2 A_2)$$

are isomorphisms, whence $[R_1{}^n, R_2{}^n, j_1 A_1] = [R^n]$ and $[R_1{}^m, R_2{}^m, j_2 A_2] = [R^m]$, and so $\delta g_1([R_1{}^n, A_1], [R_2{}^m, A_2]) = 0$, whence $\delta g_1 = 0$.

(ii) Let $[S^n, A] \in \ker \delta$, where $A \in \mathrm{GL}_n(S)$. Thus $[R_1{}^n, R_2{}^n, A] = [R^n]$, and so there is an isomorphism

$$\alpha : (R_1{}^{n+m}, R_2{}^{n+m}, A \oplus 1) \to (R_1{}^{n+m}, R_2{}^{n+m}, 1)$$

for some m. By Corollary 62, there exist $B_1 \in \mathrm{GL}_{n+m}(R_1)$ and $B_2 \in \mathrm{GL}_{n+m}(R_2)$ such that $\alpha = (B_1, B_2)$, that is, such that the diagram

$$\begin{array}{ccc} S^{n+m} & \xrightarrow{A \oplus 1} & S^{n+m} \\ {\scriptstyle j_1B_1}\downarrow & & \downarrow{\scriptstyle j_2B_2} \\ S^{n+m} & \xrightarrow{1} & S^{n+m} \end{array}$$

commutes. Thus

$$\begin{aligned} [S^n, A] &= [S^{n+m}, A \oplus 1] \\ &= [S^{n+m}, j_2B_2{}^{-1}j_1B_1] \\ &= [S^{n+m}, j_1B_1] - [S^{n+m}, j_2B_2] \\ &= g_1([R_1{}^{n+m}, B_1], [R_2{}^{n+m}, B_2]) \end{aligned}$$

which is in the image of g_1.

4. Exactness at $K_1R_1 \oplus K_1R_2$.

(i) $$\begin{aligned} g_1f_1[R^n, A] &= g_1([R_1{}^n, i_1A], [R_2{}^n, i_2A]) \\ &= [S^n, j_1i_1A] - [S^n, j_2i_2A] = 0 \end{aligned}$$

since $j_1i_1 = j_2i_2$. Thus $g_1f_1 = 0$.

(ii) Let $([R_1{}^n, A_1], [R_2{}^m, A_2]) \in \ker g_1$, where $A_1 \in \mathrm{GL}_n(R_1)$ and $A_2 \in \mathrm{GL}_m(R_2)$. So $[S^n, j_1A_1] = [S^m, j_2A_2]$. Passing to $\mathrm{GL}(S)$, we have $j_2A_2{}^{-1}j_1A_1 \in \mathrm{E}(S)$, and since j_2 is surjective we can find $B \in \mathrm{E}(R_2)$ with $j_2B = j_2A_2{}^{-1}j_1A_1$. Choose r large enough that $B \in \mathrm{E}_r(R_2)$, and also $r > n$, $r > m$, and we have, in $\mathrm{GL}_r(S)$, $j_1(A_1 \oplus 1) = j_2((A_2 \oplus 1)B)$. This means that

$$(A_1 \oplus 1, (A_2 \oplus 1)B) : (R_1{}^r, R_2{}^r, 1) \to (R_1{}^r, R_2{}^r, 1)$$

is an isomorphism. We then have

$$\begin{aligned} ([R_1{}^n, A_1], [R_2{}^m, A_2]) &= ([R_1{}^r, A_1 \oplus 1], [R_2{}^r, A_2 \oplus 1]) \\ &= ([R_1{}^r, A_1 \oplus 1], [R_2{}^r, (A_2 \oplus 1)B]) \end{aligned}$$

(since $B \in E_r(R_2)$, so that $[R_2{}^r, B] = 0$)

$$= f_1[(R_1{}^r, R_2{}^r, 1), (A_1 \oplus 1, (A_2 \oplus 1)B)] \in f_1(K_1R).$$

This completes the proof of Proposition 63.

Now let

$$\begin{array}{ccc} R & \xrightarrow{i_2} & R_2 \\ {\scriptstyle i_1}\downarrow & & \downarrow{\scriptstyle j_2} \\ R_1 & \xrightarrow{j_1} & S \end{array}$$

be a cartesian square of *commutative* rings, with j_2 surjective. Let $M = (P_1, P_2, h)$ and $M' = (P_1', P_2', h')$. Then there is an isomorphism of S-modules $h \otimes h' : (S \otimes_{j_1} P_1) \otimes_S (S \otimes_{j_1} P_1') \to (S \otimes_{j_2} P_2) \otimes_S (S \otimes_{j_2} P_2')$. If we identify

$(S \otimes_{j_1} P_1) \otimes_S (S \otimes_{j_1} P_1')$ with $S \otimes_{j_1} (P_1 \otimes_{R_1} P_1')$

and

$(S \otimes_{j_2} P_2) \otimes_S (S \otimes_{j_2} P_2')$ with $S \otimes_{j_2} (P_2 \otimes_{R_2} P_2')$

(cf. the argument on pp.16-17), then we may write

$$N = (P_1 \otimes_{R_1} P_1', P_2 \otimes_{R_2} P_2', h \otimes h').$$

We then have:

Proposition 64. If $M = (P_1, P_2, h)$, $M' = (P_1', P_2', h')$, and $N = (P_1 \otimes_{R_1} P_1', P_2 \otimes_{R_2} P_2', h \otimes h')$, then $M \otimes_R M' \simeq N$.

Proof. The map $M \times M' \to N$, $((p_1, p_2), (p_1', p_2')) \mapsto (p_1 \otimes p_1', p_2 \otimes p_2')$, is bilinear and balanced, and so induces an R-module homomorphism $f(M, M') : M \otimes_R M' \to N$. The proof now proceeds on lines similar to the proofs of Propositions 60 and 61:

(a) Let $\alpha : M \to M_0$, $\alpha' : M' \to M_0'$ be isomorphisms of R-modules, where $M_0 = (Q_1, Q_2, k)$ and $M_0' = (Q_1', Q_2', k')$. Put $N_0 = (Q_1 \otimes_{R_1} Q_1', Q_2 \otimes_{R_2} Q_2', k \otimes k')$. By Corollary 62, we may write $\alpha = (\beta, \gamma)$ and $\alpha' = (\beta', \gamma')$, where $\beta : P_1 \to Q_1$, $\gamma : P_2 \to Q_2$, $\beta' : P_1' \to Q_1'$, $\gamma' : P_2' \to Q_2'$ are isomorphisms such that the diagrams

$$\begin{array}{ccc} S \otimes_{j_1} P_1 & \xrightarrow{h} & S \otimes_{j_2} P_2 \\ {\scriptstyle 1 \otimes \beta}\downarrow & & \downarrow{\scriptstyle 1 \otimes \gamma} \\ S \otimes_{j_1} Q_1 & \xrightarrow{k} & S \otimes_{j_2} Q_2 \end{array}$$

and

$$\begin{array}{ccc} S \otimes_{j_1} P_1' & \xrightarrow{h'} & S \otimes_{j_2} P_2' \\ {\scriptstyle 1 \otimes \beta'}\downarrow & & \downarrow{\scriptstyle 1 \otimes \gamma'} \\ S \otimes_{j_1} Q_1' & \xrightarrow{k'} & S \otimes_{j_2} Q_2' \end{array}$$

commute. It follows that, making the identifications agreed above, the diagram

$$\begin{array}{ccc} S \otimes_{j_1} (P_1 \otimes_{R_1} P_1') & \xrightarrow{h \otimes h'} & S \otimes_{j_2} (P_2 \otimes_{R_2} P_2') \\ {\scriptstyle 1 \otimes (\beta \otimes \beta')}\downarrow & & \downarrow{\scriptstyle 1 \otimes (\gamma \otimes \gamma')} \\ S \otimes_{j_1} (Q_1 \otimes_{R_1} Q_1') & \xrightarrow{k \otimes k'} & S \otimes_{j_2} (Q_2 \otimes_{R_2} Q_2') \end{array}$$

commutes, and so $(\beta \otimes \beta', \gamma \otimes \gamma') : N \to N_0$ is an isomorphism of R-modules. It is clear that the diagram

$$\begin{array}{ccc} M \otimes_R M' & \xrightarrow{f(M,\ M')} & N \\ {\scriptstyle \alpha \otimes \alpha'}\downarrow & & \downarrow{\scriptstyle (\beta \otimes \beta',\ \gamma \otimes \gamma')} \\ M_0 \otimes_R M_0' & \xrightarrow{f(M_0,\ M_0')} & N_0 \end{array}$$

commutes, and so $f(M, M')$ is an isomorphism if and only if $f(M_0, M_0')$ is an isomorphism.

(b) Let $M = (P_1, P_2, h)$, $M' = (P_1', P_2', h')$, $M_0 = (Q_1, Q_2, k)$, $N = (P_1 \otimes_{R_1} Q_1, P_2 \otimes_{R_2} Q_2, h \otimes k)$,
$N' = (P_1' \otimes_{R_1} Q_1, P_2' \otimes_{R_2} Q_2, h' \otimes k)$, and
$N_0 = ((P_1 \oplus P_1') \otimes_{R_1} Q_1, (P_2 \oplus P_2') \otimes_{R_2} Q_2, (h \oplus h') \otimes k)$. Then there is an obvious isomorphism

$$(M \oplus M') \otimes_R M_0 \to (M \otimes_R M_0) \oplus (M' \otimes_R M_0).$$

Further, the obvious isomorphisms

$$(P_1 \oplus P_1') \otimes_{R_1} Q_1 \to (P_1 \otimes_{R_1} Q_1) \oplus (P_1' \otimes_{R_1} Q_1)$$

and

$$(P_2 \oplus P_2') \otimes_{R_2} Q_2 \to (P_2 \otimes_{R_2} Q_2) \oplus (P_2' \otimes_{R_2} Q_2)$$

give rise to an isomorphism $N_0 \to N \oplus N'$, where we are identifying $N \oplus N'$ with (A_1, A_2, ℓ), where $A_1 =$ $(P_1 \otimes_{R_1} Q_1) \oplus (P_1' \otimes_{R_1} Q_1)$, $A_2 = (P_2 \otimes_{R_2} Q_2) \oplus (P_2' \otimes_{R_2} Q_2)$, and $\ell = (h \otimes k) \oplus (h' \otimes k)$. If we also identify $M \oplus M'$ with $(P_1 \oplus P_1', P_2 \oplus P_2', h \oplus h')$, we obtain a commutative diagram

$$\begin{array}{ccc} (M \oplus M') \otimes_R M_0 & \xrightarrow{f(M \oplus M',\, M_0)} & N_0 \\ \simeq \downarrow & & \downarrow \simeq \\ (M \otimes_R M_0) \oplus (M' \otimes_R M_0) & \xrightarrow{f(M,\, M_0) \oplus f(M',\, M_0)} & N \oplus N' \end{array}$$

and so $f(M \oplus M', M_0)$ is an isomorphism if and only if $f(M, M_0) \oplus f(M', M_0)$ is an isomorphism. Similarly, $f(M_0, M \oplus M')$ is an isomorphism if and only if $f(M_0, M) \oplus f(M_0, M')$ is an isomorphism.

(c) Making the identifications $R = (R_1, R_2, 1)$, $R_1 \otimes_{R_1} R_1 = R_1$, and $R_2 \otimes_{R_2} R_2 = R_2$, we obtain

$$(R_1 \otimes_{R_1} R_1, R_2 \otimes_{R_2} R_2, 1 \otimes 1) = (R_1, R_2, 1) = R$$

and now it is easy to check that $f(R, R) : R \otimes_R R \to R$ is the obvious isomorphism, $r \otimes r' \mapsto rr'$. So $f(R, R)$ is an isomorphism, and by (b) it follows that $f(R^n, R^m)$ is an isomorphism.

Now, given M, M' as in the statement of the proposition, find M_0, M_0' with $M \oplus M_0 \simeq R^n$ and $M' \oplus M_0' \simeq R^m$. By (c), $f(R^n, R^m)$ is an isomorphism, so by (a), $f(M \oplus M_0, M' \oplus M_0')$ is an isomorphism. By (b), $f(M \oplus M_0, M') \oplus f(M \oplus M_0, M_0')$ is an isomorphism, so $f(M \oplus M_0, M')$ is an isomorphism. Again, by (b), $f(M, M') \oplus f(M_0, M')$ is an isomorphism, and so finally $f(M, M')$ is an isomorphism, as required.

Corollary 65. $M = (P_1, P_2, h)$ is an invertible R-module if and only if P_1 is an invertible R_1-module and P_2 is an invertible R_2-module.

Proof. First suppose $(P_1, P_2, h) \otimes_R (P_1', P_2', h') \simeq R$. By Proposition 64, it follows that

$$(P_1 \otimes_{R_1} P_1', P_2 \otimes_{R_2} P_2', h \otimes h') \simeq (R_1, R_2, 1)$$

and so $P_1 \otimes_{R_1} P_1' \simeq R_1$ and $P_2 \otimes_{R_2} P_2' \simeq R_2$, by Corollary 62, whence P_1 and P_2 are invertible.

Conversely, suppose P_1 and P_2 are invertible. Then there is an isomorphism $\lambda : P_1 \otimes_{R_1} P_1^* \to R_1$, where $P_1^* = \mathrm{Hom}_{R_1}(P_1, R_1)$ is the dual of P_1, given by $\lambda(p_1 \otimes v_1) = v_1 p_1$, where $p_1 \in P_1$ and $v_1 \in P_1^*$, that is, $v_1 : P_1 \to R_1$. Similarly, there is an isomorphism $\lambda : P_2 \otimes_{R_2} P_2^* \to R_2$, $\lambda(p_2 \otimes v_2) = v_2 p_2$, $p_2 \in P_2$, $v_2 \in P_2^*$. Given $s \in S$, there is a homomorphism $s : S \to S$ of S-modules given by $s' \mapsto ss'$. With this abuse of notation, given $s \in S$ and $v_2 \in P_2^*$, there is an S-module homomorphism $s \otimes v_2 :$ $S \otimes_{j_2} P_2 \to S \otimes_{j_2} R_2$, and if we make the identification $S \otimes_{j_2} R_2 = S$, then $s \otimes v_2 \in (S \otimes_{j_2} P_2)^*$. Thus we obtain a bilinear balanced map $S \times P_2^* \to (S \otimes_{j_2} P_2)^*$, given by $(s, v_2) \mapsto$ $s \otimes v_2$, and this induces an S-module homomorphism f_{P_2} : $S \otimes_{j_2} P_2^* \to (S \otimes_{j_2} P_2)^*$, given by $s \otimes v_2 \mapsto s \otimes v_2$ (sic). We leave it to the reader to show that f_{P_2} is an isomorphism; the proof is on similar lines to the proofs of Propositions 60, 61, and 64.

Now let $v_1 \in P_1$ and $s \in S$; then we have an S-module homomorphism $s \otimes v_1 : S \otimes_{j_1} P_1 \to S \otimes_{j_1} R_1$, and if we make the identification $S \otimes_{j_1} R_1 = S$, then $s \otimes v_1 \in (S \otimes_{j_1} P_1)^*$. Then h : $S \otimes_{j_1} P_1 \to S \otimes_{j_2} P_2$ is an S-module isomorphism, and so $(s \otimes v_1)h^{-1} \in (S \otimes_{j_2} P_2)^*$. If we now identify $(S \otimes_{j_2} P_2)^*$ with $S \otimes_{j_2} P_2^*$ via f_{P_2}, we obtain a bilinear balanced map $S \times P_1^* \to$ $S \otimes_{j_2} P_2^*$, given by $(s, v_1) \mapsto (s \otimes v_1)h^{-1}$, and this induces an S-module homomorphism $k : S \otimes_{j_1} P_1^* \to S \otimes_{j_2} P_2^*$. We leave it to the reader to show that k is an isomorphism. By Proposition 64, $(P_1, P_2, h) \otimes_R (P_1^*, P_2^*, k) \simeq (P_1 \otimes_{R_1} P_1^*, P_2 \otimes_{R_2} P_2^*, h \otimes k)$, and then the diagram

$$\begin{array}{ccc} S \otimes_{j_1} (P_1 \otimes_{R_1} P_1^*) & \xrightarrow{h \otimes k} & S \otimes_{j_2} (P_2 \otimes_{j_2} P_2^*) \\ {\scriptstyle 1 \otimes \lambda}\downarrow & & \downarrow{\scriptstyle 1 \otimes \lambda} \\ S \otimes_{j_1} R_1 = S & \xrightarrow{\quad 1 \quad} & S = S \otimes_{j_2} R_2 \end{array}$$

commutes, whence

$$(P_1 \otimes_{R_1} P_1^*, P_2 \otimes_{R_2} P_2^*, h \otimes k) \simeq (R_1, R_2, 1) \simeq R$$

and so (P_1, P_2, h) is invertible.

We shall now apply the determinant map to the Mayer-Vietoris sequence to obtain another sequence

$$R^{\cdot} \xrightarrow{f_1} R_1^{\cdot} \oplus R_2^{\cdot} \xrightarrow{g_1} S^{\cdot} \xrightarrow{\delta}$$

$$\xrightarrow{\delta} \text{Pic } R \xrightarrow{f_0} \text{Pic } R_1 \oplus \text{Pic } R_2 \xrightarrow{g_0} \text{Pic } S.$$

Explicitly, if $\alpha \in R^{\cdot}$, $f_1\alpha = (i_1\alpha, i_2\alpha)$. If $\beta_1 \in R_1^{\cdot}$ and $\beta_2 \in R_2^{\cdot}$, then $g_1(\beta_1, \beta_2) = (j_1\beta_1)(j_2\beta_2)^{-1}$. If $\gamma \in S^{\cdot}$, then $\gamma \in GL_1(S)$, and (R_1, R_2, γ) is an invertible R-module, by Corollary 65, and we define $\delta\gamma = \langle R_1, R_2, \gamma\rangle$. Then $f_0\langle P_1, P_2, h\rangle = (\langle P_1\rangle, \langle P_2\rangle)$, which makes sense, by Corollary 65, and finally $g_0(\langle P_1\rangle, \langle P_2\rangle) = \langle S \otimes_{j_1} P_1\rangle\langle S \otimes_{j_2} P_2\rangle^{-1}$.

Proposition 66. The diagram

$$\begin{array}{ccccccc} K_1R & \xrightarrow{f_1} & K_1R_1 \oplus K_1R_2 & \xrightarrow{g_1} & K_1S & \xrightarrow{\delta} \\ \det \downarrow & & \det \oplus \det \downarrow & & \det \downarrow & \\ R^{\cdot} & \xrightarrow[f_1]{} & R_1^{\cdot} \oplus R_2^{\cdot} & \xrightarrow[g_1]{} & S^{\cdot} & \xrightarrow[\delta]{} \end{array}$$

$$\begin{array}{ccccccc} \xrightarrow{\delta} & K_0R & \xrightarrow{f_0} & K_0R_1 \oplus K_0R_2 & \xrightarrow{g_0} & K_0S \\ & \det \downarrow & & \det \oplus \det \downarrow & & \det \downarrow \\ \xrightarrow[\delta]{} & \text{Pic } R & \xrightarrow[f_0]{} & \text{Pic } R_1 \oplus \text{Pic } R_2 & \xrightarrow[g_0]{} & \text{Pic } S \end{array}$$

is commutative with exact rows.

Proof. We already have that the top row is exact (Proposition 63), and the commutativity is easy, once we note that det : $K_0R \to \text{Pic } R$ is given by $\det[P_1, P_2, h] = \langle \det P_1, \det P_2, \det h\rangle$. Next, in the top row we have $g_1f_1 = 0$, and so $(\det)(g_1f_1) = 0$. By commutativity, $(g_1f_1)(\det) = 0$; but det is surjective, and so in the bottom row, $g_1f_1 = 0$. Similarly, $\delta g_1 = 0$, $f_0\delta = 0$, and

$g_0f_0 = 0$, in the bottom row. So at each place in the bottom row we have image $\subset$ kernel, and we need only show kernel $\subset$ image.

1. Exactness at Pic $R_1 \oplus$ Pic R_2.

Let $(\langle P_1\rangle, \langle P_2\rangle) \in \ker g_0$, that is, $\langle S \otimes_{j_1} P_1\rangle = \langle S \otimes_{j_2} P_2\rangle$ in Pic S. So there is an isomorphism $h : S \otimes_{j_1} P_1 \to S \otimes_{j_2} P_2$, and by Corollary 65, $\langle P_1, P_2, h\rangle \in$ Pic R, and further,

$$(\langle P_1\rangle, \langle P_2\rangle) = f_0\langle P_1, P_2, h\rangle \in f_0(\text{Pic } R)$$

as required.

2. Exactness at Pic R.

Let $\langle P_1, P_2, h\rangle \in \ker f_0$, that is, $\langle P_1\rangle = 1$ in Pic R_1 and $\langle P_2\rangle = 1$ in Pic R_2. So there are isomorphisms $\alpha : P_1 \to R_1$ and $\beta : P_2 \to R_2$, and if we define $k : S \to S$ so as to make the diagram

$$\begin{array}{ccc} S \otimes_{j_1} P_1 & \xrightarrow{\quad h \quad} & S \otimes_{j_2} P_2 \\ {\scriptstyle 1\otimes\alpha}\downarrow & & \downarrow{\scriptstyle 1\otimes\beta} \\ S \otimes_{j_1} R_1 = S & \xrightarrow{\quad k \quad} & S = S \otimes_{j_2} R_2 \end{array}$$

commute, then $(\alpha, \beta) : (P_1, P_2, h) \to (R_1, R_2, k)$ is an isomorphism of R-modules, and so $\langle P_1, P_2, h\rangle = \langle R_1, R_2, k\rangle$. Then $k \in GL_1(S) = S^{\bullet}$, and $\langle R_1, R_2, k\rangle = \delta k \in \delta(S^{\bullet})$, as required.

3. Exactness at $S^{\bullet}$.

Let $\alpha \in \ker \delta$, that is, $\langle R_1, R_2, \alpha\rangle = \langle R\rangle = \langle R_1, R_2, 1\rangle$ in Pic R. So, using Corollary 62, there is an isomorphism $(\beta, \gamma) : (R_1, R_2, \alpha) \to (R_1, R_2, 1)$. That is, $\beta : R_1 \to R_1$, or $\beta \in GL_1(R_1) = R_1^{\bullet}$, and similarly $\gamma \in R_2^{\bullet}$, and the diagram

$$\begin{array}{ccc} S \otimes_{j_1} R_1 = S & \xrightarrow{\quad \alpha \quad} & S = S \otimes_{j_2} R_2 \\ {\scriptstyle 1\otimes\beta}\downarrow & & \downarrow{\scriptstyle 1\otimes\gamma} \\ S \otimes_{j_1} R_1 = S & \xrightarrow{\quad 1 \quad} & S = S \otimes_{j_2} R_2 \end{array}$$

commutes. In other words, the diagram $\begin{array}{ccc} S & \xrightarrow{\alpha} & S \\ {\scriptstyle j_1\beta}\downarrow & & \downarrow{\scriptstyle j_2\gamma} \\ S & \xrightarrow{1} & S \end{array}$ commutes, or $\alpha = (j_1\beta)(j_2\gamma)^{-1} = g_1(\beta, \gamma)$, which is in the image of g_1, as required.

4. Exactness at $R_1^{\bullet} \oplus R_2^{\bullet}$.

Let $(\beta, \gamma) \in \ker g_1$, that is, $j_1\beta = j_2\gamma$. Since we have a cartesian square, there is a (unique) element $\alpha \in R$ with $i_1\alpha = \beta$ and $i_2\alpha = \gamma$. We must show $\alpha \in R^{\bullet}$. Now $(j_1\beta)^{-1} = (j_2\gamma)^{-1}$, or $j_1(\beta^{-1}) = j_2(\gamma^{-1})$. So there is a (unique) element $\alpha_1 \in R$ with $i_1\alpha_1 = \beta^{-1}$ and $i_2\alpha_1 = \gamma^{-1}$. Then $j_1(1) = 1 = j_2(1)$, so there is a unique element $\alpha_2 \in R$ with $i_1\alpha_2 = 1$ and $i_2\alpha_2 = 1$. Since $i_1(1) = 1$ and $i_2(1) = 1$, we must have $\alpha_2 = 1$. But then $i_1(\alpha\alpha_1) = (i_1\alpha)(i_1\alpha_1) = \beta\beta^{-1} = 1$, and $i_2(\alpha\alpha_1) = (i_2\alpha)(i_2\alpha_1) = \gamma\gamma^{-1} = 1$, so $\alpha\alpha_1 = 1$, and $\alpha \in R^{\bullet}$. Finally, $(\beta, \gamma) = (i_1\alpha, i_2\alpha) = f_1\alpha \in f_1(R^{\bullet})$, as required.

Examples. Let R_2 be any ring, and R a subring of R_2, and denote by $i_2 : R \hookrightarrow R_2$ the injection map $r \mapsto r$. Suppose $I \lhd R_2$ with $I \subset R$; then $I \lhd R$. Write $R_1 = R/I$ and $S = R_2/I$, and let $i_1 : R \to R_1$, $j_2 : R_2 \twoheadrightarrow S$ be the natural epimorphisms, and let $j_1 : R_1 \hookrightarrow S$ be the monomorphism induced by i_2. Then the diagram

$$\begin{array}{ccc} R & \xrightarrow{i_2} & R_2 \\ {\scriptstyle i_1}\downarrow & & \downarrow{\scriptstyle j_2} \\ R_1 & \xrightarrow{j_1} & S \end{array}$$

commutes, and furthermore it is a cartesian square. For if $r_2 \in R_2$ and $r_1 + I \in R/I = R_1$ $(r_1 \in R)$ with $j_1(r_1 + I) = j_2r_2$, then in $R_2/I = S$, $r_1 + I = r_2 + I$, whence $r_2 \in r_1 + I \subset R$. Then $i_2r_2 = r_2$, and $i_1r_2 = r_2 + I = r_1 + I$; uniqueness is clear, since i_2 is injective. Also note that j_2 is surjective, so Proposition 63 applies; if, further, R_2 is commutative, then so are R, R_1, and S, and then Proposition 66 applies.

Let us take a special case. Let $R_2 = \mathbb{Z}[\alpha]$, where $\alpha^2 = a\alpha + b$ for some $a, b \in \mathbb{Z}$ with $a^2 + 4b$ not a square. More formally, $R_2 = \mathbb{Z}[x]/(x^2 - ax - b)$, and we write α for the image of x in R_2. So $R_2 = \{n + m\alpha : n, m \in \mathbb{Z}\}$ is a subring of $\mathbb{C}$, the field of complex numbers. Now let $r \in \mathbb{Z}$, $r > 0$, and put $R = \mathbb{Z}[r\alpha] = \{n + mr\alpha : n, m \in \mathbb{Z}\}$; put $I = rR_2 = \{nr + mr\alpha : n, m \in \mathbb{Z}\}$. Then $I \lhd R_2$ and $I \subset R$, so we can put $R_1 = R/I$ and $S = R_2/I$ to obtain a cartesian square, as above.

Write $n \mapsto \bar{n}$ for the natural map $\mathbb{Z} \to \mathbb{Z}/r\mathbb{Z}$. Then there is an epimorphism $R \to \mathbb{Z}/r\mathbb{Z}$ given by $n + mr\alpha \mapsto \bar{n}$ and having kernel I, whence $R_1 \simeq \mathbb{Z}/r\mathbb{Z}$. There is an epimorphism $R_2 \to (\mathbb{Z}/r\mathbb{Z})[\alpha]$ given by $n + m\alpha \mapsto \bar{n} + \bar{m}\alpha$ and having kernel I, whence $S \simeq (\mathbb{Z}/r\mathbb{Z})[\alpha]$. (The latter is the ring of all expressions $\bar{n} + \bar{m}\alpha$, $\bar{n}, \bar{m} \in \mathbb{Z}/r\mathbb{Z}$, where $\alpha^2 = \bar{a}\alpha + \bar{b}$; more formally, it is the ring $(\mathbb{Z}/r\mathbb{Z})[x]/(x^2 - \bar{a}x - \bar{b})$.) If we identify R_1 with $\mathbb{Z}/r\mathbb{Z}$ and S with $(\mathbb{Z}/r\mathbb{Z})[\alpha]$, we obtain a cartesian square

$$\begin{array}{ccc} \mathbb{Z}[r\alpha] & \longrightarrow & \mathbb{Z}[\alpha] \\ \downarrow & & \downarrow \\ \mathbb{Z}/r\mathbb{Z} & \longrightarrow & (\mathbb{Z}/r\mathbb{Z})[\alpha] \end{array}$$

to which we may apply Proposition 66.

Let us first take $\alpha = \frac{1}{2}(1 + \sqrt{-d})$, where $d \in \mathbb{Z}$, $d > 0$, $d \equiv 3 \pmod 4$, and let us take $r = 2$. Then $R = \mathbb{Z}[1 + \sqrt{-d}] = \mathbb{Z}[\sqrt{-d}]$, $R_2 = \mathbb{Z}[\frac{1}{2}(1 + \sqrt{-d})]$, and $R_1 = \mathbb{Z}/2\mathbb{Z}$. Next, $\alpha^2 = \alpha - \frac{1}{4}(1 + d)$, so if $\frac{1}{4}(1 + d) \equiv 1 \pmod 2$, that is, if $d \equiv 3 \pmod 8$, then since $x^2 + x + 1$ is irreducible over R_1, which is a field, S is a field. Thus $K_0R_1 \simeq \mathbb{Z}$, $K_0S \simeq \mathbb{Z}$, Pic $R_1 = 0$, Pic $S = 0$, and since $|R_1| = 2$ and $|S| = 4$, $K_1R_1 \simeq R_1^{\bullet} = 0$, and $K_1S \simeq S^{\bullet} \simeq \mathbb{Z}/3\mathbb{Z}$ (writing all the groups additively). Now when $d = 3$, 11, or 19, R_2 is a principal ideal domain. (See pp.76-77. Actually, for $d = 3$ or 11, R_2 is Euclidean, but we do not need this fact here.) So in these cases, $K_0R_2 \simeq \mathbb{Z}$ and Pic $R_2 = 0$. When $d = 3$, $R_2^{\bullet} \simeq \mathbb{Z}/6\mathbb{Z}$, and $R^{\bullet} \simeq \mathbb{Z}/2\mathbb{Z}$; Proposition 66 gives

$$\begin{array}{ccccccccccc} K_1R & \xrightarrow{f_1} & K_1R_2 & \xrightarrow{g_1} & \mathbb{Z}/3\mathbb{Z} & \xrightarrow{\delta} & K_0R & \to & \mathbb{Z} \oplus \mathbb{Z} & \to & \mathbb{Z} \\ \Downarrow & & \Downarrow & & 1\downarrow & & \Downarrow & & \downarrow & & \downarrow \\ \mathbb{Z}/2\mathbb{Z} & \xrightarrow{f_1} & \mathbb{Z}/6\mathbb{Z} & \xrightarrow{g_1} & \mathbb{Z}/3\mathbb{Z} & \xrightarrow{\delta} & \text{Pic } R & \longrightarrow & 0 & \longrightarrow & 0. \end{array}$$

In the bottom row, $f_1 \neq 0$, so f_1 is a monomorphism, and $\ker g_1 = \mathbb{Z}/2\mathbb{Z}$, whence g_1 is surjective. It follows that, in the top row,

g_1 is surjective, and so $\delta = 0$ in each row. Thus Pic $R = 0$ and $K_0R \simeq \mathbb{Z}$. This is the alternative method of calculating $K_0\mathbb{Z}[\sqrt{-3}]$ that was promised on p.76. When $d = 11$ or 19, then $R_2^{\bullet} \simeq R^{\bullet} \simeq \mathbb{Z}/2\mathbb{Z}$, and Proposition 66 gives

$$\begin{array}{ccccccccccc} K_1R & \xrightarrow{f_1} & K_1R_2 & \xrightarrow{g_1} & \mathbb{Z}/3\mathbb{Z} & \xrightarrow{\delta} & K_0R & \to & \mathbb{Z} \oplus \mathbb{Z} & \to & \mathbb{Z} \\ \downarrow & & \downarrow & & \downarrow 1 & & \downarrow & & \downarrow & & \downarrow \\ \mathbb{Z}/2\mathbb{Z} & \xrightarrow{f_1} & \mathbb{Z}/2\mathbb{Z} & \xrightarrow{g_1} & \mathbb{Z}/3\mathbb{Z} & \xrightarrow{\delta} & \text{Pic } R & \to & 0 & \to & 0. \end{array}$$

This time, in the bottom row we must have $g_1 = 0$, so $g_1 = 0$ in the top row also. Thus in the bottom row, δ is an isomorphism, and in the top row δ is a monomorphism, giving Pic $R \simeq \mathbb{Z}/3\mathbb{Z}$ and $K_0R \simeq \mathbb{Z} \oplus \mathbb{Z}/3\mathbb{Z}$.

Now let us consider the case where $d \equiv 7 \pmod 8$. Then $\frac{1}{4}(1 + d) \equiv 0 \pmod 2$, and so $\alpha^2 = \alpha$ in S, and $S \simeq \mathbb{Z}/2\mathbb{Z} \times \mathbb{Z}/2\mathbb{Z}$. So $K_0S \simeq \mathbb{Z} \oplus \mathbb{Z}$, Pic $S = 0$, and $K_1S \simeq S^{\bullet} = 0$. When $d = 7$, 15, or 23, we have seen that R_2 is a Dedekind domain (see pp.76-77), with Pic $R_2 \simeq 0$, $\mathbb{Z}/2\mathbb{Z}$, $\mathbb{Z}/3\mathbb{Z}$ respectively, and $K_0R_2 \simeq \mathbb{Z} \oplus \text{Pic } R_2$. Proposition 66 gives

$$\begin{array}{ccccccccccc} K_1R & \to & K_1R_2 & \to & 0 & \to & K_0R & \xrightarrow{f_0} & \mathbb{Z} \oplus \mathbb{Z} \oplus \text{Pic } R_2 & \xrightarrow{g_0} & \mathbb{Z} \oplus \mathbb{Z} \\ \downarrow & & \downarrow & & \downarrow & & \downarrow & & \downarrow & & \downarrow \\ \mathbb{Z}/2\mathbb{Z} & \to & \mathbb{Z}/2\mathbb{Z} & \to & 0 & \to & \text{Pic } R & \xrightarrow{f_0} & \text{Pic } R_2 & \xrightarrow{g_0} & 0 \end{array}$$

and in the bottom row, f_0 is an isomorphism. It is easy to see that in the top row, $g_0(\mathbb{Z} \oplus \mathbb{Z} \oplus \text{Pic } R_2) \simeq \mathbb{Z}$, and so $\ker g_0 \simeq \mathbb{Z} \oplus \text{Pic } R_2$. Then f_0 is a monomorphism, so $K_0R \simeq \mathbb{Z} \oplus \text{Pic } R_2 \simeq K_0R_2$, and Pic $R \simeq$ Pic R_2.

As a further example, let us take $\alpha = i$, so $\alpha^2 = -1$, and let us take $r = p$, where p is prime, and $p \equiv 3 \pmod 4$. Then $R_1 = \mathbb{Z}/p\mathbb{Z}$ is a field, $K_0R_1 \simeq \mathbb{Z}$, Pic $R_1 = 0$, and $K_1R_1 \simeq R_1^{\bullet} \simeq \mathbb{Z}/(p - 1)\mathbb{Z}$ (writing it additively). Since $p - 1 \equiv 2 \pmod 4$ there is no element of order 4 in $R_1^{\bullet}$, or in other words $x^2 + 1$ is irreducible over R_1. Thus S is a field of order p^2, $K_0S \simeq \mathbb{Z}$, Pic $S = 0$, and $K_1S \simeq S^{\bullet} \simeq \mathbb{Z}/(p^2 - 1)\mathbb{Z}$. Next, $R_2 = \mathbb{Z}[i]$ is Euclidean, $K_0R_2 \simeq \mathbb{Z}$, Pic $R_2 = 0$, and $K_1R_2 \simeq R_2^{\bullet} \simeq \mathbb{Z}/4\mathbb{Z}$. Then

$R^{\bullet} \simeq Z/2Z$, and so Proposition 66 gives

$$\begin{array}{ccccccc}
K_1R & \xrightarrow{f_1} & Z/(p-1)Z \oplus Z/4Z & \xrightarrow{g_1} & Z/(p^2-1)Z & \xrightarrow{\delta} \\
\downarrow & & 1\downarrow & & 1\downarrow & \\
Z/2Z & \xrightarrow{f_1} & Z/(p-1)Z \oplus Z/4Z & \xrightarrow{g_1} & Z/(p^2-1)Z & \xrightarrow{\delta}
\end{array}$$

$$\begin{array}{ccccccc}
\xrightarrow{\delta} & K_0R & \to & Z \oplus Z & \to & Z \\
 & \downarrow & & \downarrow & & \downarrow \\
\xrightarrow{\delta} & \mathrm{Pic}\, R & \longrightarrow & 0 & \longrightarrow & 0.
\end{array}$$

Clearly f_1, in the bottom row, is not the zero map, so it is a monomorphism, and thus $|\ker g_1| = 2$. Therefore $|\ker \delta| = |\text{image } g_1| = 2(p - 1)$, and since δ is surjective, and a quotient of a cyclic group is always cyclic, we deduce $\mathrm{Pic}\, R \simeq Z/\tfrac{1}{2}(p + 1)Z$, and from the top row, $K_0R \simeq Z \oplus Z/\tfrac{1}{2}(p + 1)Z$.

Exercise: Show that $f(r) = 2|\mathrm{Pic}\, Z[ri]|$ is a multiplicative function, that is, $f(rs) = f(r)f(s)$ when r, s are coprime. Calculate $|\mathrm{Pic}\, Z[p^n i]|$, where p is prime. Give an example of an integral domain R such that $\mathrm{Pic}\, R$ is not cyclic.

Finally in this section, we show the connection between the exact sequences of Corollary 57 and Proposition 63. Let R, S be rings, and let $f : R \twoheadrightarrow S$ be a ring epimorphism. Write $D = \{(r_1, r_2) \in R \times R : f(r_1) = f(r_2)\}$. Then D is a subring of $R \times R$, and there are homomorphisms $p_1 : D \to R$, $p_2 : D \to R$ given by $p_1(r_1, r_2) = r_1$, $p_2(r_1, r_2) = r_2$ such that the diagram

$$\begin{array}{ccc}
D & \xrightarrow{p_2} & R \\
p_1\downarrow & & \downarrow f \\
R & \overset{f}{\twoheadrightarrow} & S
\end{array}$$

commutes. This is obviously a cartesian square, and since f is surjective we may apply Proposition 63 to obtain an exact sequence

$$K_1D \to K_1R \oplus K_1R \to K_1S \xrightarrow{\delta} K_0D \to K_0R \oplus K_0R \to K_0S.$$

Let $A = \ker f$: so $A \lhd R$. The ring $D = D(R, A)$ is called the *double* of R over A. Define $K_i(R, A) = \ker(K_ip_2)$: so $K_i(R, A) \subset K_iD$, $i = 0, 1$. Note that $\delta(K_1S) \subset K_0(R, A)$. We obtain:

Proposition 67. The sequence

$$K_1(R, A) \xrightarrow{K_1p_1} K_1R \xrightarrow{K_1f} K_1S \xrightarrow{\delta}$$

$$\xrightarrow{\delta} K_0(R, A) \xrightarrow{K_0p_1} K_0R \xrightarrow{K_0f} K_0S$$

is exact.

Proof. The diagram

$$\begin{array}{ccccccc} K_1(R, A) & \xrightarrow{K_1p_1} & K_1R & \xrightarrow{K_1f} & K_1S & \xrightarrow{\delta} \\ \downarrow & & \downarrow i_1 & & \downarrow 1 & \\ K_1D & \xrightarrow{f_1} & K_1R \oplus K_1R & \xrightarrow{g_1} & K_1S & \xrightarrow{\delta} \end{array}$$

$$\begin{array}{ccccccc} \xrightarrow{\delta} & K_0(R, A) & \xrightarrow{K_0p_1} & K_0R & \xrightarrow{K_0f} & K_0S \\ & \downarrow & & \downarrow i_0 & & \downarrow 1 \\ \xrightarrow{\delta} & K_0D & \xrightarrow{f_0} & K_0R \oplus K_0R & \xrightarrow{g_0} & K_0S \end{array}$$

commutes, where i_0, i_1 are injections to the first component, $x \mapsto (x, 0)$. The result now follows by a simple diagram chase, and details are left to the reader.

Now there is a map $\pi : K_0D \to K_0\Phi f$ suggested by the notation, that is, $\pi[P_1, P_2, h] = [P_1, P_2, h]$. However, Φf is a category with product and composition, so there are relations in $K_0\Phi f$, of the form

$$[P_1, P_2, h] + [P_2, P_3, k] = [P_1, P_3, kh]$$

not present in K_0D, so π is not an isomorphism. Nevertheless:

Corollary 68. The restriction $\pi : K_0(R, A) \to K_0\Phi f$ is an isomorphism.

Proof. The diagram

$$\begin{array}{ccccccccc}
K_1R & \xrightarrow{K_1f} & K_1S & \xrightarrow{\delta} & K_0(R, A) & \xrightarrow{K_0p_1} & K_0R & \xrightarrow{K_0f} & K_0S \\
1\downarrow & & 1\downarrow & & \pi\downarrow & & 1\downarrow & & 1\downarrow \\
K_1R & \xrightarrow[K_1f]{} & K_1S & \xrightarrow[d']{} & K_0\Phi f & \xrightarrow[d]{} & K_0R & \xrightarrow[K_0f]{} & K_0S
\end{array}$$

commutes, and the rows are exact, by Corollary 57 and Proposition 67. The result now follows by a simple diagram chase, and details are left to the reader.

Thus if we identify $K_0(R, A)$ with $K_0\Phi f$, the sequence of Proposition 67 is the same as the sequence of Corollary 57, except that we have extended the latter by one term on the left, $K_1(R, A)$.

Exercise: (i) Show that $K_1D \simeq K_1\Phi f$.
(ii) By taking $R = \mathbb{Z}/4\mathbb{Z}$ and $A = 2R$, show that $K_1\Phi f \not\simeq K_1(R, A)$ in general.
(Compare the exercise on p.152.)

We give an alternative description of the group $K_1(R, A)$ in terms of matrices. Let $\mathrm{GL}(R, A)$ be the kernel of $\mathrm{GL}(f)$: $\mathrm{GL}(R) \to \mathrm{GL}(S)$, that is, $\mathrm{GL}(R, A)$ is the subgroup of $\mathrm{GL}(R)$ consisting of all matrices congruent to the identity matrix modulo A. Let $\mathrm{E}(R, A)$ be the *normal* subgroup of $\mathrm{E}(R)$ generated by all elementary matrices $B_{ij}(x)$ with $x \in A$. It is clear that $\mathrm{E}(R, A) \subset \mathrm{GL}(R, A)$.

Proposition 69. $\mathrm{E}(R, A)$ is a normal subgroup of $\mathrm{GL}(R, A)$, and $K_1(R, A) \simeq \mathrm{GL}(R, A)/\mathrm{E}(R, A)$.
Proof. We have a map g : $\mathrm{GL}(R) \to \mathrm{GL}(R) \oplus \mathrm{GL}(R)$ given by injection to the first component, $M \mapsto (M, I)$. Then, identifying $\mathrm{GL}(R) \oplus \mathrm{GL}(R)$ with $\mathrm{GL}(R \times R)$, and noting that $\mathrm{GL}(D) \subset \mathrm{GL}(R \times R)$, where $D = D(R, A)$, we see that g restricts to give a map g : $\mathrm{GL}(R, A) \to \mathrm{GL}(D)$, and hence induces $\bar{g}$: $\mathrm{GL}(R, A) \to \mathrm{GL}(D)/\mathrm{E}(D) = K_1D$. Further, it is clear that $\bar{g}(\mathrm{GL}(R, A)) = K_1(R, A)$, so it

remains to show that $\ker \bar{g} = E(R, A)$. Obviously $E(R, A) \subset \ker \bar{g}$. Then let $M \in \ker \bar{g}$, so $M \equiv I \pmod{A}$, and

$$g(M) = (M, I) \in GL(R) \oplus GL(R) = GL(R \times R).$$

Since $\bar{g}(M) = 1$, $g(M) \in E(D)$, that is $g(M) = N_1 N_2 \ldots N_n$, say, where for each i, $N_i = B_{jk}(x_i, y_i)$ (j, k depending on i), where $x_i, y_i \in R$ with $x_i - y_i \in A$. Put $P_i = B_{jk}(x_i)$ and $Q_i = B_{jk}(y_i)$, each i, so that $M = P_1 P_2 \ldots P_n$ and $I = Q_1 Q_2 \ldots Q_n$. So

$$M = (P_1 Q_1^{-1})(Q_1(P_2 Q_2^{-1})Q_1^{-1})(Q_1 Q_2(P_3 Q_3^{-1})Q_2^{-1} Q_1^{-1})\ldots$$
$$\ldots(Q_1 Q_2 \ldots Q_{n-1}(P_n Q_n^{-1})Q_{n-1}^{-1} \ldots Q_2^{-1} Q_1^{-1})$$

and since $P_i Q_i^{-1} = B_{jk}(x_i - y_i)$, each i, we have $M \in E(R, A)$, as required.

If H, K are subgroups of the group G, we write $[H, K]$ for the subgroup of G generated by all commutators $[h, k] = hkh^{-1}k^{-1}$ ($h \in H$, $k \in K$).

Proposition 70. $E(R, A) = [E(R), E(R, A)] = [GL(R), GL(R, A)]$.

Proof. Let $x \in A$. Then $B_{ij}(x) = [B_{ik}(1), B_{kj}(x)]$, where $k \neq i, j$, so $E(R, A) \subset [E(R), E(R, A)]$. The reverse inclusion follows since $E(R, A)$ is normal in $E(R)$. Next, $E(R) \subset GL(R)$ and $E(R, A) \subset GL(R, A)$, so $[E(R), E(R, A)] \subset [GL(R), GL(R, A)]$. Finally, let $M \in GL(R)$ and $N \in GL(R, A)$, and choose n large enough so that $M, N \in GL_n(R)$. Working in $GL_{3n}(R)$, we have

$$\begin{pmatrix} [M, N] & 0 & 0 \\ 0 & I & 0 \\ 0 & 0 & I \end{pmatrix} = \left[\begin{pmatrix} M & 0 & 0 \\ 0 & M^{-1} & 0 \\ 0 & 0 & I \end{pmatrix}, \begin{pmatrix} N & 0 & 0 \\ 0 & I & 0 \\ 0 & 0 & N^{-1} \end{pmatrix}\right].$$

In $GL_{2n}(R)$, we have

$$\begin{pmatrix} M & 0 \\ 0 & M^{-1} \end{pmatrix} = \begin{pmatrix} I & 0 \\ M^{-1}-I & I \end{pmatrix}\begin{pmatrix} I & I \\ 0 & I \end{pmatrix}\begin{pmatrix} I & 0 \\ M-I & I \end{pmatrix}\begin{pmatrix} I & -M^{-1} \\ 0 & I \end{pmatrix}$$

which is in $E(R)$, and

$$\begin{pmatrix} N & 0 \\ 0 & N^{-1} \end{pmatrix} = \begin{pmatrix} I & 0 \\ N^{-1}-I & I \end{pmatrix}\begin{pmatrix} I & I \\ 0 & I \end{pmatrix}\begin{pmatrix} I & 0 \\ N-I & I \end{pmatrix}\begin{pmatrix} I & I \\ 0 & I \end{pmatrix}^{-1}\begin{pmatrix} I & I-N^{-1} \\ 0 & I \end{pmatrix}$$

which is in $E(R, A)$, since the entries of $N - I$ and $N^{-1} - I$ are all in A. Permuting rows and columns, we see

$$\begin{pmatrix} N & 0 & 0 \\ 0 & I & 0 \\ 0 & 0 & N^{-1} \end{pmatrix} \in E(R, A)$$

and hence $[M, N] \in [E(R), E(R, A)]$, as required.

From Propositions 69 and 70,

$$K_1(R, A) \simeq GL(R, A)/[GL(R), GL(R, A)].$$

Note that, when $A = R$, this is just Proposition 35.

Proposition 71. (i) Let H be a subgroup of $GL(R)$ with

$$E(R, A) \subset H \subset GL(R, A)$$

for some $A \lhd R$. Then H is normal in $GL(R)$.

(ii) Let H be a normal subgroup of $GL(R)$. Then there is a unique $A \lhd R$ with

$$E(R, A) \subset H \subset GL(R, A).$$

Proof. (i) Let $M \in GL(R)$, $N \in H$. Then, using Proposition 70,

$$[M, N] \in [GL(R), GL(R, A)] = E(R, A) \subset H$$

and so $MNM^{-1} \in H$.

(ii) Let A be the ideal of R generated by all the entries of $N - I$, for all $N \in H$. Then clearly $H \subset GL(R, A)$, and if also $B \lhd R$ with $H \subset GL(R, B)$, then $A \subset B$, and $E(R, A) \subset E(R, B)$. We must show $E(R, A) \subset H$; then if $E(R, B) \subset H$, and if $b \in B$, then $B_{12}(b) \in H$, so $B_{12}(b) \in GL(R, A)$, whence $b \in A$, and $A = B$.

So let $N \in H$, and choose n large enough so that $N \in GL_n(R)$. Put $N - I = X = (x_{ij})$. To show $E(R, A) \subset H$, we must show $B_{k\ell}(x_{ij}) \in H$, all $k \neq \ell$, all $i, j \leq n$. Let Y, Z be any $n \times n$ matrices over R. Working in $GL_{2n}(R)$, we have $\begin{pmatrix} I & Y \\ 0 & I \end{pmatrix} \in E_{2n}(R)$, so

$$\begin{pmatrix} I & Y \\ 0 & I \end{pmatrix}^{-1} \begin{pmatrix} N & 0 \\ 0 & I \end{pmatrix} \begin{pmatrix} I & Y \\ 0 & I \end{pmatrix} = \begin{pmatrix} N & XY \\ 0 & I \end{pmatrix} \in H$$

and so

$$\begin{pmatrix} N & XY \\ 0 & I \end{pmatrix} \begin{pmatrix} N^{-1} & 0 \\ 0 & I \end{pmatrix} = \begin{pmatrix} I & XY \\ 0 & I \end{pmatrix} \in H.$$

Passing to $\mathrm{GL}_{3n}(R)$, we have

$$\begin{pmatrix} I & 0 & 0 \\ 0 & I & 0 \\ Z & 0 & I \end{pmatrix} \in \mathrm{E}_{3n}(R)$$

and so

$$\begin{pmatrix} I & 0 & 0 \\ 0 & I & 0 \\ Z & 0 & I \end{pmatrix} \begin{pmatrix} I & XY & 0 \\ 0 & I & 0 \\ 0 & 0 & I \end{pmatrix} \begin{pmatrix} I & 0 & 0 \\ 0 & I & 0 \\ Z & 0 & I \end{pmatrix}^{-1} = \begin{pmatrix} I & XY & 0 \\ 0 & I & 0 \\ 0 & ZXY & I \end{pmatrix} \in H$$

and hence

$$\begin{pmatrix} I & XY & 0 \\ 0 & I & 0 \\ 0 & 0 & I \end{pmatrix}^{-1} \begin{pmatrix} I & XY & 0 \\ 0 & I & 0 \\ 0 & ZXY & I \end{pmatrix} = \begin{pmatrix} I & 0 & 0 \\ 0 & I & 0 \\ 0 & ZXY & I \end{pmatrix} \in H.$$

Write E_{ij} for the $n \times n$ matrix with 1 in the i, j position and 0 elsewhere. Choosing $Y = E_{j1}$ and $Z = E_{1i}$, we obtain

$$\begin{pmatrix} I & 0 & 0 \\ 0 & I & 0 \\ 0 & ZXY & I \end{pmatrix} = B_{2n+1\ n+1}(x_{ij}) \in H.$$

If $k \neq 2n + 1$, $n + 1$, we have

$$B_{k\ n+1}(x_{ij}) = [B_{k\ 2n+1}(1), B_{2n+1\ n+1}(x_{ij})] \in H$$

and if $\ell \neq k$, $n + 1$, then

$$B_{k\ell}(x_{ij}) = [B_{k\ n+1}(x_{ij}), B_{n+1\ \ell}(1)] \in H.$$

Finally,

$$B_{n+1\ \ell}(x_{ij}) = [B_{n+1\ k}(1), B_{k\ell}(x_{ij})] \in H$$

and so $B_{k\ell}(x_{ij}) \in H$ for all k, ℓ with $k \neq \ell$, and the proof is complete.

Proposition 71 shows that there is a one-to-one correspondence between normal subgroups $H \subset \mathrm{GL}(R)$ and subgroups

$$H/\mathrm{E}(R, A) \subset \mathrm{GL}(R, A)/\mathrm{E}(R, A) \simeq K_1(R, A)$$

for various ideals $A \lhd R$. For example, if R is a field, we must have $A = 0$ or R, and then $K_1(R, 0)$ is trivial and $K_1(R, R) = K_1R = R^{\bullet}$. So for instance if $|R| = 2$, $\mathrm{GL}(R)$ is a simple group, and if $|R| = 3$, then the only proper non-trivial normal subgroup of $\mathrm{GL}(R)$ is $\mathrm{E}(R)$.

Exercise: Let $A \lhd R$, where R is commutative, and write

$$(R, A)^{\bullet} = \{\alpha \in R^{\bullet} : \alpha \equiv 1 \pmod{A}\}.$$

Show that $\det : K_1R \to R^{\bullet}$ restricts to give $\det : K_1(R, A) \to (R, A)^{\bullet}$, and that if the kernel of this map is denoted by $SK_1(R, A)$, then

$$K_1(R, A) \cong (R, A)^{\bullet} \oplus SK_1(R, A).$$

Show that the normal subgroups of $\mathrm{SL}(R)$ are in one-to-one correspondence with the subgroups of $SK_1(R, A)$, for various ideals A.

The problem of finding the normal subgroups of $\mathrm{GL}_n(R)$, for various special rings R, is of interest outside algebraic K-theory, and attempts to solve this problem have had a strong developing influence on algebraic K-theory. Given $A \lhd R$, we define the *congruence subgroups*

$$\mathrm{GL}_n(R, A) = \ker(\mathrm{GL}_n(R) \to \mathrm{GL}_n(R/A))$$

and

$$\mathrm{E}_n(R, A) = \ker(\mathrm{E}_n(R) \to \mathrm{E}_n(R/A)).$$

The plan then is to prove a *non-stable* version of Proposition 71 (that is, with GL_n, E_n in place of GL, E), at any rate for non-central subgroups H, and also to show that the natural map $\mathrm{GL}_n(R, A) \hookrightarrow \mathrm{GL}(R, A)$ induces an isomorphism

$\mathrm{GL}_n(R, A)/\mathrm{E}_n(R, A) \to K_1(R, A)$ for large enough n. Once this is done, what remains is to calculate $K_1(R, A)$, each A, and also to settle what happens when n is small. The carrying out of this program for a particular ring R is the solution of the *congruence subgroup problem* for R. For more details, see [2], Chapter V, and also the references given in [11], preface p.xi. In the next chapter, we shall carry out a small part of the program for $R = \mathbb{Z}$, namely the computation of $SK_1(\mathbb{Z}, n\mathbb{Z})$, all n (see Corollary 11l).

CHAPTER EIGHT

Steinberg Groups and K_2R

Let R be a ring. Recall that the elementary group $E(R)$ is generated by the elementary matrices $B_{ij}(a)$, $a \in R$, $i \neq j$, and that these matrices satisfy certain relations, for instance,

$$B_{ij}(a)B_{ij}(b) = B_{ij}(a + b) \qquad \ldots(1)$$

$$[B_{ij}(a), B_{k\ell}(b)] = \begin{cases} 1 & (i \neq \ell,\ j \neq k) \qquad \ldots(2) \\ B_{i\ell}(ab) & (i \neq \ell,\ j = k) \qquad \ldots(3) \end{cases}$$

where $[u, v] = uvu^{-1}v^{-1}$, and $a, b \in R$. These are not the only relations between the generators of $E(R)$; indeed, we always have

$$[B_{ij}(a), B_{k\ell}(b)] = B_{kj}(-ba) \quad (i = \ell,\ j \neq k)$$

and this is a consequence of the relations (1) - (3). For we have

$$\begin{aligned} [B_{ij}(a), B_{ki}(b)] &= [B_{ki}(b), B_{ij}(a)]^{-1} \\ &= B_{kj}(ba)^{-1}, \text{ by (3)} \\ &= B_{kj}(-ba), \text{ by (1).} \end{aligned}$$

There is in general no easy formula for $[B_{ij}(a), B_{ji}(b)]$. The relations (1) - (3) do not in general form a set of defining relations for $E(R)$.

Definition: The *Steinberg group of R*, ST(R), is the (non-abelian) group with the following presentation:

Generators: $x_{ij}(a)$ ($i, j = 1, 2, 3, \ldots, i \neq j, a \in R$),

Relations:

$$x_{ij}(a)x_{ij}(b) = x_{ij}(a + b) \qquad \ldots(1)$$

$$[x_{ij}(a), x_{k\ell}(b)] = \begin{cases} 1 & (i \neq \ell,\ j \neq k) \qquad \ldots(2) \\ x_{i\ell}(ab) & (i \neq \ell,\ j = k) \qquad \ldots(3). \end{cases}$$

These relations (1) - (3) are called the *Steinberg relations*; their similarity with (1) - (3) in E(R) is obvious. So, for instance, it is immediate that

$$[x_{ij}(a), x_{k\ell}(b)] = x_{kj}(-ba) \quad (i = \ell,\ j \neq k)$$

the calculation being exactly as in E(R), as above. Then there is a natural epimorphism ϕ : ST(R) $\rightarrow$ E(R) given on generators by $\phi(x_{ij}(a)) = B_{ij}(a)$.

Definition: $K_2R = \ker \phi$.

So we have an exact sequence

$$0 \rightarrow K_2R \rightarrow \mathrm{ST}(R) \xrightarrow{\phi} \mathrm{E}(R) \rightarrow 0$$

or indeed,

$$0 \rightarrow K_2R \rightarrow \mathrm{ST}(R) \xrightarrow{\phi} \mathrm{GL}(R) \rightarrow K_1R \rightarrow 0.$$

Of course, if $K_2R = 0$, then ST(R) $\simeq$ E(R), and then the Steinberg relations are a set of defining relations for E(R). We may think of K_2R, in general, as measuring the extent to which the Steinberg relations fail to be a set of defining relations for E(R).

Given a ring homomorphism $f : R \rightarrow S$, there is an induced group homomorphism ST(f) : ST(R) $\rightarrow$ ST(S), given on generators by $x_{ij}(a) \mapsto x_{ij}(f(a))$. This makes ST into a functor, and it restricts to a homomorphism $K_2f : K_2R \rightarrow K_2S$, so that the diagram

$$\begin{array}{ccccccccc} 0 \rightarrow & K_2R & \longrightarrow & \mathrm{ST}(R) & \longrightarrow & \mathrm{E}(R) & \rightarrow 0 \\ & K_2f \downarrow & & \mathrm{ST}(f) \downarrow & & \mathrm{E}(f) \downarrow & \\ 0 \rightarrow & K_2S & \longrightarrow & \mathrm{ST}(S) & \longrightarrow & \mathrm{E}(S) & \rightarrow 0 \end{array}$$

commutes, and making K_2 into a functor also.

For any group G, we write $\zeta(G)$ for the centre of G.

Proposition 72. For any ring R, $\zeta(\mathrm{E}(R)) = 1$, and $\zeta(\mathrm{ST}(R)) = K_2R$. In particular, K_2R is an abelian group.

Proof. Let $N \in \zeta(\mathrm{E}(R))$, and choose n large enough so that $N \in \mathrm{E}_n(R)$, Working in $\mathrm{E}_{2n}(R)$, we have

$$\begin{pmatrix} N & 0 \\ 0 & I \end{pmatrix}\begin{pmatrix} I & I \\ 0 & I \end{pmatrix} = \begin{pmatrix} I & I \\ 0 & I \end{pmatrix}\begin{pmatrix} N & 0 \\ 0 & I \end{pmatrix}$$

or

$$\begin{pmatrix} N & N \\ 0 & I \end{pmatrix} = \begin{pmatrix} N & I \\ 0 & I \end{pmatrix}$$

whence $N = I$, and so $\zeta(\mathrm{E}(R)) = 1$. Next, let $M \in \zeta(\mathrm{ST}(R))$, and put $\phi(M) = N \in \mathrm{E}(R)$. Since ϕ is surjective, $N \in \zeta(\mathrm{E}(R))$, so $N = 1$ and $M \in K_2R$. Thus $\zeta(\mathrm{ST}(R)) \subset K_2R$.

Let us write C_n for the subgroup of $\mathrm{ST}(R)$ generated by all $x_{in}(a)$, with $i \neq n$ and $a \in R$, where n is fixed. Then

$$x_{in}(a)x_{in}(b) = x_{in}(a + b)$$

and

$$[x_{in}(a), x_{jn}(b)] = 1$$

so C_n is abelian, and any element $y \in C_n$ can be written as a finite product $y = \Pi_{i \neq n} x_{in}(a_i)$. But then $\phi(y) = \Pi_{i\neq n} B_{in}(a_i)$, and this matrix differs from the identity matrix only in the nth column, where the entries are $a_1, a_2, \ldots, a_{n-1}, 1, a_{n+1}, \ldots$. It follows that ϕ restricted to C_n is a monomorphism. Next, if $p \neq n$, then

$$x_{pq}(a)x_{in}(b)x_{pq}(a)^{-1} = \begin{cases} x_{in}(b) \text{ if } q \neq i \\ x_{pn}(ab)x_{in}(b) \text{ if } q = i \end{cases}$$

and thus $x_{pq}(a)$ normalizes C_n when $p \neq n$. Let $\alpha \in K_2R$: so $\phi(\alpha) = 1$. Choose an expression for α in terms of the generators of $\mathrm{ST}(R)$, and choose n larger than any subscript occurring in this expression. By the above argument, α normalizes C_n, that is, $\alpha\beta\alpha^{-1} \in C_n$, all $\beta \in C_n$. But $\phi(\alpha) = 1$, and so $\phi(\alpha\beta\alpha^{-1}) =$

$\phi(\beta)$; then ϕ restricted to C_n is a monomorphism, so $\alpha\beta\alpha^{-1} = \beta$, all $\beta \in C_n$, that is, α centralizes C_n. Similarly, we can write R_n for the subgroup of ST(R) generated by all $x_{nj}(a)$, with $j \neq n$, and $a \in R$, where n is as above, and then by a similar argument α centralizes R_n also. But C_n and R_n together generate ST(R), for if $p \neq q$,

$$x_{pq}(a) \in C_n \text{ if } q = n$$

$$x_{pq}(a) \in R_n \text{ if } p = n$$

and

$$x_{pq}(a) = [x_{pn}(a), x_{nq}(1)] \in [C_n, R_n] \text{ if } p \neq n \text{ and } q \neq n.$$

It follows that α centralizes ST(R), that is, $\alpha \in \zeta(\text{ST}(R))$, or $K_2R \subset \zeta(\text{ST}(R))$. Thus $K_2R = \zeta(\text{ST}(R))$, and the proof is complete.

An epimorphism $\phi : G \to H$ of groups is called a *central extension* if ker $\phi \subset \zeta(G)$. So Proposition 72 shows that ϕ : ST$(R) \to$ E(R) is a central extension. In fact, it is a *universal central extension*, that is, it is a central extension satisfying the following universal property:

Proposition 73. Given any central extension $\theta : G \to \text{E}(R)$, there is a unique homomorphism ψ : ST$(R) \to G$ such that the diagram

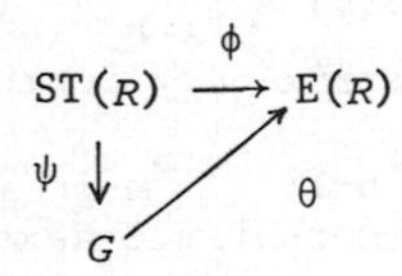

commutes.

Proof. First we make some preliminary statements.

(A) $[u, [v, w]] = [uv, w][w, u][w, v]$, any $u, v, w \in G$.

The proof is immediate. Remembering that $[w, u] = [u, w]^{-1}$ and $[w, v] = [v, w]^{-1}$, we have:

(B) $[uv, w] = [u, [v, w]][v, w][u, w]$, any $u, v, w \in G$.

Next, for $u \in G$ write $\bar{u} = \theta(u) \in \text{E}(R)$, and for $u, v \in G$ write $u \sim v$ if $uv^{-1} \in \zeta(G)$. Since θ is a central extension, we have in particular that if $u, v \in G$ with $\bar{u} = \bar{v}$, then $u \sim v$.

(C) If $u \sim u'$ and $v \sim v'$, then $[u, v] = [u', v']$.

The proof of this is immediate on making the substitutions $u = au'$, $v = bv'$, where $a, b \in \zeta(G)$.

(D) If $u, v, w \in G$ with $[\bar{u}, \bar{v}] = [\bar{u}, \bar{w}] = 1$, then

$$[u, [v, w]] = 1.$$

For if $a = [u, v]$ and $b = [u, w]$, we have $\bar{a} = \bar{b} = 1$, whence $a, b \in \zeta(G)$. Then $uvu^{-1} = av$ and $uwu^{-1} = bw$, and $[av, bw] = [v, w]$, by (C), whence

$$[u, [v, w]] = u[v, w]u^{-1}[v, w]^{-1} = [uvu^{-1}, uwu^{-1}][v, w]^{-1}$$
$$= [av, bw][v, w]^{-1} = [v, w][v, w]^{-1} = 1.$$

Next, for each i, j with $i \neq j$, and each $a \in R$, choose $y_{ij}(a) \in \theta^{-1}(B_{ij}(a))$.

(E) If $i \neq j$, $k \neq \ell$, $i \neq \ell$, $j \neq k$, then $[y_{ij}(a), y_{k\ell}(b)] = 1$.

For, choosing n different from i, j, k, and ℓ, we have $y_{k\ell}(b) \sim [y_{kn}(b), y_{n\ell}(1)]$, and so

$$[y_{ij}(a), y_{k\ell}(b)] = [y_{ij}(a), [y_{kn}(b), y_{n\ell}(1)]], \text{ by (C)}$$
$$= 1, \text{ by (D)}.$$

Now if i, j, n are distinct, and $a \in R$, write

$$z^n_{ij}(a) = [y_{in}(a), y_{nj}(1)].$$

Note that $\theta(z^n_{ij}(a)) = B_{ij}(a)$, so $z^n_{ij}(a) \sim y_{ij}(a)$, all n, and also $z^n_{ij}(a)$ depends only on i, j, n, and a, and not on the choice of $y_{in}(a)$ and $y_{nj}(1)$, by (C). We show next that $z^n_{ij}(a)$ does not depend on n.

(F) $[y_{ij}(a), y_{jk}(b)] = z^n_{ik}(ab)$, where i, j, k, n are distinct.

For

$$[y_{ij}(a), y_{jk}(b)]z^n_{jk}(b)$$
$$= [y_{ij}(a), [y_{jn}(b), y_{nk}(1)]]z^n_{jk}(b), \text{ by (C)}$$
$$= [y_{ij}(a)y_{jn}(b), y_{nk}(1)], \text{ by (A) and (E)}$$
$$= [y_{jn}(b)y_{ij}(a)y_{in}(ab), y_{nk}(1)], \text{ by (C)}$$

$= [y_{jn}(b)y_{ij}(a),\ z^n_{ik}(ab)]z^n_{ik}(ab)[y_{jn}(b)y_{ij}(a),\ y_{nk}(1)]$, by (B)

$= [y_{jn}(b)y_{ij}(a),\ y_{ik}(ab)]z^n_{ik}(ab)[y_{jn}(b)y_{ij}(a),\ y_{nk}(1)]$, by (C)

$= z^n_{ik}(ab)[y_{jn}(b)y_{ij}(a),\ y_{nk}(1)]$, by (E)

$= z^n_{ik}(ab)[y_{jn}(b),\ 1]z^n_{jk}(b)$, by (B) and (E)

$= z^n_{ik}(ab)z^n_{jk}(b)$

and (F) follows.

Putting $b = 1$ in (F), we see $z^j_{ik}(a) = z^n_{ik}(a)$, and so $z^n_{ik}(a)$ is independent of n, and we can write $z_{ik}(a) = z^n_{ik}(a)$. Let ψ : $\mathrm{ST}(R) \to G$ be any homomorphism with $\theta\psi = \phi$. For any $i \neq j$ and any $a \in R$, $\psi(x_{ij}(a)) \in \theta^{-1}(B_{ij}(a))$, so $\psi(x_{ij}(a)) \sim y_{ij}(a)$. Then if $k \neq i, j$, we have

$$\begin{aligned}\psi(x_{ij}(a)) &= \psi([x_{ik}(a),\ x_{kj}(1)])\\ &= [\psi(x_{ik}(a)),\ \psi(x_{kj}(1))]\\ &= [y_{ik}(a),\ y_{kj}(1)], \text{ by (C)}\\ &= z^k_{ij}(a) = z_{ij}(a).\end{aligned}$$

Since the elements $x_{ij}(a)$ generate $\mathrm{ST}(R)$, we have shown that ψ, if it exists, is unique. To complete the proof of Proposition 73, we must show that ψ, defined on the generators of $\mathrm{ST}(R)$ by $\psi(x_{ij}(a)) = z_{ij}(a)$, as the above argument shows it must be, extends to give a homomorphism $\psi : \mathrm{ST}(R) \to G$. Since $\theta(z_{ij}(a)) = B_{ij}(a)$, it will follow that $\theta\psi = \phi$. We need only show that the elements $z_{ij}(a) \in G$ satisfy relations corresponding to the Steinberg relations. We have $z_{ij}(a) \sim y_{ij}(a)$ and $z_{k\ell}(b) \sim y_{k\ell}(b)$, so

$$[z_{ij}(a),\ z_{k\ell}(b)] = [y_{ij}(a),\ y_{k\ell}(b)], \text{ by (C)}$$
$$= \begin{cases} 1 \text{ if } i \neq \ell \text{ and } j \neq k, \text{ by (E)} \\ z_{i\ell}(ab) \text{ if } i \neq \ell \text{ and } j = k, \text{ by (F).}\end{cases}$$

Finally,

$$\begin{aligned} z_{ij}(a+b) &= z_{ij}(b+a) \\ &= [z_{ik}(b+a), z_{kj}(1)], \text{ where } k \neq i, j \\ &= [z_{ik}(b)z_{ik}(a), z_{kj}(1)], \text{ by (C)} \\ &= [z_{ik}(b), z_{ij}(a)]z_{ij}(a)z_{ij}(b), \text{ by (B)} \\ &= z_{ij}(a)z_{ij}(b) \end{aligned}$$

and the proof is complete.

Now let $f : R \to S$ be an epimorphism of rings, and let $A = \ker f$. Recall that the inclusion $\mathrm{GL}(R, A) \hookrightarrow \mathrm{GL}(R)$ induces a homomorphism $K_1(R, A) \to K_1R$ so that the sequence

$$K_1(R, A) \to K_1R \xrightarrow{K_1f} K_1S$$

is exact (Proposition 67). Recall also that $\mathrm{GL}(R, A)$ is the kernel of $\mathrm{GL}(f) : \mathrm{GL}(R) \to \mathrm{GL}(S)$, and that $K_1(R, A) = \mathrm{GL}(R, A)/\mathrm{E}(R, A)$, where $\mathrm{E}(R, A)$ is the kernel of $\mathrm{E}(f) : \mathrm{E}(R) \to \mathrm{E}(S)$, or equivalently $\mathrm{E}(R, A)$ is the normal subgroup of $\mathrm{E}(R)$ generated by all elementary matrices $B_{ij}(a)$ with $a \in A$. If we define $\mathrm{ST}(R, A)$ to be the kernel of $\mathrm{ST}(f) : \mathrm{ST}(R) \to \mathrm{ST}(S)$, then $\mathrm{ST}(R, A)$ is the normal subgroup of $\mathrm{ST}(R)$ generated by all $x_{ij}(a)$ with $a \in A$. It is clear that $\phi : \mathrm{ST}(R) \to \mathrm{GL}(R)$ restricts to $\phi : \mathrm{ST}(R, A) \to \mathrm{GL}(R, A)$, with image $\mathrm{E}(R, A)$. In other words, there is an exact sequence

$$\mathrm{ST}(R, A) \xrightarrow{\phi} \mathrm{GL}(R, A) \to K_1(R, A).$$

We also have exact sequences

$$0 \to K_2R \to \mathrm{ST}(R) \xrightarrow{\phi} \mathrm{GL}(R) \to K_1R \to 0$$

and

$$0 \to K_2S \to \mathrm{ST}(S) \xrightarrow{\phi} \mathrm{GL}(S) \to K_1S \to 0.$$

Finally, since f is surjective, so is $\mathrm{ST}(f) : \mathrm{ST}(R) \to \mathrm{ST}(S)$. Putting all this together, we obtain a diagram

$$\begin{array}{ccccccccc}
 & & 0 & & 0 & & & & \\
 & & \downarrow & & \downarrow & & & & \\
 & & \mathrm{ST}(R, A) & \xrightarrow{\phi} & \mathrm{GL}(R, A) & \to & K_1(R, A) & \to & 0 \\
 & & \downarrow & & \downarrow & & \downarrow & & \\
0 \to K_2R & \longrightarrow & \mathrm{ST}(R) & \xrightarrow{\phi} & \mathrm{GL}(R) & \longrightarrow & K_1R & \longrightarrow & 0 \\
K_2f \downarrow & & \mathrm{ST}(f) \downarrow & & \mathrm{GL}(f) \downarrow & & K_1f \downarrow & & \\
0 \to K_2S & \longrightarrow & \mathrm{ST}(S) & \xrightarrow[\phi]{} & \mathrm{GL}(S) & \longrightarrow & K_1S & \longrightarrow & 0 \\
 & & \downarrow & & & & & & \\
 & & 0 & & & & & &
\end{array}$$

which is commutative and has exact rows and columns. We use this diagram to construct a homomorphism $\delta' : K_2S \to K_1(R, A)$.

Let $\alpha \in K_2S$. Then $\alpha \in \mathrm{ST}(S)$, so α is the image of β, say, where $\beta \in \mathrm{ST}(R)$. Then $\phi(\beta) \in \mathrm{GL}(R)$, and the image of $\phi(\beta)$ in $\mathrm{GL}(S)$ is the same as the image of α in $\mathrm{GL}(S)$, which is trivial. So $\phi(\beta) \in \mathrm{GL}(R, A)$, and thus $\phi(\beta)$ maps to $\overline{\phi(\beta)}$, say, in $K_1(R, A)$. We write $\delta'(\alpha) = \overline{\phi(\beta)}$. This is well defined, for if α is also the image of $\beta_1 \in \mathrm{ST}(R)$, then $\beta\beta_1^{-1}$ has trivial image in $\mathrm{ST}(S)$. Thus $\beta\beta_1^{-1} \in \mathrm{ST}(R, A)$, so

$$\overline{\phi(\beta\beta_1^{-1})}$$

is trivial, or $\overline{\phi(\beta)} = \overline{\phi(\beta_1)}$. It is also easy to see that δ' is a homomorphism. For if $\alpha_0, \alpha_1, \alpha_2 \in K_2S$ with $\alpha_0 = \alpha_1\alpha_2$, we can choose $\beta_1, \beta_2 \in \mathrm{ST}(R)$ with images $\alpha_1, \alpha_2 \in \mathrm{ST}(S)$, respectively, and if we put $\beta_0 = \beta_1\beta_2$, then β_0 has image α_0 in $\mathrm{ST}(S)$. Then

$$\begin{aligned}
\delta'(\alpha_1\alpha_2) &= \delta'(\alpha_0) = \overline{\phi(\beta_0)} = \overline{\phi(\beta_1\beta_2)} \\
&= \overline{\phi(\beta_1)\phi(\beta_2)} = \overline{\phi(\beta_1)}\,\overline{\phi(\beta_2)} = \delta'(\alpha_1)\delta'(\alpha_2).
\end{aligned}$$

Proposition 74. The sequence

$$K_2R \xrightarrow{K_2f} K_2S \xrightarrow{\delta'} K_1(R, A) \to K_1R \to \dots \to K_0S$$

is exact.

Proof. This is the sequence of Proposition 67, with two extra terms on the left, so we need only check exactness at $K_1(R, A)$

and K_2S.

1. Exactness at $K_1(R, A)$.

(i) Let $\alpha \in K_2S$. Then $\delta'(\alpha) = \overline{\phi(\beta)}$ for some $\beta \in \mathrm{ST}(R)$, as above. The image of $\delta'(\alpha)$ in K_1R is thus equal to the image of β in K_1R, and this is trivial.

(ii) Let $\gamma \in K_1(R, A)$. Thus $\gamma = \bar{M}$ for some $M \in \mathrm{GL}(R, A)$, and if γ has trivial image in K_1R, then M, regarded as belonging to $\mathrm{GL}(R)$, has trivial image in K_1R, so $M = \phi(\beta)$ for some $\beta \in \mathrm{ST}(R)$. Suppose β maps to $\alpha \in \mathrm{ST}(S)$; then the image of α in $\mathrm{GL}(S)$ is equal to the image of M in $\mathrm{GL}(S)$, which is trivial. So $\alpha \in K_2S$, and clearly $\delta'(\alpha) = \gamma$.

2. Exactness at K_2S.

(i) Let $\beta \in K_2R$, and suppose β maps to $\alpha \in K_2S$. Then $\delta'(\alpha) = \overline{\phi(\beta)}$, and $\phi(\beta)$ is trivial, so $\delta'(\alpha)$ is trivial also.

(ii) Let $\alpha \in K_2S$, and suppose $\delta'(\alpha)$ is trivial. Thus, if $\beta \in \mathrm{ST}(R)$ maps to $\alpha \in \mathrm{ST}(S)$, then $\overline{\phi(\beta)}$ is trivial. So $\phi(\beta) = \phi(\beta_1)$ for some $\beta_1 \in \mathrm{ST}(R, A)$, and then $\beta\beta_1^{-1} \in K_2R$. The image of β_1 in $\mathrm{ST}(S)$ is trivial, so $\beta\beta_1^{-1} \in K_2R$ maps to $\alpha \in K_2S$, and the proof is complete.

Now suppose R, S are commutative, and that $A = \ker f$, where $f : R \to S$ is a surjective ring homomorphism. We have $K_1R = R^{\bullet} \oplus SK_1R$, $K_1S = S^{\bullet} \oplus SK_1S$, and $K_1(R, A) = (R, A)^{\bullet} \oplus SK_1(R, A)$. The exact sequence $K_1(R, A) \to K_1R \to K_1S$ is the direct sum of the exact sequences $(R, A)^{\bullet} \to R^{\bullet} \to S^{\bullet}$ and $SK_1(R, A) \to SK_1R \to SK_1S$; but $(R, A)^{\bullet} \to R^{\bullet}$ is injective, so $\ker(K_1(R, A) \to K_1R) \subset SK_1(R, A)$, or, in the notation of Proposition 74, $\delta'(K_2S) \subset SK_1(R, A)$. Thus we have proved:

Corollary 75. The sequence

$$K_2R \xrightarrow{K_2f} K_2S \xrightarrow{\delta'} SK_1(R, A) \to SK_1R \xrightarrow{SK_1f} SK_1S$$

is exact.

Consider the special case $R = \mathbb{Z}$, $A = n\mathbb{Z}$. Since $\mathbb{Z}$ is Euclidean, $SK_1\mathbb{Z} = 0$, and so we obtain an exact sequence

$$K_2\mathbb{Z} \to K_2\mathbb{Z}/n\mathbb{Z} \to SK_1(\mathbb{Z}, n\mathbb{Z}) \to 0.$$

We shall use this sequence to give a proof of the Mennicke-Bass-Lazard-Serre theorem that $SK_1(\mathbb{Z}, n\mathbb{Z})$ is trivial, for all n. (See Corollary 111, and also the remarks at the end of Chapter 7.) To do this, we do not need to compute $K_2\mathbb{Z}$ or $K_2\mathbb{Z}/n\mathbb{Z}$, but merely to find a set of generators for $K_2\mathbb{Z}/n\mathbb{Z}$ and show that they come from $K_2\mathbb{Z}$, thus giving that $K_2\mathbb{Z} \to K_2\mathbb{Z}/n\mathbb{Z}$ is surjective. (Compare [11], p.92.) Recall that the Chinese remainder theorem says that $\mathbb{Z}/n\mathbb{Z}$ is a product of local rings of type $\mathbb{Z}/p^r\mathbb{Z}$. We have:

Proposition 76. Let R, S be rings. Then $\mathrm{ST}(R \times S) \simeq \mathrm{ST}(R) \oplus \mathrm{ST}(S)$ and $K_2(R \times S) \simeq K_2R \oplus K_2S$.

Proof. We can map the generators of $\mathrm{ST}(R)$ to the generators of $\mathrm{ST}(R \times S)$ by $x_{ij}(r) \mapsto x_{ij}(r, 0)$, and it is easy to see that this map extends to give a group homomorphism $\mathrm{ST}(R) \to \mathrm{ST}(R \times S)$. (Of course, the map $R \to R \times S$, $r \mapsto (r, 0)$, is not a ring homomorphism.) This group homomorphism is injective, since it has a left inverse $\mathrm{ST}(R \times S) \to \mathrm{ST}(R)$ induced by the ring homomorphism $R \times S \to R$, $(r, s) \mapsto r$. So we can identify $\mathrm{ST}(R)$ with the subgroup of $\mathrm{ST}(R \times S)$ generated by all $x_{ij}(r, 0)$; similarly, we can identify $\mathrm{ST}(S)$ with the subgroup of $\mathrm{ST}(R \times S)$ generated by all $x_{ij}(0, s)$. These two subgroups generate $\mathrm{ST}(R \times S)$, since

$$x_{ij}(r, s) = x_{ij}(r, 0)x_{ij}(0, s).$$

Next we must show that the two subgroups centralize each other; it is enough to check this for generators. We have:

$$[x_{ij}(r, 0), x_{k\ell}(0, s)] = 1 \quad (i \neq \ell,\ j \neq k)$$

$$[x_{ij}(r, 0), x_{jk}(0, s)] = x_{ik}((r, 0)(0, s))$$

$$= x_{ik}(0, 0) = 1 \quad (i \neq k)$$

$$[x_{ij}(r, 0), x_{ki}(0, s)] = x_{kj}(-(0, s)(r, 0))$$

$$= x_{kj}(0, 0) = 1 \quad (k \neq j)$$

and finally

$$[x_{ij}(r, 0), x_{ji}(0, s)]$$

$$= [x_{ij}(r, 0), [x_{jk}(0, s), x_{ki}(0, 1)]] \quad (k \neq i, j)$$

$$= 1, \text{ by the previous cases.}$$

Lastly, we have homomorphisms $\mathrm{ST}(R \times S) \to \mathrm{ST}(R)$, $\alpha \mapsto \alpha_1$ say, and $\mathrm{ST}(R \times S) \to \mathrm{ST}(S)$, $\alpha \mapsto \alpha_2$ say, giving $\alpha = \alpha_1\alpha_2$, since $x_{ij}(r, s) = x_{ij}(r, 0)x_{ij}(0, s)$, and furthermore, if $\alpha \in \mathrm{ST}(R)$, then $\alpha = \alpha_1$, and if $\alpha \in \mathrm{ST}(S)$, then $\alpha = \alpha_2$ (again, just look at the generators); so if $\alpha \in \mathrm{ST}(R) \cap \mathrm{ST}(S)$, then $\alpha = \alpha_1 = \alpha_2$, so $\alpha = \alpha_1\alpha_2 = \alpha^2$, or $\alpha = 1$. Thus $\mathrm{ST}(R) \cap \mathrm{ST}(S) = 1$, and $\mathrm{ST}(R \times S) = \mathrm{ST}(R) \oplus \mathrm{ST}(S)$. We now have a commutative diagram

$$\begin{array}{ccc} \mathrm{ST}(R \times S) & \xrightarrow{\simeq} & \mathrm{ST}(R) \oplus \mathrm{ST}(S) \\ \phi \downarrow & & \downarrow \phi \oplus \phi \\ \mathrm{GL}(R \times S) & \xrightarrow{\simeq} & \mathrm{GL}(R) \oplus \mathrm{GL}(S) \end{array}$$

giving an isomorphism between the kernels of the vertical arrows, that is, $K_2(R \times S) \simeq K_2R \oplus K_2S$, as required.

8.1 *Generators for K_2R: the symbols* $\{\alpha, \beta\}$

Our next task is to construct elements of K_2R, enough to generate K_2R in certain special cases, e.g., when R is a field or a commutative local ring. In order to do this, we must first perform some computations in $\mathrm{ST}(R)$; we follow [11], section 9 fairly closely.

Let R be a ring; for each $\alpha \in R^{\bullet}$, write

$$w_{ij}(\alpha) = x_{ij}(\alpha)x_{ji}(-\alpha^{-1})x_{ij}(\alpha) \in \mathrm{ST}(R).$$

Note that $w_{ij}(\alpha)^{-1} = w_{ij}(-\alpha)$. Note also that

$$\phi(w_{12}(\alpha)) = \begin{pmatrix} 1 & \alpha \\ 0 & 1 \end{pmatrix} \begin{pmatrix} 1 & 0 \\ -\alpha^{-1} & 1 \end{pmatrix} \begin{pmatrix} 1 & \alpha \\ 0 & 1 \end{pmatrix} = \begin{pmatrix} 0 & \alpha \\ -\alpha^{-1} & 0 \end{pmatrix}$$

$$= \begin{pmatrix} 0 & 1 \\ 1 & 0 \end{pmatrix} \begin{pmatrix} -\alpha^{-1} & 0 \\ 0 & \alpha \end{pmatrix}.$$

Let us write $h_{ij}(\alpha) = w_{ij}(\alpha)w_{ij}(-1)$; note that $h_{ij}(1) = 1$, and that

$$\phi(h_{12}(\alpha)) = \begin{pmatrix} \alpha & 0 \\ 0 & \alpha^{-1} \end{pmatrix}.$$

A matrix is called a *monomial matrix* if it is the product of a permutation matrix and a diagonal matrix. If $W = W(R)$ is the subgroup of $\mathrm{ST}(R)$ generated by all the $w_{ij}(\alpha)$, then $\phi(W)$ is contained in the subgroup of all monomial matrices in $\mathrm{E}(R)$; since $\phi(w_{ij}(1))$ is the product of a permutation matrix corresponding to the 2-cycle (ij) and a diagonal matrix, we shall have that $\phi(W)$ is equal to the subgroup of all monomial matrices in $\mathrm{E}(R)$ provided every diagonal matrix in $\mathrm{E}(R)$ can be expressed in terms of the diagonal matrices $\phi(h_{ij}(\alpha))$; this will be the case if R is commutative, or more generally if the kernel of the natural map $R^{\bullet} \to K_1R$ is precisely the derived subgroup $(R^{\bullet})'$. (Compare Proposition 42 and the proof of Proposition 38(i).)

Let us write $P(\pi)$ for the permutation matrix corresponding to the permutation π, that is, the i, j entry of $P(\pi)$ is 1 if $i = \pi(j)$, and 0 otherwise. Let us write $\mathrm{diag}(\alpha_1, \alpha_2, \ldots)$ for the diagonal matrix with diagonal entries $\alpha_1, \alpha_2, \ldots$. Then if $w \in W$, we have $\phi(w) = P(\pi)\mathrm{diag}(\alpha_1, \alpha_2, \ldots)$ for some permutation π and some $\alpha_i \in R^{\bullet}$. For instance, if $w = w_{k\ell}(\alpha)$, then $\pi = (k\ell)$, $\alpha_k = -\alpha^{-1}$, $\alpha_\ell = \alpha$, and $\alpha_m = 1$ for $m \neq k, \ell$. Also, if $w = h_{k\ell}(\alpha)$, then $\pi = 1$, $\alpha_k = \alpha$, $\alpha_\ell = \alpha^{-1}$, and $\alpha_m = 1$ for $m \neq k, \ell$.

Proposition 77. Let $w \in W(R)$, $\phi(w) = P(\pi)\mathrm{diag}(\alpha_1, \alpha_2, \ldots)$. Then if $\pi(i) = k$ and $\pi(j) = \ell$, we have

(i) $wx_{ij}(a)w^{-1} = x_{k\ell}(\alpha_i a \alpha_j^{-1})$

(ii) $ww_{ij}(\beta)w^{-1} = w_{k\ell}(\alpha_i \beta \alpha_j^{-1})$

(iii) $wh_{ij}(\beta)w^{-1} = h_{k\ell}(\alpha_i\beta\alpha_j^{-1})h_{k\ell}(\alpha_i\alpha_j^{-1})^{-1}$.

Proof. (i) We have

$$\begin{aligned}
&\phi(wx_{ij}(a)w^{-1})\\
&= P(\pi)\operatorname{diag}(\alpha_1, \alpha_2, \ldots)B_{ij}(a)\operatorname{diag}(\alpha_1, \alpha_2, \ldots)^{-1}P(\pi)^{-1}\\
&= P(\pi)B_{ij}(\alpha_i a\alpha_j^{-1})P(\pi)^{-1}\\
&= B_{k\ell}(\alpha_i a\alpha_j^{-1}).
\end{aligned}$$

Thus it is sufficient to prove that conjugation by w carries the generator $x_{ij}(a)$ to some generator $x_{st}(b)$, for then

$$B_{k\ell}(\alpha_i a\alpha_j^{-1}) = \phi(x_{st}(b)) = B_{st}(b)$$

whence the result. It is sufficient to consider the case $w = w_{pq}(\alpha)$, that is, to show that conjugation by generators of W carries generators of ST(R) to generators of ST(R).

Case (a). Suppose p, q, i, j are distinct. Then clearly

$$w_{pq}(\alpha)x_{ij}(a) = x_{ij}(a)w_{pq}(\alpha).$$

Case (b). Suppose p, i, j are distinct and $q = i$. Then

$$\begin{aligned}
w_{pi}(\alpha)x_{ij}(a) &= x_{pi}(\alpha)x_{ip}(-\alpha^{-1})x_{pi}(\alpha)x_{ij}(a)\\
&= x_{pi}(\alpha)x_{ip}(-\alpha^{-1})x_{ij}(a)x_{pj}(\alpha a)x_{pi}(\alpha)\\
&= x_{pi}(\alpha)x_{pj}(\alpha a)x_{ip}(-\alpha^{-1})x_{pi}(\alpha)\\
&= x_{pj}(\alpha a)w_{pi}(\alpha).
\end{aligned}$$

Case (c). Suppose p, i, j are distinct and $q = j$. Then

$$\begin{aligned}
w_{pj}(\alpha)x_{ij}(a) &= x_{pj}(\alpha)x_{jp}(-\alpha^{-1})x_{pj}(\alpha)x_{ij}(a)\\
&= x_{pj}(\alpha)x_{jp}(-\alpha^{-1})x_{ij}(a)x_{pj}(\alpha)\\
&= x_{pj}(\alpha)x_{ij}(a)x_{ip}(a\alpha^{-1})x_{jp}(-\alpha^{-1})x_{pj}(\alpha)\\
&= x_{ip}(a\alpha^{-1})w_{pj}(\alpha).
\end{aligned}$$

Case (d). Suppose q, i, j are distinct and $p = i$. Then by case (b),

$$w_{iq}(-\alpha)x_{qj}(-\alpha^{-1}a) = x_{ij}(a)w_{iq}(-\alpha)$$

whence

$$w_{iq}(\alpha)x_{ij}(a) = x_{qj}(-\alpha^{-1}a)w_{iq}(\alpha).$$

Case (e). Suppose q, i, j are distinct and $p = j$. Then by case (c),

$$w_{jq}(-\alpha)x_{iq}(-a\alpha) = x_{ij}(a)w_{jq}(-\alpha)$$

whence

$$w_{jq}(\alpha)x_{ij}(a) = x_{iq}(-a\alpha)w_{jq}(\alpha).$$

Case (f). Suppose $p = i$ and $q = j$. Then, choosing $h \neq i, j$ and using cases (c) and (d), we have

$$\begin{aligned} w_{ij}(\alpha)x_{ij}(a) &= w_{ij}(\alpha)[x_{ih}(a), x_{hj}(1)] \\ &= [x_{jh}(-\alpha^{-1}a), x_{hi}(\alpha^{-1})]w_{ij}(\alpha) \\ &= x_{ji}(-\alpha^{-1}a\alpha^{-1})w_{ij}(\alpha). \end{aligned}$$

Case (g). Suppose $p = j$ and $q = i$. Then, choosing $h \neq i, j$ and using cases (b) and (e), we have

$$\begin{aligned} w_{ji}(\alpha)x_{ij}(a) &= w_{ji}(\alpha)[x_{ih}(a), x_{hj}(1)] \\ &= [x_{jh}(\alpha a), x_{hi}(-\alpha)]w_{ji}(\alpha) \\ &= x_{ji}(-\alpha a\alpha)w_{ji}(\alpha). \end{aligned}$$

This completes the proof of statement (i) of the proposition.

(ii) Using part (i), we have

$$\begin{aligned} ww_{ij}(\beta)w^{-1} &= wx_{ij}(\beta)x_{ji}(-\beta^{-1})x_{ij}(\beta)w^{-1} \\ &= x_{k\ell}(\alpha_i\beta\alpha_j^{-1})x_{\ell k}(-\alpha_j\beta^{-1}\alpha_i^{-1})x_{k\ell}(\alpha_i\beta\alpha_j^{-1}) \\ &= w_{k\ell}(\alpha_i\beta\alpha_j^{-1}). \end{aligned}$$

(iii) Using part (ii), we have

$$\begin{aligned} wh_{ij}(\beta)w^{-1} &= ww_{ij}(\beta)w_{ij}(-1)w^{-1} \\ &= w_{k\ell}(\alpha_i\beta\alpha_j^{-1})w_{k\ell}(-\alpha_i\alpha_j^{-1}) \\ &= w_{k\ell}(\alpha_i\beta\alpha_j^{-1})w_{k\ell}(-1)w_{k\ell}(1)w_{k\ell}(-\alpha_i\alpha_j^{-1}) \end{aligned}$$

$$= h_{k\ell}(\alpha_i\beta\alpha_j^{-1})h_{k\ell}(\alpha_i\alpha_j^{-1})^{-1}.$$

Corollary 78. (i) $w_{ij}(\alpha) = w_{ji}(-\alpha^{-1})$.

(ii) $[h_{ij}(\alpha), h_{ik}(\beta)] = h_{ik}(\alpha\beta)h_{ik}(\alpha)^{-1}h_{ik}(\beta)^{-1}$.

Proof. (i) Apply Proposition 77(ii) with $w = w_{ij}(\alpha)$; so $\alpha_i = -\alpha^{-1}$ and $\alpha_j = \alpha$. We obtain

$$w_{ij}(\alpha) = w_{ij}(\alpha)w_{ij}(\alpha)w_{ij}(\alpha)^{-1} = w_{ji}(-\alpha^{-1}\alpha\alpha^{-1}) = w_{ji}(-\alpha^{-1}).$$

(ii) By Proposition 77(iii),

$$h_{ij}(\alpha)h_{ik}(\beta)h_{ij}(\alpha)^{-1} = h_{ik}(\alpha\beta)h_{ik}(\alpha)^{-1}$$

(remembering that if $w = h_{ij}(\alpha)$, then $\pi = 1$, $\alpha_i = \alpha$, and $\alpha_j = \alpha^{-1}$); the result follows.

Now we are in a position to construct certain elements in K_2R. Suppose $\alpha, \beta \in R^{\bullet}$, and suppose $\alpha\beta = \beta\alpha$. Then

$$\phi(h_{12}(\alpha)) = \begin{pmatrix} \alpha & 0 & 0 \\ 0 & \alpha^{-1} & 0 \\ 0 & 0 & 1 \end{pmatrix}, \quad \phi(h_{13}(\beta)) = \begin{pmatrix} \beta & 0 & 0 \\ 0 & 1 & 0 \\ 0 & 0 & \beta^{-1} \end{pmatrix}$$

and since $\alpha\beta = \beta\alpha$, these diagonal matrices commute. Thus

$$\phi([h_{12}(\alpha), h_{13}(\beta)]) = 1$$

or in other words

$$[h_{12}(\alpha), h_{13}(\beta)] \in K_2R.$$

Using Corollary 78(ii), we may write

$$\{\alpha, \beta\} = h_{13}(\alpha\beta)h_{13}(\alpha)^{-1}h_{13}(\beta)^{-1} \in K_2R$$

and more generally

$$\{\alpha, \beta\}_{ik} = h_{ik}(\alpha\beta)h_{ik}(\alpha)^{-1}h_{ik}(\beta)^{-1} \in K_2R$$

(so that $\{\alpha, \beta\} = \{\alpha, \beta\}_{13}$); we shall show that in fact

$$\{\alpha, \beta\}_{ik} = \{\alpha, \beta\}$$

for all i, k with $i \neq k$. For if $p \neq i, k$, then

$$w_{pi}(1)h_{ik}(\gamma)w_{pi}(1)^{-1} = h_{pk}(\gamma)h_{pk}(1)^{-1} = h_{pk}(\gamma)$$

for any $\gamma \in R^{\bullet}$, by Proposition 77(iii). Then if $q \neq p, k$ we obtain similarly

$$w_{qk}(1)h_{pk}(\gamma)w_{qk}(1)^{-1} = h_{pq}(\gamma).$$

So if we choose distinct p, q, different from all of i, k, 1, 3, and put $w = w_{3q}(1)w_{1p}(1)w_{qk}(1)w_{pi}(1)$, then

$$w\{\alpha, \beta\}_{ik}w^{-1} = \{\alpha, \beta\}.$$

But $\{\alpha, \beta\}_{ik} \in K_2R = \zeta(\mathrm{ST}(R))$ (Proposition 72), so we conclude that $\{\alpha, \beta\}_{ik} = \{\alpha, \beta\}$.

We now restrict attention to the case where R is commutative, so that we can define a symbol $\{\alpha, \beta\} \in K_2R$ for all $\alpha, \beta \in R^{\bullet}$. The use of the word *symbol* to describe $\{\alpha, \beta\}$ is quite deliberate, and goes back to the classical Legendre and Hilbert quadratic residue symbols; the latter obey identities like those in Propositions 79 and 80, below. The connection between these classical symbols and K_2 is more than superficial; for more details, see [11], section 11 and [2], Chapter VI.

Proposition 79. Let R be a commutative ring. Then for all α, α_1, α_2, β, β_1, $\beta_2 \in R^{\bullet}$, we have:

(i) $\{\beta, \alpha\} = \{\alpha, \beta\}^{-1}$

(ii) $\{\alpha_1\alpha_2, \beta\} = \{\alpha_1, \beta\}\{\alpha_2, \beta\}$

(iii) $\{\alpha, \beta_1\beta_2\} = \{\alpha, \beta_1\}\{\alpha, \beta_2\}$.

Proof. (i) $\{\beta, \alpha\} = [h_{12}(\beta), h_{13}(\alpha)]$

$$= [h_{13}(\alpha), h_{12}(\beta)]^{-1} = \{\alpha, \beta\}^{-1}$$

using the fact that $\{\alpha, \beta\} = \{\alpha, \beta\}_{12}$.

(ii) By Corollary 78(ii), we have

$$\begin{aligned}\{\alpha_1\alpha_2, \beta\} &= [h_{12}(\alpha_1\alpha_2), h_{13}(\beta)] \\ &= [\{\alpha_1, \alpha_2\}h_{12}(\alpha_2)h_{12}(\alpha_1), h_{13}(\beta)] \\ &= [h_{12}(\alpha_2)h_{12}(\alpha_1), h_{13}(\beta)]\end{aligned}$$

since $\{\alpha_1, \alpha_2\} \in \zeta(\mathrm{ST}(R))$. So, using the commutator identity (B) on p.186, we have

$$\{\alpha_1\alpha_2,\ \beta\} = [h_{12}(\alpha_2),\ \{\alpha_1,\ \beta\}]\{\alpha_1,\ \beta\}\{\alpha_2,\ \beta\}$$
$$= \{\alpha_1,\ \beta\}\{\alpha_2,\ \beta\}$$

since $\{\alpha_1,\ \beta\} \in \zeta(\mathrm{ST}(R))$.

(iii) $\{\alpha,\ \beta_1\beta_2\} = \{\beta_1\beta_2,\ \alpha\}^{-1}$, by (i)

$= (\{\beta_1,\ \alpha\}\{\beta_2,\ \alpha\})^{-1}$, by (ii)

$= \{\beta_1,\ \alpha\}^{-1}\{\beta_2,\ \alpha\}^{-1}$, since K_2R is abelian

$= \{\alpha,\ \beta_1\}\{\alpha,\ \beta_2\}$, by (i).

Exercise: Show that Proposition 79 holds even if R is not commutative, whenever the symbols are defined, that is, whenever α commutes with β, β_1, β_2, and also β commutes with α_1, α_2. (The only difficulty is in the proof of (ii), where $\{\alpha_1,\ \alpha_2\}$ may not be defined if $\alpha_1\alpha_2 \neq \alpha_2\alpha_1$. Modify the second line of the proof to read

$$\ldots = [[h_{14}(\alpha_1),\ h_{12}(\alpha_2)]h_{12}(\alpha_2)h_{12}(\alpha_1),\ h_{13}(\beta)]$$

and show that $[h_{14}(\alpha_1),\ h_{12}(\alpha_2)]$ commutes with $h_{13}(\beta)$; the proof then proceeds much as before.)

Proposition 80. Let $\alpha,\ \beta \in R^{\bullet}$ with $\alpha + \beta = 0$ or 1. Then

$$\{\alpha,\ \beta\} = 1.$$

Proof. Note that $\beta = -\alpha$ or $1 - \alpha$, so that $\alpha\beta = \beta\alpha$, and the symbol $\{\alpha,\ \beta\}$ is defined. First let $\alpha + \beta = 0$. Then

$h_{12}(\alpha)h_{12}(\beta) = h_{12}(\alpha)h_{12}(-\alpha)$

$= w_{12}(\alpha)w_{12}(-1)w_{12}(-\alpha)w_{12}(-1)$

$= w_{12}(\alpha)w_{12}(-1)w_{12}(\alpha)^{-1}w_{12}(-1)$

$= w_{21}(\alpha^{-2})w_{12}(-1)$, by Proposition 77(ii)

$= w_{12}(-\alpha^2)w_{12}(-1)$, by Corollary 78(i)

$= h_{12}(-\alpha^2) = h_{12}(\alpha\beta)$

and the result follows. Now let $\alpha + \beta = 1$. Then

$h_{12}(\alpha)h_{12}(\beta)w_{12}(-1)^{-1}$

$= w_{12}(\alpha)w_{12}(-1)w_{12}(\beta)$

$= w_{12}(\alpha)w_{21}(1)w_{12}(\beta)$, by Corollary 78(i)

$= w_{12}(\alpha)x_{21}(1)x_{12}(-1)x_{21}(1)w_{12}(\beta)$

$= x_{12}(-\alpha^2)w_{12}(\alpha)x_{12}(-1)w_{12}(\beta)x_{12}(-\beta^2)$, by Proposition 77

$= x_{12}(-\alpha^2 + \alpha)x_{21}(-\alpha^{-1})x_{12}(\alpha - 1 + \beta)x_{21}(-\beta^{-1})x_{12}(\beta - \beta^2)$

$= x_{12}(\alpha - \alpha^2)x_{21}(-\alpha^{-1} - \beta^{-1})x_{12}(\beta - \beta^2)$

$= x_{12}(\alpha\beta)x_{21}(-(\alpha\beta)^{-1})x_{12}(\alpha\beta)$

$= w_{12}(\alpha\beta) = h_{12}(\alpha\beta)w_{12}(-1)^{-1}$

so $h_{12}(\alpha)h_{12}(\beta) = h_{12}(\alpha\beta)$, as required.

Recall that for $w \in W(R)$ (any ring R) we can write

$$\phi(w) = P(\pi)\mathrm{diag}(\alpha_1, \alpha_2, \ldots)$$

where $P(\pi)$ is the permutation matrix corresponding to the permutation π (p.194). Write $S = \cup_{n=1}^{\infty} S_n$, where S_n is the symmetric group on $\{1, 2, \ldots, n\}$, regarded as a subgroup of S_{n+1} in the obvious way. It is clear that we have a group epimorphism $\psi : W \to S$ given by $w \mapsto \pi$. If we write $H = H(R)$ for the subgroup of W generated by all $h_{ij}(\alpha)$, it is clear that $H \subset \ker \psi$; furthermore H is a normal subgroup of W, by Proposition 77(iii).

Proposition 81. $H = \ker \psi$.

Proof. By the preceding remarks, it suffices to prove that the map $\bar{\psi} : W/H \to S$ induced by ψ is injective. Write $w \mapsto \bar{w}$ for the natural map $W \to W/H$. Since $h_{ij}(\alpha) = w_{ij}(\alpha)w_{ij}(1)^{-1}$, we have $\overline{w_{ij}(\alpha)} = \overline{w_{ij}(1)}$, all $\alpha \in R^{\bullet}$. So if we write w_{ij} for $\overline{w_{ij}(1)}$, we have, by Proposition 77(ii),

$$\bar{w}w_{ij} = w_{k\ell}\bar{w}$$

where $w \in W$, $\psi(w) = \pi$, $\pi(i) = k$, and $\pi(j) = \ell$. Also, since $w_{ij}(1)w_{ij}(-1) = 1$, we have $w_{ij}{}^2 = 1$, and by Corollary 78(i) we also have $w_{ij} = w_{ji}$.

So suppose we are given $\bar{w} \in W/H$, say

$$\bar{w} = w_{i_1 j_1} w_{i_2 j_2} \ldots w_{i_m j_m} \qquad \ldots(1)$$

and suppose $n = \max\{i_1, i_2, \ldots, i_m, j_1, j_2, \ldots, j_m\}$. Since $w_{ij} = w_{ji}$, we may as well assume $i_k < j_k$, all k. Suppose $r = \min\{k : j_k = n\}$. If $r < m$, we can use the relations

$$w_{in} w_{jk} = w_{jk} w_{pn}, \text{ some } p \ (k \neq n,\ j \neq n)$$

$$w_{in} w_{jn} = w_{ji} w_{in} \quad (i \neq j)$$

$$w_{in} w_{in} = 1$$

to replace the terms

$$w_{i_k j_k} w_{i_{k+1} j_{k+1}}$$

in (1), giving a similar expression for $\bar{w}$ with either the same values of m and n but a larger value of r, or else with m replaced by $m - 2$. After a finite number of repetitions of this argument we obtain either $\bar{w} = 1$ or else we obtain an expression (1) for $\bar{w}$ with $r = m$, that is, if n is the largest subscript occurring, it only occurs in the last term. Passing to S, this latter case gives $\bar{\psi}(\bar{w}) = \pi_1\pi_2$, where

$$\pi_1 = (i_1 j_1)(i_2 j_2)\ldots(i_{m-1} j_{m-1})$$

and

$$\pi_2 = (i_m j_m) = (i_m n).$$

But then $\pi_1 \in S_{n-1}$ and $\pi_2 \notin S_{n-1}$, whence $\pi_1\pi_2 \neq 1$, or $\bar{\psi}(\bar{w}) \neq 1$. We have shown that $\ker \bar{\psi} = 1$, or in other words, $H = \ker \psi$.

Next, let $C = C(R)$ be the subgroup of $\mathrm{ST}(R)$ defined by

$$C = W \cap K_2R = W \cap \ker \phi.$$

Corollary 82. $C \subset H$.

Proof. Immediate, since if $c \in C$, we have $\phi(c) = 1$, whence $\psi(c) = 1$ also. (Here we are using the equation $\phi(c) = P(\psi(c))\mathrm{diag}(\alpha_1, \alpha_2, \ldots)$.)

Lemma 83. H is generated by all $h_{1j}(\alpha)$.

Proof. We have

$$\begin{aligned} h_{jk}(\alpha) &= w_{jk}(\alpha)w_{jk}(-1) \\ &= h_{ik}(\alpha)w_{jk}(1)h_{ik}(\alpha)^{-1}w_{jk}(-1) \text{ by Proposition 77(ii)} \\ &= h_{ik}(\alpha)h_{ij}(\alpha)^{-1} \end{aligned}$$

since

$$w_{jk}(1)h_{ik}(\alpha)w_{jk}(1)^{-1} = h_{ij}(\alpha)h_{ij}(1)^{-1} = h_{ij}(\alpha)$$

by Proposition 77(iii). Putting $i = 1$, we obtain

$$h_{jk}(\alpha) = h_{1k}(\alpha)h_{1j}(\alpha)^{-1}$$

provided $j, k \neq 1$. We also have $h_{kj}(\alpha) = h_{jk}(\alpha)^{-1}$, for

$$h_{jk}(\alpha) = h_{ik}(\alpha)h_{ij}(\alpha)^{-1}$$

and similarly

$$h_{kj}(\alpha) = h_{ij}(\alpha)h_{ik}(\alpha)^{-1}.$$

So in particular, $h_{j1}(\alpha) = h_{1j}(\alpha)^{-1}$, and the result follows.

Now let us return to the case when R is commutative. For each $\alpha, \beta \in R^{\bullet}$, we constructed the symbol $\{\alpha, \beta\} \in W \cap K_2R$, that is, $\{\alpha, \beta\} \in C(R)$.

Proposition 84. If R is commutative (or more generally if $R^{\bullet}$ is abelian), $C(R)$ is generated by all symbols $\{\alpha, \beta\}$ $(\alpha, \beta \in R^{\bullet})$.

Proof. Let us write (temporarily) C^* for the subgroup of C generated by all symbols $\{\alpha, \beta\}$. We shall show that $C^* = C$. First note that since $C^* \subset K_2R = \zeta(\mathrm{ST}(R))$, we have $C^* \subset \zeta(H)$, so that C^* is a normal subgroup of H. Writing $h \mapsto \bar{h}$ for the natural map $H \to H/C^*$, we have by Lemma 83 that H/C^* is generated by all $\overline{h_{1j}(\alpha)}$; further, since

$$[h_{1j}(\alpha), h_{1k}(\beta)] = \{\alpha, \beta\} \in C^*$$

we have

$$[\overline{h_{1j}(\alpha)}, \overline{h_{1k}(\beta)}] = 1.$$

Next,

$$h_{1k}(\alpha\beta)h_{1k}(\alpha)^{-1}h_{1k}(\beta)^{-1} = \{\alpha, \beta\} \in C^*$$

so

$$\overline{h_{1k}(\beta)}\,\overline{h_{1k}(\alpha)} = \overline{h_{1k}(\alpha\beta)}.$$

Thus any element $\bar{h} \in H/C^*$ can be written in the form

$$\bar{h} = \overline{h_{12}(\alpha_2)}\,\overline{h_{13}(\alpha_3)}\cdots\overline{h_{1n}(\alpha_n)}$$

for some $\alpha_2, \alpha_3, \ldots, \alpha_n \in R^{\bullet}$. Writing $\bar{\phi} : H/C^* \to \mathrm{E}(R)$ for the map induced by the restriction of $\phi : \mathrm{ST}(R) \to \mathrm{E}(R)$ to H, we see

$$\bar{\phi}(\bar{h}) = \mathrm{diag}(\alpha, \alpha_2^{-1}, \alpha_3^{-1}, \ldots, \alpha_n^{-1}, 1, 1, \ldots)$$

where $\alpha = \alpha_2\alpha_3\ldots\alpha_n$. Thus if $\bar{h} \in C/C^*$, we have $\bar{\phi}(\bar{h}) = 1$, whence $\alpha_2 = \alpha_3 = \ldots = \alpha_n = 1$, and so $\bar{h} = 1$, or $C^* = C$.

Corollary 85. Let R be a commutative ring. Then the symbols $\{\alpha, \beta\}$ $(\alpha, \beta \in R^{\bullet})$ are a set of generators for K_2R if and only if $K_2R \subset W(R)$.

Proof. Immediate from Proposition 84 and the definition

$$C(R) = W(R) \cap K_2R.$$

Before we can check that the condition $K_2R \subset W(R)$ holds in certain special cases, we need some more general results. For any ring R, write $T = T(R)$ for the subgroup of $\mathrm{ST}(R)$ generated by all $x_{ij}(a)$ with $i < j$.

Proposition 86. For any ring R, the restriction $\phi : T(R) \to \mathrm{E}(R)$ is a monomorphism (or, $T(R) \cap K_2R = 1$).

Proof. Let

$$\underline{a} = (a_1, a_2, \ldots, a_{r-1}) \in R^{r-1}$$

and put

$$c_r(\underline{a}) = x_{1r}(a_1)x_{2r}(a_2)\ldots x_{r-1\,r}(a_{r-1}) \in T.$$

It is clear that T is generated by all such $c_r(\underline{a})$, for various r; we leave it to the reader to verify the relations

$$c_r(\underline{a})c_r(\underline{b}) = c_r(\underline{a} + \underline{b}) \quad (\underline{a}, \underline{b} \in R^{r-1})$$

and

$$c_r(\underline{a})c_s(\underline{b}) = c_s(\underline{b} + \underline{a}b_r)c_r(\underline{a}) \quad (\underline{a} \in R^{r-1}, \underline{b} \in R^{s-1})$$

where $r < s$, b_r is the rth coordinate of $\underline{b}$, and R^{r-1} is embedded in R^{s-1} in the obvious way. It follows from these relations that any element $t \in T$ can be written in the form

$$t = c_n(\underline{a}_n)c_{n-1}(\underline{a}_{n-1})\ldots c_2(\underline{a}_2)$$

for some n and $\underline{a}_r \in R^{r-1}$, $1 \le r \le n$. It is easy to check that $\phi(t)$ is an upper triangular matrix whose entries above the main diagonal in the nth column are the coordinates of $\underline{a}_n$. So if $\phi(t) = 1$, we must have $\underline{a}_n = \underline{0}$, and since $c_n(\underline{0}) = 1$, induction gives $t = 1$.

Of course, a similar result holds for the subgroup $T'(R)$ of $\mathrm{ST}(R)$ generated by all $x_{ij}(a)$ with $i > j$.

Proposition 87. If R is a field or skew field, $K_2R \subset W(R)$.

Proof. Let us write X_{ij} for the set of all $x_{ij}(a)$ $(a \in R)$, and $W_{ij} = \{w_{ij}(\alpha) : \alpha \in R^{\bullet}\}$. Now $\mathrm{ST}(R)$ is generated by all the X_{ij}, and since by Proposition 77(i) we have

$$X_{ji} \subset W_{ij}X_{ij}W_{ij}$$

it follows that $\mathrm{ST}(R)$ is generated by T together with all the W_{ij} with $i < j$. Next, by Proposition 77(ii), we have

$$W_{i\,i+2} \subset W_{i+1\,i+2}W_{i\,i+1}W_{i+1\,i+2}$$

and

$$W_{i\,i+r+1} \subset W_{i+r\,i+r+1}W_{i\,i+r}W_{i+r\,i+r+1}$$

whence $\mathrm{ST}(R)$ is generated by T together with all W_{ij} with $j = i + 1$.

Consider the subset $TWT \subset \mathrm{ST}(R)$. Now $(TWT)T \subset TWT$, since $TT \subset T$; if we can show that $WTW_{ij} \subset TWT$ for $j = i + 1$, then it will follow that $(TWT)W_{ij} \subset T(TWT) \subset TWT$, and hence by the previous remarks that $(TWT)G \subset TWT$, where $G = \mathrm{ST}(R)$, and since $1 \in TWT$, it will follow that $TWT = \mathrm{ST}(R)$. So let $w \in W$ and $t \in T$:

we shall show that $wtW_{ij} \subset TWT$, where i, j are now fixed, with $j = i + 1$. Using the notation of p.203, we can write

$$t = c_n(\underline{a}_n)c_{n-1}(\underline{a}_{n-1})\ldots c_2(\underline{a}_2)$$

for some n and $\underline{a}_r \in R^{r-1}$, $1 \leq r \leq n$. Since $c_r(\underline{0}) = 1$, all r, we may assume $n > j$. The relations on p.204 allow us to write

$$t = c_j(\underline{a}_j)c_n(\underline{a}'_n)\ldots c_{j+1}(\underline{a}'_{j+1})c_i(\underline{a}_i)\ldots c_2(\underline{a}_2)$$

for some $\underline{a}'_r \in R^{r-1}$, $j < r \leq n$. Thus we can write $t = x_{ij}(a)t'$ for some $a \in R$ and $t' \in T$, where t' can be expressed in terms of generators $x_{k\ell}(b)$ of T *other* than generators of the form $x_{ij}(b)$, that is, with $k \neq i$ or $\ell \neq j$ or both, and also $k < \ell$. For such $x_{k\ell}(b)$, it follows from Proposition 77(i) that for $w' \in W_{ij}$, we have $w'^{-1}x_{k\ell}(b)w' \in T$ (remembering that $j = i + 1$). So

$$tw' = x_{ij}(a)w'(w'^{-1}t'w') \in X_{ij}W_{ij}T$$

and thus it is enough to show $wX_{ij}W_{ij} \subset TWT$. Let

$$\phi(w) = P(\pi)\mathrm{diag}(\alpha_1, \alpha_2, \ldots)$$

and let $\pi(i) = p$, $\pi(j) = q$. Then $wX_{ij}w^{-1} = X_{pq}$, so if $p < q$,

$$wX_{ij}W_{ij} = X_{pq}wW_{ij} \subset TW \subset TWT.$$

It remains to consider the case when $p > q$. Given $a \in R$, we either have $a = 0$, when

$$wx_{ij}(a)W_{ij} = wW_{ij} \subset W \subset TWT$$

or else $a \in R^{\bullet}$, since R is a field or skew field, when

$$x_{ij}(a) = w_{ij}(a)x_{ij}(-a)x_{ji}(a^{-1}) \in W_{ij}X_{ij}X_{ji}$$

so

$$wx_{ij}(a)W_{ij} \subset (wW_{ij}X_{ij})(X_{ji}W_{ij}).$$

By Proposition 77(i),

$$wW_{ij}X_{ij} = wX_{ji}W_{ij} = X_{qp}wW_{ij} \subset TW$$

and

$$X_{ji}W_{ij} = W_{ij}X_{ij} \subset WT$$

so

$$wx_{ij}(a)W_{ij} \subset (TW)(WT) \subset TWT.$$

This completes the proof that $\mathrm{ST}(R) = TWT$.

To complete the proof of the proposition, let $x \in K_2R$; by the above argument we can find t_1, $t_2 \in T$ and $w \in W$ with $x = t_1wt_2$. Thus

$$1 = \phi(x) = \phi(t_1)\phi(w)\phi(t_2)$$

or

$$\phi(t_1^{-1}t_2^{-1}) = \phi(w).$$

But $\phi(t_1^{-1}t_2^{-1})$ is an upper unitriangular matrix, whereas $\phi(w)$ is a monomial matrix, so

$$\phi(t_1^{-1}t_2^{-1}) = 1 = \phi(w).$$

By Proposition 86, $t_1^{-1}t_2^{-1} = 1$, or $t_2 = t_1^{-1}$; also $w \in K_2R = \zeta(\mathrm{ST}(R))$, so

$$x = t_1wt_1^{-1} = w \in W(R)$$

as required.

Corollary 88. If R is a field, K_2R is generated by all symbols $\{\alpha, \beta\}$ $(\alpha, \beta \in R^{\bullet})$.

Proof. Immediate from Proposition 87 and Corollary 85.

Corollary 89. If R is a finite field, then $K_2R = 1$.

Proof. If $|R| = q$, then $R^{\bullet}$ is a cyclic group of order $q - 1$, generated by γ, say. Given $\alpha, \beta \in R^{\bullet}$, we have $\alpha = \gamma^n$, $\beta = \gamma^m$, for some $n, m \in \mathbb{Z}$, and by Proposition 79,

$$\{\alpha, \beta\} = \{\gamma^n, \gamma^m\} = \{\gamma, \gamma^m\}^n = \{\gamma, \gamma\}^{nm}.$$

So by Corollary 88, K_2R is cyclic, generated by $\{\gamma, \gamma\}$. By Proposition 79(i), $\{\gamma, \gamma\}^2 = 1$, so it suffices to find an odd integer r such that $\{\gamma, \gamma\}^r = 1$. Now if q is even, $q - 1$ is odd, and

$$\{\gamma, \gamma\}^{q-1} = \{\gamma, \gamma^{q-1}\} = \{\gamma, 1\} = 1$$

and the proof is complete. On the other hand, if q is odd, consider the map $R^{\bullet} - \{1\} \to R^{\bullet} - \{1\}$, $\alpha \mapsto 1 - \alpha$. This is a bijection, and since $R^{\bullet}$ contains $\frac{1}{2}(q - 1)$ squares and $\frac{1}{2}(q - 1)$ non-squares, and 1 is a square, $R^{\bullet} - \{1\}$ contains $\frac{1}{2}(q - 3)$

squares and $\frac{1}{2}(q - 1)$ non-squares. Thus we can find $\alpha \in R^{\cdot} - \{1\}$ such that α and $1 - \alpha$ are both non-squares. So $\alpha = \gamma^n$, $1 - \alpha = \gamma^m$, say, where n and m are odd, and by Proposition 80,

$$1 = \{\alpha, 1 - \alpha\} = \{\gamma^n, \gamma^m\} = \{\gamma, \gamma\}^{nm}$$

and since nm is odd, the proof is complete.

We now quote, without proof, the following theorem of Matsumoto:

Theorem 90. If R is a field, K_2R has the following presentation as an abelian group:

Generators: $\{\alpha, \beta\}$ $(\alpha, \beta \in R^{\cdot})$

Relations:
$$\{\alpha_1\alpha_2, \beta\} = \{\alpha_1, \beta\}\{\alpha_2, \beta\}$$
$$\{\alpha, \beta_1\beta_2\} = \{\alpha, \beta_1\}\{\alpha, \beta_2\}$$
$$\{\alpha, 1 - \alpha\} = 1 \ (\alpha \neq 1).$$

The proof can be found in [11], section 12. Note that the relations $\{\alpha, -\alpha\} = 1$ and $\{\alpha, \beta\}^{-1} = \{\beta, \alpha\}$ are consequences of the above relations. For $\{1, -1\} = 1$, and if $\alpha \neq 1$,

$$\begin{aligned}
\{\alpha, -\alpha\} &= \{\alpha, (1 - \alpha)(1 - \alpha^{-1})^{-1}\} \\
&= \{\alpha, 1 - \alpha\}\{\alpha, (1 - \alpha^{-1})^{-1}\} \\
&= \{\alpha, (1 - \alpha^{-1})^{-1}\} \\
&= \{\alpha, 1 - \alpha^{-1}\}^{-1} \\
&= \{\alpha^{-1}, 1 - \alpha^{-1}\} = 1
\end{aligned}$$

and then

$$\begin{aligned}
\{\alpha, \beta\} &= \{\alpha, \beta\}\{\alpha, -\alpha\} \\
&= \{\alpha, -\alpha\beta\} \\
&= \{\alpha, -\alpha\beta\}\{\alpha\beta, -\alpha\beta\}^{-1} \\
&= \{\alpha, -\alpha\beta\}\{\alpha^{-1}\beta^{-1}, -\alpha\beta\} \\
&= \{\beta^{-1}, -\alpha\beta\} \\
&= \{\beta^{-1}, -\alpha\beta\}\{\beta^{-1}, -\beta^{-1}\} \\
&= \{\beta^{-1}, \alpha\} \\
&= \{\beta, \alpha\}^{-1}.
\end{aligned}$$

So in proving Corollary 89, we have in effect proved Matsumoto's theorem in the case where the field R is finite.

8.2 *Generators for K_2R: the symbols $\langle a, b\rangle$*

We now introduce a new notation for the generators of K_2R, when R is a field. Let $a, b \in R$ with $1 + ab \in R^{\bullet}$ (that is, $ab \neq -1$, since R is a field), and define

$$\langle a, b\rangle = \begin{cases} 1 \text{ if } b = 0 \\ \{1 + ab, b\} \text{ if } b \in R^{\bullet}. \end{cases}$$

Note that $\langle 0, b\rangle = 1$, any $b \in R$. Note also that, if a, b, $1 + ab \in R^{\bullet}$, then

$$\{1 + ab, b\}\{1 + ab, -a\} = \{1 + ab, -ab\} = 1$$

so $\langle a, b\rangle = \{-a, 1 + ab\}$; this also holds if $a \in R^{\bullet}$ and $b = 0$.

Theorem 91. If R is a field, K_2R has the following presentation as an abelian group:

Generators: $\langle a, b\rangle$ $(a, b \in R, 1 + ab \in R^{\bullet})$

Relations:

$$\langle a, b\rangle^{-1} = \langle -b, -a\rangle$$

$$\langle a, b\rangle\langle a, c\rangle = \langle a, b + c + bac\rangle$$

$$\langle a, bc\rangle\langle b, ac\rangle\langle c, ab\rangle = 1.$$

Proof. We use Matsumoto's theorem: take K_2R to be the group whose presentation is given in Theorem 90, and define the symbols $\langle a, b\rangle$ as above. Then given $\alpha, \beta \in R^{\bullet}$, put $a = -\alpha$ and choose b such that $1 + ab = \beta$. Then the equation $\langle a, b\rangle = \{-a, 1 + ab\}$, obtained above, can be rewritten $\{\alpha, \beta\} = \langle -\alpha, (1 - \beta)\alpha^{-1}\rangle$, and it follows that K_2R is generated by the symbols $\langle a, b\rangle$. Next,

$$\langle a, b\rangle\langle -b, -a\rangle = \{1 + ab, b\}\{b, 1 + ab\} = 1$$

if $a, b, 1 + ab \in R^{\bullet}$ (the relation being trivial if $a = 0$ or $b = 0$). Also,

$$\begin{aligned} \langle a, b\rangle\langle a, c\rangle &= \{-a, 1 + ab\}\{-a, 1 + ac\} \\ &= \{-a, (1 + ab)(1 + ac)\} \\ &= \{-a, 1 + a(b + c + bac)\} \end{aligned}$$

$$= \langle a, b + c + bac\rangle$$

if a, $1 + ab$, $1 + ac \in R^{\bullet}$ (the relation being trivial if $a = 0$). Finally,

$$\langle a, bc\rangle\langle b, ac\rangle\langle c, ab\rangle$$

$$= \{-a, 1 + abc\}\{-b, 1 + abc\}\{-c, 1 + abc\}$$

$$= \{-abc, 1 + abc\} = 1$$

if a, b, c, $1 + abc \in R^{\bullet}$ (the relation being trivial if $a = 0$ or $b = 0$ or $c = 0$).

Now let G be the group whose presentation is given in the statement of the theorem. What we have shown is that it is possible to define a homomorphism $f : G \to K_2R$ whose effect on generators is given by $\langle a, 0\rangle \mapsto 1$ and $\langle a, b\rangle \mapsto \{1 + ab, b\}$ $(b \neq 0)$, or equivalently by $\langle 0, b\rangle \mapsto 1$ and $\langle a, b\rangle \mapsto \{-a, 1 + ab\}$ $(a \neq 0)$. We show that f is an isomorphism by constructing its inverse $g : K_2R \to G$; this is given on generators by $\{\alpha, \beta\} \mapsto \langle -\alpha, (1 - \beta)\alpha^{-1}\rangle$, and it is immediate that fg and gf preserve the generators of K_2R and G respectively. So for α, $\beta \in R^{\bullet}$ we define $\{\alpha, \beta\} \in G$ by $\{\alpha, \beta\} = \langle -\alpha, (1 - \beta)\alpha^{-1}\rangle$, and we need only check that these elements of G satisfy the relations of Theorem 90. So if α, β, $\gamma \in R^{\bullet}$ we have, working in G,

$$\{\alpha, \beta\}\{\alpha, \gamma\}$$

$$= \langle -\alpha, (1 - \beta)\alpha^{-1}\rangle\langle -\alpha, (1 - \gamma)\alpha^{-1}\rangle$$

$$= \langle -\alpha, (1 - \beta)\alpha^{-1} + (1 - \gamma)\alpha^{-1} + (1 - \beta)\alpha^{-1}(-\alpha)(1 - \gamma)\alpha^{-1}\rangle$$

$$= \langle -\alpha, (1 - \beta\gamma)\alpha^{-1}\rangle$$

$$= \{\alpha, \beta\gamma\}.$$

Next,

$$\langle -\alpha, (1 - \gamma)\alpha^{-1}\rangle\langle -\beta, (1 - \gamma)\beta^{-1}\rangle\langle -(1 - \gamma)\alpha^{-1}\beta^{-1}, \alpha\beta\rangle = 1$$

that is,

$$\{\alpha, \gamma\}\{\beta, \gamma\} = \{\alpha\beta, \gamma\}.$$

Then if $a \neq 1$,

$$\langle a, -1\rangle\langle 1, -a\rangle\langle -1, a\rangle = 1$$

but $\langle a, -1\rangle\langle 1, -a\rangle = 1$, so $\langle -1, a\rangle = 1$. If $b \neq -1$, we then have

$$\langle b, 1\rangle = \langle -1, -b\rangle^{-1} = 1.$$

Now let $\alpha \in R^{\bullet}$, $\alpha \neq 1$, so $1 - \alpha \in R^{\bullet}$. We have

$$\langle -\alpha^{-1}, \alpha^{2}\rangle\langle -\alpha, 1\rangle\langle -\alpha, 1\rangle = 1$$

whence $\langle -\alpha^{-1}, \alpha^{2}\rangle = 1$, or $\{\alpha^{-1}, 1 - \alpha\} = 1$. But now

$$\{\alpha, 1 - \alpha\} = \{\alpha^{-1}, 1 - \alpha\}^{-1} = 1$$

and the proof is complete.

Remarks: (i) In view of the first relation, the third can be rewritten

$$\langle a, bc\rangle = \langle ab, c\rangle\langle ac, b\rangle.$$

There is also a 'left-handed' version:

$$\langle -ab, c\rangle\langle -ac, b\rangle\langle -bc, a\rangle = 1.$$

There is a 'left-handed' version of the second relation:

$$\langle a, c\rangle\langle b, c\rangle = \langle a + b + acb, c\rangle.$$

(ii) The symbols $\langle a, b\rangle$ suffer from a curious lack of symmetry or skew symmetry: witness the relation $\langle a, b\rangle = 1$ if $a = -1$ or $b = 1$. It might have been easier to define $\langle a, b\rangle' = \langle -a, b\rangle$ for $1 - ab \in R^{\bullet}$; the defining relations for K_2R (when R is a field) would then read

$$\langle a, b\rangle'^{-1} = \langle b, a\rangle'$$

$$\langle a, b\rangle'\langle a, c\rangle' = \langle a, b + c - bac\rangle'$$

$$\langle a, bc\rangle'\langle b, ac\rangle'\langle c, ab\rangle' = 1$$

and the corresponding 'left-handed' relations are then obtained just by switching the order in each symbol; also $\langle a, b\rangle' = 1$ if $a = 1$ or $b = 1$.

We shall now generalize some of the foregoing calculations, starting with Proposition 77. Let R be a ring, not necessarily commutative, and let $w \in \mathrm{ST}(R)$. We say w is *monomial* if $\phi(w)$ is a monomial matrix (see p.194), and w is *diagonal* if $\phi(w)$ is a diagonal matrix.

Proposition 92. Let w be a monomial element of ST(R). Then

$$wx_{ij}(a)w^{-1} = x_{k\ell}(\alpha_i a \alpha_j^{-1})$$

where $\phi(w) = P(\pi)\mathrm{diag}(\alpha_1, \alpha_2, \ldots)$, $\pi(i) = k$, and $\pi(j) = \ell$. (Compare Proposition 77(i).)

Proof. First suppose w is diagonal, that is $\pi = 1$, $k = i$, and $\ell = j$. We can choose n so large that there is an expression for w in terms of the generators $x_{st}(b)$ with all $s, t \le n$, and also $i, j \le n$. So $\alpha_s = 1$ for $s > n$. Put $w_1 = w_{i\ n+1}(1)w_{j\ n+2}(1)$, and $w_2 = w_1 w w_1^{-1}$. By Proposition 77 there is an expression for w_2 in terms of the generators $x_{st}(b)$ with all s, t different from i, j, and so w_2 commutes with $x_{ij}(a)$. Further, using Proposition 77(i) again,

$$\phi(w_2) = \mathrm{diag}(\alpha_1', \alpha_2', \ldots, \alpha_n', \alpha_i, \alpha_j, 1, 1, \ldots)$$

where $\alpha_i' = 1 = \alpha_j'$, and $\alpha_s' = \alpha_s$ for $s \ne i, j$. Thus

$$\phi(ww_2^{-1}) = \phi(h_{i\ n+1}(\alpha_i)h_{j\ n+2}(\alpha_j))$$

whence $ww_2^{-1} \equiv h \pmod{K_2R}$, where $h = h_{i\ n+1}(\alpha_i)h_{j\ n+2}(\alpha_j)$. So

$$\begin{aligned} wx_{ij}(a)w^{-1} &= (ww_2^{-1})x_{ij}(a)(ww_2^{-1})^{-1} \\ &= hx_{ij}(a)h^{-1} \\ &= x_{ij}(\alpha_i a \alpha_j^{-1}), \text{ by Proposition 77(i).} \end{aligned}$$

For the general case, choose $w_0 \in W(R)$ such that $\phi(w_0) = P(\pi)\mathrm{diag}(\beta_1, \beta_2, \ldots)$, for some $\beta_1, \beta_2, \ldots \in R^{\cdot}$. This can be done, since π can be written as a product of 2-cycles (st), and $\phi(w_{st}(1)) = P(st)\mathrm{diag}(\ldots)$. Thus $w_0^{-1}w$ is diagonal, say $\phi(w_0^{-1}w) = \mathrm{diag}(\gamma_1, \gamma_2, \ldots)$. By the first part of the proof,

$$w_0^{-1}wx_{ij}(a)w^{-1}w_0 = x_{ij}(\gamma_i a \gamma_j^{-1})$$

so

$$\begin{aligned} wx_{ij}(a)w^{-1} &= w_0 x_{ij}(\gamma_i a \gamma_j^{-1})w_0^{-1} \\ &= x_{k\ell}(\beta_i \gamma_i a \gamma_j^{-1} \beta_j^{-1}), \text{ by Proposition 77(i)} \\ &= x_{k\ell}(\alpha_i a \alpha_j^{-1}), \text{ as required.} \end{aligned}$$

Now let $a, b \in R$ with $1 + ab = \alpha \in R^{\bullet}$. Put $\beta = 1 + ba$ and note that $\beta \in R^{\bullet}$; explicitly, $\beta^{-1} = 1 - b\alpha^{-1}a$. In ST($R$), define

$$H_{ij}(a, b) = x_{ji}(-b\alpha^{-1})x_{ij}(a)x_{ji}(b)x_{ij}(-a\beta^{-1}).$$

Proposition 93. $H_{ij}(a, b)$ is diagonal; explicitly,

$$\phi(H_{ij}(a, b)) = \mathrm{diag}(\alpha_1, \alpha_2, \dots)$$

where $\alpha_i = \alpha$, $\alpha_j = \beta^{-1}$, and $\alpha_k = 1$ for $k \neq i, j$.

Proof. We have

$$\begin{pmatrix} 1 & a \\ 0 & 1 \end{pmatrix}\begin{pmatrix} 1 & 0 \\ b & 1 \end{pmatrix}\begin{pmatrix} 1 & 0 \\ 0 & \beta \end{pmatrix} = \begin{pmatrix} \alpha & 0 \\ 0 & 1 \end{pmatrix}\begin{pmatrix} 1 & 0 \\ b & 1 \end{pmatrix}\begin{pmatrix} 1 & a \\ 0 & 1 \end{pmatrix}$$

(compare p.135), whence

$$\begin{pmatrix} 1 & a \\ 0 & 1 \end{pmatrix}\begin{pmatrix} 1 & 0 \\ b & 1 \end{pmatrix} = \begin{pmatrix} \alpha & 0 \\ 0 & 1 \end{pmatrix}\begin{pmatrix} 1 & 0 \\ b & 1 \end{pmatrix}\begin{pmatrix} 1 & a \\ 0 & 1 \end{pmatrix}\begin{pmatrix} 1 & 0 \\ 0 & \beta^{-1} \end{pmatrix}$$

$$= \begin{pmatrix} 1 & 0 \\ b\alpha^{-1} & 1 \end{pmatrix}\begin{pmatrix} \alpha & 0 \\ 0 & \beta^{-1} \end{pmatrix}\begin{pmatrix} 1 & a\beta^{-1} \\ 0 & 1 \end{pmatrix}.$$

This shows that $\phi(H_{12}(a, b)) = \mathrm{diag}(\alpha, \beta^{-1}, 1, 1, \dots)$, and the general result follows by a similar calculation (or by conjugating by a suitable $w \in W(R)$).

Proposition 94. The following relations hold in ST(R), any R:

(i) $H_{ij}(a, b) = H_{ji}(-b, -a)^{-1}$ $(a, b \in R,\ 1 + ab \in R^{\bullet})$

(ii) $H_{ij}(a, b + c + bac) = H_{ij}(a, b)H_{ij}(a\beta^{-1}, \beta c)$ $(a, b, c \in R,\ 1 + ab,\ 1 + ac \in R^{\bullet},\ \beta = 1 + ba)$

(iii) $H_{ij}(a, bc)^{-1}H_{jk}(b, ca)^{-1}H_{ki}(c, ab)^{-1} = 1$ $(a, b, c \in R,\ 1 + abc \in R^{\bullet})$.

Proof. (i) Immediate, noting that $1 + (-b)(-a) = 1 + ba \in R^{\bullet}$.

(ii) Note first that $1 + a(b + c + bac) = (1 + ab)(1 + ac) \in R^{\bullet}$ and $1 + (a\beta^{-1})(\beta c) = 1 + ac \in R^{\bullet}$, so all the terms are defined. Put $\alpha = 1 + ab$, $\beta = 1 + ba$, $\gamma = 1 + ac$, $\delta = 1 + ca$. Then

$1 + a(b + c + bac) = \alpha\gamma$ and $1 + (b + c + bac)a = \beta\delta$. So

$$x_{ji}((b + c + bac)\gamma^{-1}\alpha^{-1})H_{ij}(a, b + c + bac)x_{ij}(a\delta^{-1}\beta^{-1})$$
$$= x_{ij}(a)x_{ji}(b + c + bac)$$
$$= x_{ij}(a)x_{ji}(b)x_{ji}(\beta c)$$
$$= x_{ji}(b\alpha^{-1})H_{ij}(a, b)x_{ij}(a\beta^{-1})x_{ji}(\beta c)$$
$$= x_{ji}(b\alpha^{-1})H_{ij}(a, b)x_{ji}(\beta c\gamma^{-1})H_{ij}(a\beta^{-1}, \beta c)x_{ij}(a\delta^{-1}\beta^{-1})$$

(since $1 + (a\beta^{-1})(\beta c) = \gamma$ and $1 + (\beta c)(a\beta^{-1}) = \beta\delta\beta^{-1}$)

$$= x_{ji}(b\alpha^{-1} + c\gamma^{-1}\alpha^{-1})H_{ij}(a, b)H_{ij}(a\beta^{-1}, \beta c)x_{ij}(a\delta^{-1}\beta^{-1})$$

by Propositions 92 and 93, and the result follows, since

$$b\alpha^{-1} + c\gamma^{-1}\alpha^{-1} = (b\gamma + c)\gamma^{-1}\alpha^{-1} = (b + c + bac)\gamma^{-1}\alpha^{-1}.$$

(iii) Note first that, since $1 + abc \in R^{\bullet}$, we have $1 + bca \in R^{\bullet}$ and $1 + cab \in R^{\bullet}$, so the terms are defined. The relation is proved by using Philip Hall's commutator identity

$$({}^{x}[[x^{-1}, z], y])({}^{y}[[y^{-1}, x], z])({}^{z}[[z^{-1}, y], x]) = 1$$

where ${}^{u}v$ means uvu^{-1}. We put $x = x_{ij}(a)$, $y = x_{ki}(c)$, and $z = x_{jk}(b)$. Then, putting $d = (1 + abc)^{-1} - 1$, $e = (1 + cab)^{-1} - 1$, and $f = (1 + bca)^{-1} - 1$, we have

$$[[x^{-1}, z], y] = [[x_{ij}(-a), x_{jk}(b)], x_{ki}(c)]$$
$$= [x_{ik}(-ab), x_{ki}(c)]$$
$$= x_{ik}(-ab)x_{ki}(c)x_{ik}(ab)x_{ki}(-c)$$
$$= x_{ik}(abe)H_{ki}(c, ab)x_{ki}(cd)$$

and similarly

$$[[y^{-1}, x], z] = x_{kj}(caf)H_{jk}(b, ca)x_{jk}(be)$$

and

$$[[z^{-1}, y], x] = x_{ji}(bcd)H_{ij}(a, bc)x_{ij}(af).$$

Next,

$$x_{ki}(cd)x_{ij}(-a)x_{ki}(c)x_{kj}(caf) = x_{ij}(-a)x_{ki}(cd + c)$$

since $cda = caf$, and then similarly

$$x_{jk}(be)x_{ki}(-c)x_{jk}(b)x_{ji}(bcd) = x_{ki}(-c)x_{jk}(be + b)$$

and

$$x_{ij}(af)x_{jk}(-b)x_{ij}(a)x_{ik}(abe) = x_{jk}(-b)x_{ij}(af + a).$$

Putting all this into the commutator identity, we obtain

$$H_{ki}(c, ab)x_{ij}(-a)x_{ki}(cd + c)H_{jk}(b, ca)x_{ki}(-c)x_{jk}(be + b)$$
$$\times H_{ij}(a, bc)x_{jk}(-b)x_{ij}(af + a) = 1.$$

Then

$$H_{ki}(c, ab)x_{ij}(-a) = x_{ij}((1 + abc)^{-1}(-a))H_{ki}(c, ab)$$

by Propositions 92 and 93. But $(1 + abc)a = a(1 + bca)$, whence $(1 + abc)^{-1}a = af + a$, and so

$$x_{ij}(af + a)H_{ki}(c, ab)x_{ij}(-a) = H_{ki}(c, ab).$$

Similarly,

$$x_{ki}(cd + c)H_{jk}(b, ca)x_{ki}(-c) = H_{jk}(b, ca)$$

and

$$x_{jk}(be + b)H_{ij}(a, bc)x_{jk}(-b) = H_{ij}(a, bc).$$

Substituting back, we obtain

$$H_{ki}(c, ab)H_{jk}(b, ca)H_{ij}(a, bc) = 1$$

and the result follows.

We are now in a position to introduce a more general definition of the symbols $\langle a, b\rangle$. Let R be any ring, and let $a, b \in R$ with $1 + ab = \alpha \in R^{\cdot}$, and suppose $ab = ba$, so $\alpha = 1 + ba$. Then it follows from Proposition 93 that

$$\phi(H_{ij}(a, b)) = \phi(h_{ij}(\alpha))$$

so $H_{ij}(a, b)h_{ij}(\alpha)^{-1} \in K_2R$. Since $K_2R = \zeta(\mathrm{ST}(R))$, it follows on conjugation by suitable elements in $W(R)$ that $H_{ij}(a, b)h_{ij}(\alpha)^{-1}$ is independent of the choice of i, j (compare the argument on pp.197-198). We define

$$\langle a, b\rangle = H_{ij}(a, b)h_{ij}(\alpha)^{-1}.$$

Proposition 95. Let $a, b \in R$ with $1 + ab = \alpha \in R^{\cdot}$, $ab = ba$, and let $\beta \in R^{\cdot}$ with $\alpha\beta = \beta\alpha$. Then $\langle a\beta^{-1}, \beta b\rangle = \langle a, b\rangle\{\alpha, \beta\}$.

Proof. Using Proposition 92,

$$\begin{aligned}
&h_{23}(\beta)H_{12}(a, b)h_{23}(\beta)^{-1} \\
&= h_{23}(\beta)x_{21}(-b\alpha^{-1})x_{12}(a)x_{21}(b)x_{12}(-a\alpha^{-1})h_{23}(\beta)^{-1} \\
&= x_{21}(-\beta b\alpha^{-1})x_{12}(a\beta^{-1})x_{21}(\beta b)x_{12}(-a\beta^{-1}\alpha^{-1}) \\
&= H_{12}(a\beta^{-1}, \beta b).
\end{aligned}$$

Also, from Proposition 77(iii),

$$h_{23}(\beta)h_{12}(\alpha)h_{23}(\beta)^{-1} = h_{12}(\alpha\beta^{-1})h_{12}(\beta^{-1})^{-1}.$$

Since $\langle a, b\rangle$ is central, we then have

$$\begin{aligned}
\langle a, b\rangle &= h_{23}(\beta)\langle a, b\rangle h_{23}(\beta)^{-1} \\
&= h_{23}(\beta)H_{12}(a, b)h_{12}(\alpha)^{-1}h_{23}(\beta)^{-1} \\
&= H_{12}(a\beta^{-1}, \beta b)h_{12}(\beta^{-1})h_{12}(\alpha\beta^{-1})^{-1} \\
&= \langle a\beta^{-1}, \beta b\rangle h_{12}(\alpha)h_{12}(\beta^{-1})h_{12}(\alpha\beta^{-1})^{-1} \\
&= \langle a\beta^{-1}, \beta b\rangle\{\beta^{-1}, \alpha\}^{-1} \\
&= \langle a\beta^{-1}, \beta b\rangle\{\beta, \alpha\} \\
&= \langle a\beta^{-1}, \beta b\rangle\{\alpha, \beta\}^{-1}, \text{ as required.}
\end{aligned}$$

Proposition 96. Let $a, b \in R$, $1 + ab = \alpha \in R^{\bullet}$, $ab = ba$. Then:

(i) $\langle a, b\rangle = 1$ if $b = 0$ or 1;

(ii) $\langle a, b\rangle = 1$ if $a = 0$ or -1;

(iii) $\langle a, b\rangle = \{\alpha, b\}$ if $b \in R^{\bullet}$;

(iv) $\langle a, b\rangle = \{-a, \alpha\}$ if $a \in R^{\bullet}$.

Proof. (i) If $b = 0$, then $\alpha = 1$, $h_{12}(1) = 1$ and $H_{12}(a, 0) = 1$. If $b = 1$, then $a = \alpha - 1$. Now

$$\begin{aligned}
h_{12}(\alpha) &= w_{12}(\alpha)w_{12}(-1) \\
&= x_{12}(\alpha)x_{21}(-\alpha^{-1})x_{12}(\alpha - 1)x_{21}(1)x_{12}(-1).
\end{aligned}$$

Then

$$\begin{aligned}
H_{12}(\alpha - 1, 1) &= x_{21}(-\alpha^{-1})x_{12}(\alpha - 1)x_{21}(1)x_{12}(\alpha^{-1} - 1) \\
&= x_{12}(-\alpha)h_{12}(\alpha)x_{12}(\alpha^{-1})
\end{aligned}$$

$= h_{12}(\alpha)$, by Proposition 77(iii).

Thus $H_{12}(\alpha - 1, 1)h_{12}(\alpha)^{-1} = 1$, as required.

(ii) Proposition 94(i) says $H_{12}(a, b) = H_{21}(-b, -a)^{-1}$, so

$$\langle a, b\rangle h_{12}(\alpha) = h_{21}(\alpha)^{-1}\langle -b, -a\rangle^{-1}.$$

But $\langle a, b\rangle$ is central, and $h_{12}(\alpha) = h_{21}(\alpha)^{-1}$, so $\langle a, b\rangle = \langle -b, -a\rangle^{-1}$. The result now follows from (i).

(iii) Using Proposition 95 with $\beta = b^{-1}$,

$$\langle a, b\rangle\{\alpha, b^{-1}\} = \langle ab, 1\rangle = 1, \text{ by (i)}.$$

So

$$\langle a, b\rangle = \{\alpha, b^{-1}\}^{-1} = \{\alpha, b\}.$$

(iv) $\langle a, b\rangle = \langle -b, -a\rangle^{-1}$, as in (ii)

$= \{\alpha, -a\}^{-1}$, by (iii)

$= \{-a, \alpha\}$.

Note that the above proposition shows that our new (and more general) definition of the symbol $\langle a, b\rangle$ is compatible with the previous definition (p.208). Next we show that our new symbols satisfy relations like those of Theorem 91:

Proposition 97. Let R be any ring. Then:

(i) $\langle a, b\rangle = \langle -b, -a\rangle^{-1}$ ($a, b \in R$, $1 + ab \in R^{\bullet}$, $ab = ba$);

(ii) $\langle a, b\rangle\langle a, c\rangle = \langle a, b + c + bac\rangle$ ($a, b, c \in R$, $1 + ab$, $1 + ac \in R^{\bullet}$, $ab = ba$, $ac = ca$, and also $abac = acab$ (but see the exercise, below));

(iii) $\langle a, bc\rangle\langle b, ca\rangle\langle c, ab\rangle = 1$ ($a, b, c \in R$, $1 + abc \in R^{\bullet}$, $abc = bca = cab$).

Proof. (i) This was proved in the course of proving Proposition 96(ii).

(ii) From Proposition 94(ii), putting $\alpha = 1 + ab$ and $\beta = 1 + ac$,

$$H_{12}(a, b + c + bac) = H_{12}(a, b)H_{12}(a\alpha^{-1}, \alpha c)$$

whence

$$\langle a, b + c + bac\rangle h_{12}(\alpha\beta) = \langle a, b\rangle h_{12}(\alpha)\langle a\alpha^{-1}, \alpha c\rangle h_{12}(\beta)$$
$$= \langle a, b\rangle\langle a\alpha^{-1}, \alpha c\rangle h_{12}(\alpha)h_{12}(\beta).$$

Now $h_{12}(\alpha\beta)h_{12}(\beta)^{-1}h_{12}(\alpha)^{-1} = \{\alpha, \beta\}$, so

$$\langle a, b + c + bac\rangle = \langle a, b\rangle\langle a\alpha^{-1}, \alpha c\rangle\{\beta, \alpha\}^{-1}$$
$$= \langle a, b\rangle\langle a, c\rangle, \text{ by Proposition 95.}$$

(iii) From Proposition 94(iii),

$$H_{31}(c, ab)H_{23}(b, ca)H_{12}(a, bc) = 1.$$

Putting $\alpha = 1 + abc$, we have

$$\langle c, ab\rangle h_{31}(\alpha)\langle b, ca\rangle h_{23}(\alpha)\langle a, bc\rangle h_{12}(\alpha) = 1$$

or

$$\langle a, bc\rangle\langle b, ca\rangle\langle c, ab\rangle h_{31}(\alpha)h_{23}(\alpha)h_{12}(\alpha) = 1$$

using the centrality of K_2R. But $h_{31}(\alpha)h_{23}(\alpha)h_{12}(\alpha) = 1$ (see the proof of Lemma 83), and the result follows.

Exercise: The hypothesis $abac = acab$ in Proposition 97(ii) was needed to ensure that $\{\beta, \alpha\}$ and $\langle a\alpha^{-1}, \alpha c\rangle$ were defined. Show that this hypothesis is unnecessary. (Show that

$$H_{12}(a\alpha^{-1}, \alpha c) = h_{32}(\alpha)^{-1}H_{12}(a, c)h_{32}(\alpha)$$
$$= h_{32}(\alpha)^{-1}\langle a, c\rangle h_{12}(\beta)h_{32}(\alpha)$$

and that

$$h_{12}(\alpha\beta) = h_{12}(\alpha)h_{32}(\alpha)^{-1}h_{12}(\beta)h_{32}(\alpha)$$

even if $\alpha\beta \neq \beta\alpha$.)

Now let R be any ring and let $J \lhd R$. Write $H(R, J)$ for the subgroup of $\mathrm{ST}(R)$ generated by all $H_{ij}(a, b)$ with $a, b \in R$, $1 + ab \in R^{\bullet}$, and $a \in J$ or $b \in J$. Write $H(R) = H(R, R)$. Note that if J is a *radical* ideal, that is $J \subset \mathrm{rad}(R)$, then if $a \in J$ and $b \in R$, or if $a \in R$ and $b \in J$, it follows automatically that $1 + ab \in R^{\bullet}$.

Recall that $\mathrm{ST}(R, J) = \ker(\mathrm{ST}(R) \to \mathrm{ST}(R/J))$ is the *normal* subgroup of $\mathrm{ST}(R)$ generated by all $x_{ij}(a)$ with $a \in J$. Note that $H(R, J) \subset \mathrm{ST}(R, J)$. Write (temporarily) $S(R, J)$ for the subgroup of ST(R) generated by all $x_{ij}(a)$ with $a \in J$. So $S(R, J) \subset$ ST(R, J), and also $H(R, J)$ normalizes $S(R, J)$, by Propositions 92 and 93. Thus $S(R, J)H(R, J)$ is a subgroup of ST(R, J).

Proposition 98. Let J be a radical ideal of R. Then

$$\mathrm{ST}(R, J) = S(R, J)H(R, J).$$

Proof. Since $S(R, J)$ contains the normal generators of ST(R, J), it is sufficient to show that $S(R, J)H(R, J)$ is a normal subgroup of ST(R). Let $a \in R$ and $b \in J$. Then

$$x_{ij}(a)x_{k\ell}(b)x_{ij}(-a) = x_{k\ell}(b) \text{ if } k \neq j \text{ and } \ell \neq i$$

$$x_{ij}(a)x_{j\ell}(b)x_{ij}(-a) = x_{j\ell}(b)x_{i}(ab)$$

$$x_{ij}(a)x_{ki}(b)x_{ij}(-a) = x_{ki}(b)x_{kj}(-ba)$$

and

$$x_{ij}(a)x_{ji}(b)x_{ij}(-a) = x_{ji}(b\alpha^{-1})H_{ij}(a, b)x_{ij}(a\beta^{-1} - a)$$

where $\alpha = 1 + ab$ and $\beta = 1 + ba$. Now b, ab, $-ba$, $b\alpha^{-1}$, and $a\beta^{-1} - a$ all belong to J (the last because $\beta^{-1} - 1 \in J$), and so we have shown that if $u \in S(R, J)$, then $x_{ij}(a)ux_{ij}(-a) \in S(R, J)H(R, J)$.

Next, let $w \in H(R, J)$; so w is diagonal, indeed $\phi(w) = \mathrm{diag}(\alpha_1, \alpha_2, \ldots)$ say, where $\alpha_i - 1 \in J$, all i. Then

$$wx_{ij}(-a)w^{-1} = x_{ij}(-\alpha_i a\alpha_j^{-1})$$

by Proposition 92, whence

$$[x_{ij}(a), w] = x_{ij}(a - \alpha_i a\alpha_j^{-1}) \in S(R, J)$$

since $\alpha_i - 1 \in J$ and $\alpha_j^{-1} - 1 \in J$.

Finally, given $u \in S(R, J)$ and $w \in H(R, J)$, and $a \in R$, we have

$$x_{ij}(a)uwx_{ij}(-a) = (x_{ij}(a)ux_{ij}(-a))[x_{ij}(a), w]w$$
$$\in S(R, J)H(R, J)$$

using the fact that $S(R, J)H(R, J)$ is a subgroup; since this subgroup is normalized by the generators of ST(R), it must be a normal subgroup of ST(R), and the result follows.

Now write $T(R, J)$ for the subgroup of ST(R) generated by all $x_{ij}(a)$ with $i < j$ and $a \in J$, and write $T'(R, J)$ for the subgroup of ST(R) generated by all $x_{ij}(a)$ with $i > j$ and $a \in J$. If we put $T(R) = T(R, R)$ and $T'(R) = T'(R, R)$, then this agrees with the previous notation (pp.203-204). Note that for any $I, J \lhd R$, $H(R, J)$ normalizes $T(R, I)$ and $T'(R, I)$, by Propositions 92 and 93.

Now $T(R, J)$, $T'(R, J)$, and $H(R, J)$ are all subgroups of ST(R, J), so ST(R, J) $\supset T'(R, J)T(R, J)H(R, J)$. We have:

Proposition 99. Let J be a radical ideal of R. Then

$$\text{ST}(R, J) = T'(R, J)T(R, J)H(R, J).$$

Proof. The set $T'(R, J)T(R, J)$ contains every $x_{ij}(a)$, $a \in J$, and so contains a generating set for $S(R, J)$. Thus, by Proposition 98, the set $T'(R, J)T(R, J)H(R, J)$ contains a generating set for ST(R, J). So it suffices to prove that this set is closed under multiplication on the right by generators (remembering that here inverses of generators are also generators). In fact we shall show first that the larger set $X = T'(R)T(R, J)H(R, J)$ is closed under multiplication on the right by generators of $T'(R)$, $T(R, J)$, and $H(R, J)$.

Since $[x_{i\ i-1}(a), x_{i-1\ i-r}(1)] = x_{i\ i-r}(a)$, $i > r \geq 2$, the subgroup $T'(R)$ is generated by all $x_{i\ i-1}(a)$, $a \in R$, $i \geq 2$. (Note that $T'(R, J)$ is not necessarily generated by all $x_{i\ i-1}(a)$ with $a \in J$; this is why we choose to work with $T'(R)$ here.)

Now $H(R, J)$ is a subgroup, so X is closed under multiplication on the right by elements of $H(R, J)$. So now let $t' \in T'(R)$, $t \in T(R, J)$, and $h \in H(R, J)$; we must show $t'thx_{ij}(a) \in X$, where $a \in J$ if $i < j$ and $j = i - 1$ if $i > j$. Now $T'(R)$ is a subgroup, so $t'^{-1}X = X$; also $hx_{ij}(a)h^{-1} = x_{ij}(b)$, say, where $b \in J$ if $a \in J$, by Propositions 92 and 93. So $t'thx_{ij}(a) \in X$ provided $tx_{ij}(b) \in t'^{-1}Xh^{-1} = X$. Thus we must show that $tx_{ij}(b) \in X$, where $b \in J$ if $i < j$ and $j = i - 1$ if $i > j$.

The subgroup $T(R, J)$ is generated by all elements

$$c_r(\underline{a}) = x_{1r}(a_1)x_{2r}(a_2)\dots x_{r-1\,r}(a_{r-1})$$

where $\underline{a} = (a_1, a_2, \dots, a_{r-1}) \in J^{r-1}$, $r \geq 2$. Now if $r < i$,

$$c_r(\underline{a})x_{i\,i-1}(b) = x_{i\,i-1}(b)c_r(\underline{a})$$

and if $r > i$,

$$\begin{aligned}
&c_r(\underline{a})x_{i\,i-1}(b)\\
&= x_{1r}(a_1)\dots x_{i-1\,r}(a_{i-1})x_{ir}(a_i)\dots x_{r-1\,r}(a_{r-1})x_{i\,i-1}(b)\\
&= x_{i\,i-1}(b)x_{1r}(a_1)\dots x_{i-1\,r}(a_{i-1})x_{ir}(a_i - ba_{i-1})\dots\\
&\dots x_{r-1\,r}(a_{r-1})\\
&= x_{i\,i-1}(b)c_r(\underline{a}'), \text{ say, where } \underline{a}' \in J^{r-1}.
\end{aligned}$$

Then if $r = i$,

$$\begin{aligned}
&c_i(\underline{a})x_{i\,i-1}(b)\\
&= x_{1i}(a_1)\dots x_{i-2\,i}(a_{i-2})x_{i-1\,i}(a_{i-1})x_{i\,i-1}(b)\\
&= x_{1i}(a_1)\dots x_{i-2\,i}(a_{i-2})x_{i\,i-1}(b\alpha^{-1})H_{i-1\,i}(a_{i-1}, b)\\
&\times x_{i-1\,i}(a_{i-1}\beta^{-1}) \text{ (where } \alpha = 1 + a_{i-1}b \text{ and } \beta = 1 + ba_{i-1})\\
&= x_{i\,i-1}(b\alpha^{-1})(x_{1i}(a_1)x_{1\,i-1}(a_1b\alpha^{-1}))\dots\\
&\dots(x_{i-2\,i}(a_{i-2})x_{i-2\,i-1}(a_{i-2}b\alpha^{-1}))x_{i-1\,i}(\alpha a_{i-1})\\
&\times H_{i-1\,i}(a_{i-1}, b)\\
&= x_{i\,i-1}(b\alpha^{-1})c_{i-1}(\underline{a}')c_i(\underline{a}'')H_{i-1\,i}(a_{i-1}, b), \text{ say,}
\end{aligned}$$

where $\underline{a}' \in J^{i-2}$ and $\underline{a}'' \in J^{i-1}$. Thus

$$tx_{i\,i-1}(b) = x_{i\,i-1}(b')t_0$$

for some $b' \in R$ and $t_0 \in T(R, J)H(R, J)$, whence $tx_{i\,i-1}(b) \in X$. In the other case, when $i < j$ and $b \in J$, we have $x_{ij}(b) \in T(R, J)$, so $tx_{ij}(b) \in T(R, J) \subset X$, and we have proved that $\mathrm{ST}(R, J) \subset X$.

Let $u \in \mathrm{ST}(R, J)$; so $u = t'th$ for some $t' \in T'(R)$, $t \in$

$T(R, J)$ and $h \in H(R, J)$. But $T(R, J)$ and $H(R, J)$ are subgroups of $\mathrm{ST}(R, J)$, so $t' = uh^{-1}t^{-1} \in \mathrm{ST}(R, J) \cap T'(R)$. But

$$\mathrm{ST}(R, J) \cap T'(R) = \ker(T'(R) \to T'(R/J)) = T'(R, J)$$

(details are left to the reader), so $t' \in T'(R, J)$, whence $u \in T'(R, J)T(R, J)H(R, J)$, and the proof is complete.

Corollary 100. Let J be a radical ideal of R, and let $f : R \to R/J$ be the natural map. Then $\ker(K_2f) \subset H(R, J)$.

Proof. We have $\ker(K_2f) = K_2R \cap \mathrm{ST}(R, J)$. So if $u \in \ker(K_2f)$, we can write $u = t'th$ for some $t' \in T'(R, J)$, $t \in T(R, J)$, and $h \in H(R, J)$, by Proposition 99. Passing to $\mathrm{E}(R)$, we have $1 = \phi(u) = \phi(t'th)$, whence $\phi(t')^{-1} = \phi(t)\phi(h)$. But $\phi(t')^{-1}$ is lower unitriangular, $\phi(t)$ is upper unitriangular, and $\phi(h)$ is diagonal. So we must have $\phi(t') = \phi(t) = \phi(h) = 1$. Thus $t \in T(R) \cap K_2R = 1$ (Proposition 86), and similarly $t' = 1$, whence $u = h$, and the result is proved.

Corollary 101. Let J be a radical ideal of R, and let $S = R/J$. If $K_2S \subset H(S)$, then $K_2R \subset H(R)$.

Proof. Write $f : R \to S$ for the natural map. We claim that the induced map $\mathrm{ST}(f) : \mathrm{ST}(R) \to \mathrm{ST}(S)$ restricts to an epimorphism $H(f) : H(R) \to H(S)$. For if $x, y \in S$ with $1 + xy \in S^\bullet$, we can find $a, b \in R$ with $f(a) = x$ and $f(b) = y$; thus $f(1 + ab) = 1 + xy \in S^\bullet$, and since $\ker f = J \subset \mathrm{rad}(R)$, it follows that $1 + ab \in R^\bullet$. So $H(f) : H_{ij}(a, b) \mapsto H_{ij}(x, y)$, and we have an epimorphism.

Now let $u \in K_2R$; so $K_2f : u \mapsto \bar{u}$, say, where $\bar{u} \in K_2S \subset H(S)$, and therefore there exists $w \in H(R)$ such that $H(f) : w \mapsto \bar{u}$. Thus $\mathrm{ST}(f)u = \mathrm{ST}(f)w$, or $uw^{-1} \in \ker(\mathrm{ST}(f)) = \mathrm{ST}(R, J)$. By Proposition 99, we can find $t' \in T'(R, J)$, $t \in T(R, J)$, and $h \in H(R, J)$ such that $uw^{-1} = t'th$. So $u = t't(hw)$, and since $H(R, J) \subset H(R)$, we have $hw \in H(R)$. Passing to $\mathrm{E}(R)$, we have $1 = \phi(u) = \phi(t'thw)$, whence $\phi(t')^{-1} = \phi(t)\phi(hw)$. But $\phi(t')^{-1}$ is lower unitriangular, $\phi(t)$ is upper unitriangular, and $\phi(hw)$ is diagonal, so we must have $\phi(t') = \phi(t) = \phi(hw) = 1$. Thus $t \in T(R) \cap K_2R = 1$ (Proposition 86), and similarly $t' = 1$, whence $u = hw \in \mathcal{H}(R)$, and the result is proved.

Corollary 102. Let R be a local ring. Then $K_2R \subset \mathcal{H}(R)$.

Proof. Let $J = \text{rad}(R)$; so $S = R/J$ is a field or skew field. The result follows from Corollary 101 provided we can show $K_2S \subset \mathcal{H}(S)$.

Now $K_2S \subset W(S)$ (Proposition 87), or in other words $K_2S = C(S)$. Next, $C(S) \subset H(S)$ (Corollary 82), and finally $H(S) \subset \mathcal{H}(S)$, since $\langle \alpha - 1, 1 \rangle = 1$, or $h_{ij}(\alpha) = H_{ij}(\alpha - 1, 1)$, any $\alpha \in S^{\bullet}$ (Proposition 96(i)). This completes the proof.

Recall that, for any ring R, we defined $C(R) = W(R) \cap K_2R$. Since $C(R) \subset H(R) \subset W(R)$ (Corollary 82), we have $C(R) = H(R) \cap K_2R$. More generally, if $J \lhd R$, we can define $H(R, J)$ to be the subgroup of $\text{ST}(R)$ generated by all $h_{ij}(\alpha)$ with $\alpha \in (R, J)^{\bullet}$, that is, $\alpha \in R^{\bullet}$ and $\alpha \equiv 1 \bmod J$; we then define $C(R, J) = H(R, J) \cap K_2R$. Note that $H(R) = H(R, R)$ and $C(R) = C(R, R)$. Writing $f : R \to R/J$ for the natural map, we also have $C(R, J) = H(R, J) \cap \ker(K_2f)$. Clearly, $H(R, J) \subset H(R) \cap \text{ST}(R, J)$ and $C(R, J) \subset C(R) \cap \text{ST}(R, J)$, and in general these inclusions may be strict: for example, $K_2(\mathbb{Z}/3\mathbb{Z}) = 1$ (Corollary 89), so working in $\text{ST}(\mathbb{Z})$ we have

$$\{-1, -1\} \in C(\mathbb{Z}) \cap \text{ST}(\mathbb{Z}, 3\mathbb{Z}) \subset H(\mathbb{Z}) \cap \text{ST}(\mathbb{Z}, 3\mathbb{Z}).$$

But $(\mathbb{Z}, 3\mathbb{Z})^{\bullet} = 1$, so $C(\mathbb{Z}, 3\mathbb{Z}) = H(\mathbb{Z}, 3\mathbb{Z}) = 1$, and as we shall see later, $\{-1, -1\} \neq 1$ in $\text{ST}(\mathbb{Z})$.

Exercise: Show that $H(R, J)$ is generated by all $h_{1i}(\alpha)$ with $\alpha \in (R, J)^{\bullet}$, and that when R is commutative, $C(R, J)$ is generated by all symbols $\{\alpha, \beta\}$ with $\alpha, \beta \in (R, J)^{\bullet}$. (Compare Lemma 83 and Proposition 84.)

Now let us define $C(R, J) = H(R, J) \cap K_2R$; note that $C(R, J) = H(R, J) \cap \ker(K_2f)$, where $f : R \to R/J$ is the natural map. We write $C(R) = C(R, R)$. In this terminology, Corollaries 100 and 101 can be restated thus: if J is a radical ideal of R and $f : R \to R/J = S$ is the natural map, then $\ker(K_2f) = C(R, J)$, and if $K_2S = C(S)$ then $K_2R = C(R)$. Corollary 102 can be restated: if R is a local ring then $K_2R = C(R)$.

We now find generators for $C(R, J)$ in the case when R is commutative (compare the above exercise).

Proposition 103. Let R be a commutative ring and let $J \lhd R$. Then $C(R, J)$ is generated by all symbols $\langle a, b\rangle$ with $a, b \in R$, $1 + ab \in R^{\bullet}$, and $a \in J$ or $b \in J$ (so that in fact $1 + ab \in (R, J)^{\bullet}$).

Proof. Let $a, b \in R$ with $1 + ab \in R^{\bullet}$ and $a \in J$ or $b \in J$. Thus $H_{12}(a, b) \in H(R, J)$, and if $\alpha = 1 + ab$, then $h_{12}(\alpha) = H_{12}(\alpha - 1, 1)$ (Proposition 96(i)), and $\alpha - 1 \in J$, so $h_{12}(\alpha) \in H(R, J)$. Thus $\langle a, b\rangle \in H(R, J) \cap K_2R = C(R, J)$. Next, $H_{ij}(a, b) = \langle a, b\rangle h_{ij}(\alpha)$, and so if $w \in H(R, J)$, we may write $w = uh$ where u is a product of symbols $\langle a, b\rangle$ with $a \in J$ or $b \in J$, and $h \in H(R, J)$. If $w \in K_2R$, that is $w \in C(R, J)$, then $h = wu^{-1} \in H(R, J) \cap K_2R = C(R, J)$, and so by the above exercise h can be written as a product of symbols $\{\alpha, \beta\}$ with $\alpha, \beta \in (R, J)^{\bullet}$. But if $\alpha, \beta \in (R, J)^{\bullet}$, then putting $a = (\alpha - 1)\beta^{-1}$, we have $1 + a\beta = \alpha$, $a \in J$ and $\beta \in R^{\bullet}$, whence $\{\alpha, \beta\} = \langle a, \beta\rangle$, by Proposition 96(iii), and the result follows.

Corollary 104. Let J be a radical ideal of R, where R is commutative, and let $f : R \to R/J$ be the natural map. Then $\ker(K_2f)$ is generated by all symbols $\langle a, b\rangle$ ($a, b \in R$, $1 + ab \in R^{\bullet}$, $a \in J$ or $b \in J$).

Proof. Immediate from Proposition 103 and Corollary 100.

Corollary 105. Let J be a radical ideal of R, where R is commutative, and let $S = R/J$. If K_2S is generated by all symbols $\{\alpha, \beta\}$ $(\alpha, \beta \in S^{\bullet})$, or if K_2S is generated by all symbols $\langle a, b\rangle$ $(a, b \in S,\ 1 + ab \in S^{\bullet})$, then K_2R is generated by all symbols $\langle a, b\rangle$ $(a, b \in R,\ 1 + ab \in R^{\bullet})$.

Proof. Since $\{\alpha, \beta\} = \langle(\alpha - 1)\beta^{-1}, \beta\rangle$ (Proposition 96(iii)) we may assume that K_2S is generated by all symbols $\langle a, b\rangle$ $(a, b \in S,\ 1 + ab \in S^{\bullet})$. By Proposition 103, $K_2S = C(S)$, so by Corollary 101, $K_2R = C(R)$. The result now follows by Proposition 103.

Corollary 106. Let R be a commutative local ring. Then K_2R is generated by all symbols $\langle a, b\rangle$ $(a, b \in R,\ 1 + ab \in R^{\bullet})$.

Proof. Immediate from Proposition 103 and Corollary 102, or alternatively from Corollary 105 and Corollary 88.

8.3 *The calculation of* $K_2(\mathbb{Z}/n\mathbb{Z})$ *and* $SK_1(\mathbb{Z}, n\mathbb{Z})$

We shall now apply the above results to the commutative local ring $\mathbb{Z}/p^k\mathbb{Z}$, where p is a prime number,

Proposition 107. Let p be a prime number, and let $k \in \mathbb{Z}$, $k \geq 1$. Let $f : \mathbb{Z}/p^{k+1}\mathbb{Z} \to \mathbb{Z}/p^k\mathbb{Z}$ be the natural map. If $p > 2$ or if $k > 1$, then $\ker(K_2f) = 1$; if $p = 2$ and $k = 1$ then $|\ker(K_2f)| \leq 2$.

Proof. By abuse of notation we shall not distinguish between elements of $\mathbb{Z}$ and their images in $\mathbb{Z}/p^{k+1}\mathbb{Z}$. Now $\ker f = p^k\mathbb{Z}/p^{k+1}\mathbb{Z}$ is a radical ideal of $\mathbb{Z}/p^{k+1}\mathbb{Z}$, so by Corollary 104, $\ker(K_2f)$ is generated by all symbols $\langle a, b\rangle$ with $a \in \ker f$ or $b \in \ker f$; in view of the relation $\langle a, b\rangle = \langle -b, -a\rangle^{-1}$, $\ker(K_2f)$ is generated by all symbols $\langle ap^k, b\rangle$.

We claim $\ker(K_2f)$ is generated by the single symbol $\langle p^k, p^k\rangle$. Firstly, we show $\langle ap^k, bp^k\rangle = \langle p^k, p^k\rangle^{ab}$. For

$$\langle ap^k, bp^k\rangle\langle ap^k, cp^k\rangle = \langle ap^k, bp^k + cp^k + abcp^{3k}\rangle$$

$$= \langle ap^k, (b + c)p^k \rangle$$

whence $\langle ap^k, bp^k \rangle = \langle ap^k, p^k \rangle^b$ (remembering $\langle ap^k, 0 \rangle = 1$). A similar argument shows $\langle -p^k, -ap^k \rangle = \langle -p^k, -p^k \rangle^a$; taking inverses, $\langle ap^k, p^k \rangle = \langle p^k, p^k \rangle^a$, so $\langle ap^k, bp^k \rangle = \langle p^k, p^k \rangle^{ab}$. Next,

$$\langle ap^k, b \rangle \langle ap^k, c \rangle = \langle ap^k, b + c + abcp^k \rangle$$

and

$$\langle ap^k, b + c \rangle \langle ap^k, abcp^k \rangle$$

$$= \langle ap^k, (b + c) + abcp^k + (b + c)a^2bcp^{2k} \rangle$$

$$= \langle ap^k, b + c + abcp^k \rangle.$$

Since $\langle ap^k, abcp^k \rangle = \langle p^k, p^k \rangle^{a^2bc}$, we deduce

$$\langle ap^k, b \rangle \langle ap^k, c \rangle \equiv \langle ap^k, b + c \rangle \pmod{\langle p^k, p^k \rangle}$$

and so

$$\langle ap^k, b \rangle \equiv \langle ap^k, 1 \rangle^b \pmod{\langle p^k, p^k \rangle}.$$

But $\langle ap^k, 1 \rangle = 1$, so $\ker(K_2f)$ is generated by $\langle p^k, p^k \rangle$.

First suppose $k > 1$. Then

$$\langle p^k, (p)(p^{k-1}) \rangle \langle p, (p^k)(p^{k-1}) \rangle \langle p^{k-1}, (p)(p^k) \rangle = 1.$$

But $p^{2k-1} = 0 = p^{k+1}$, so $\langle p^k, p^k \rangle = 1$, and $\ker(K_2f) = 1$.

Now suppose $k = 1$. Then

$$\langle p, p \rangle \langle p, -p \rangle = \langle p, p - p - p^3 \rangle = \langle p, 0 \rangle = 1$$

and

$$\langle p, -p \rangle = \langle -(-p), -p \rangle^{-1} = \langle p, -p \rangle^{-1}$$

so $\langle p, p \rangle = \langle p, -p \rangle$ has order 1 or 2, and $|\ker(K_2f)| \le 2$. Then

$$\langle p, p \rangle^p = \langle p, p^2 \rangle = \langle p, 0 \rangle = 1$$

so if p is odd we deduce $\langle p, p \rangle = 1$, and $\ker(K_2f) = 1$.

Corollary 108. Let p, k, f be as in Proposition 107. Then if $p > 2$ or if $k > 1$, K_2f is an isomorphism.

Proof. First suppose $p > 2$. Then K_2f embeds $K_2(\mathbb{Z}/p^{k+1}\mathbb{Z})$ as a subgroup of $K_2(\mathbb{Z}/p^k\mathbb{Z})$; repeating the argument, after k steps we can embed $K_2(\mathbb{Z}/p^{k+1}\mathbb{Z})$ as a subgroup of $K_2(\mathbb{Z}/p\mathbb{Z})$; but this group is trivial, by Corollary 89, so all the groups $K_2(\mathbb{Z}/p^k\mathbb{Z})$ are trivial, and the result follows.

Now suppose $p = 2$. By a similar argument, we can embed $K_2(\mathbb{Z}/2^k\mathbb{Z})$ as a subgroup of $K_2(\mathbb{Z}/4\mathbb{Z})$. The proof of Proposition 107 (with $p = 2$ and $k = 1$) shows that, since $K_2(\mathbb{Z}/2\mathbb{Z})$ is trivial (Corollary 89), $K_2(\mathbb{Z}/4\mathbb{Z})$ is generated by

$$\langle 2, 2\rangle = \langle 2, -2\rangle = \langle 2, 1\rangle\langle 2, -1\rangle = \langle 2, -1\rangle.$$

But in $\mathbb{Z}$, $1 + 2(-1) = -1$ is a unit, so $\langle 2, -1\rangle$ is defined in $K_2(\mathbb{Z}/2^k\mathbb{Z})$, and under the above embedding it is identified with $\langle 2, -1\rangle$ in $K_2(\mathbb{Z}/4\mathbb{Z})$, and the result follows.

Corollary 109. Let $n \in \mathbb{Z}$, $n > 1$. Then $K_2(\mathbb{Z}/n\mathbb{Z})$ has order 1 or 2, and is generated by $\langle 2, -1\rangle$. If $n \not\equiv 0 \pmod 4$, then $K_2(\mathbb{Z}/n\mathbb{Z}) = 1$.

Proof. We have seen in the proof of Corollary 108 that $K_2(\mathbb{Z}/p^k\mathbb{Z}) = 1$ if p is an odd prime (or if $p = 2$ and $k = 1$), and $K_2(\mathbb{Z}/2^k\mathbb{Z})$ is generated by $\langle 2, -1\rangle$, whose order is at most 2. The Chinese remainder theorem, together with Proposition 76, now shows that if $n = 2^k m$, where m is odd, then the natural map $\mathbb{Z}/n\mathbb{Z} \to \mathbb{Z}/2^k\mathbb{Z}$ induces an isomorphism $K_2(\mathbb{Z}/n\mathbb{Z}) \to K_2(\mathbb{Z}/2^k\mathbb{Z})$, and the result follows.

Corollary 110. For every $n \in \mathbb{Z}$, $n > 1$, the natural map $\mathbb{Z} \to \mathbb{Z}/n\mathbb{Z}$ induces an epimorphism $K_2\mathbb{Z} \to K_2(\mathbb{Z}/n\mathbb{Z})$.

Proof. We have by Corollary 109 that $K_2(\mathbb{Z}/n\mathbb{Z})$ is generated by $\langle 2, -1\rangle$, and this comes from $K_2\mathbb{Z}$.

So at last we are able to prove the theorem of Mennicke-Bass-Lazard-Serre (see pp.181-182, 192):

Corollary 111. For every $n \in \mathbb{Z}$, $n > 1$, $SK_1(\mathbb{Z}, n\mathbb{Z}) = 1$.

Proof. Immediate from Corollary 110 and Corollary 75.

Let R be a field. A *Steinberg symbol* on R is a map c :

$R^{\bullet} \times R^{\bullet} \to G$, where G is an abelian group, satisfying:

$$c(\alpha_1\alpha_2, \beta) = c(\alpha_1, \beta)c(\alpha_2, \beta)$$
$$c(\alpha, \beta_1\beta_2) = c(\alpha, \beta_1)c(\alpha, \beta_2)$$
$$c(\alpha, 1 - \alpha) = 1 \ (\alpha \neq 1).$$

Matsumoto's theorem (Theorem 90) may be restated: given a Steinberg symbol c, as above, there is a unique homomorphism f : $K_2R \to G$ such that $f\{\alpha, \beta\} = c(\alpha, \beta)$ $(\alpha, \beta \in R^{\bullet})$, or in other words, $\{\ ,\ \} : R^{\bullet} \times R^{\bullet} \to K_2R$ is the *universal* Steinberg symbol on R.

The *2-adic Steinberg symbol* on the field $\mathbb{Q}$ of rational numbers is defined as follows. For $\alpha \in \mathbb{Q}^{\bullet}$, write $\alpha = (-1)^i 2^j 5^k n/m$, with $n, m \in 1 + 8\mathbb{Z}$; then i, j, k are determined modulo 2 by α. For $\beta \in \mathbb{Q}^{\bullet}$, write $\beta = (-1)^I 2^J 5^K N/M$, similarly, where $N, M \in 1 + 8\mathbb{Z}$, and put $c(\alpha, \beta) = (-1)^{iI+jK+kJ}$. We claim that this is a Steinberg symbol on $\mathbb{Q}$; indeed, the first two equations (above) are obviously satisfied, and then if $\alpha + \beta = 1$, we have

$$(-1)^i 2^j 5^k + (-1)^I 2^J 5^K \equiv 1 \pmod 8$$

so $2^j + 2^J \equiv 1 \pmod 2$, and $j = 0$, $J > 0$ (or vice versa). Next,

$$(-1)^i + (-1)^I 2^J \equiv 1 \pmod 4$$

so $i = 0$ and $J > 1$, or else $i = 1$ and $J = 1$. In the first case, we then have either $J = 2$, or else $J > 2$ and $k = 0$, and in the second case we have $I = 1$ and $k = 1$; in each case, $iI + jK + kJ$ is even, and so $c(\alpha, 1 - \alpha) = 1$.

Now $c(-1, -1) = -1$, so by Matsumoto's theorem (as restated above), it follows that $\{-1, -1\} \neq 1$ in $K_2\mathbb{Q}$; indeed, if R is any ring with a homomorphic image in $\mathbb{Q}$, it follows that $\{-1, -1\} \neq 1$ in K_2R. Let us take R to be a subring of $\mathbb{Q}$, specifically, take $R = \{n/m \in \mathbb{Q},\ m \text{ odd}\}$. R is a local ring, indeed $R = \mathbb{Z}_{(2)}$, the localization of $\mathbb{Z}$ at the prime ideal $2\mathbb{Z}$. Moreover, $R/8R \simeq \mathbb{Z}/8\mathbb{Z}$. We shall show that $\{-1, -1\} \neq 1$ in $K_2(\mathbb{Z}/8\mathbb{Z})$; to do this it is sufficient to show $\{-1, -1\} \notin \ker(K_2R \to K_2(R/8R))$.

Lemma 112. Let R be a commutative local ring, and let $I \lhd R$. Then $\ker(K_2R \to K_2(R/I))$ is generated by all symbols $\{\alpha, \beta\}$ $(\alpha \in 1 + I,\ \beta \in R^{\bullet})$.

Proof. By Corollary 104, together with the relation $\langle a, b\rangle = \langle -b, -a\rangle^{-1}$, $\ker(K_2R \to K_2(R/I))$ is generated by all symbols $\langle a, b\rangle$ with $a \in I$, $b \in R$. If $b \in R^{\bullet}$, then $\langle a, b\rangle = \{1 + ab, b\}$, by Proposition 96, and if $b \notin R^{\bullet}$, then $a, b \in \mathrm{rad}(R)$, and

$$\langle a, b\rangle\langle a, 1\rangle = \langle a, 1 + b + ba\rangle$$

by Proposition 97; then $\langle a, 1\rangle = 1$, by Proposition 96, and $\langle a, b\rangle = \langle a, c\rangle$, where $c = 1 + b + ba \in 1 + I$, so $\langle a, b\rangle = \{1 + ac, c\}$, by Proposition 96, and the lemma is proved.

Now let $R = \mathbb{Z}_{(2)}$, and let c be the 2-adic Steinberg symbol on $\mathbb{Q}$. Let $\alpha \in 1 + 8R$; thus $\alpha = 1 + 8n/m$, where m is odd, and indeed we may take $m \equiv 1 \pmod 8$, whence $\alpha = (m + 8n)/m$, with $m + 8n$, $m \in 1 + 8\mathbb{Z}$, and therefore $\alpha = (-1)^0 2^0 5^0 (m + 8n)/m$, so that $c(\alpha, \beta) = 1$, all $\beta \in R^{\bullet}$. Lemma 112, together with the fact that $c(-1, -1) = -1$, shows that $\{-1, -1\} \notin \ker(K_2R \to K_2(R/8R))$, and we have proved:

Proposition 113. $\{-1, -1\} \neq 1$ in $K_2(\mathbb{Z}/8\mathbb{Z})$.

Corollary 114. $K_2(\mathbb{Z}/n\mathbb{Z})$ is cyclic of order 2, generated by $\{-1, -1\}$, if $n \equiv 0 \pmod 4$, and is trivial otherwise.

Proof. For $n = 2^m$, $m > 1$, the result follows from Corollary 108 and Corollary 109, noting that $\langle 2, -1\rangle = \{-1, -1\}$. The general case follows from the Chinese remainder theorem and Corollary 109.

Exercise: Prove directly that $\{-1, -1\} \neq 1$ in $K_2(\mathbb{Z}/4\mathbb{Z})$ by showing that $c(\alpha, \beta) = 1$ when $\alpha \in 1 + 4\mathbb{Z}_{(2)}$ and $\beta \in \mathbb{Z}_{(2)}^{\bullet}$.

8.4 *The non-triviality of* $K_2\mathbb{R}$

Here is another Steinberg symbol, this time on $\mathbb{R}$. Define c : $\mathbb{R}^{\cdot} \times \mathbb{R}^{\cdot} \to \{\pm 1\}$ by

$$c(\alpha, \beta) = \begin{cases} 1 & \text{if } \alpha > 0 \text{ or } \beta > 0 \\ -1 & \text{if } \alpha < 0 \text{ and } \beta < 0. \end{cases}$$

Verification that this is a Steinberg symbol is trivial, and it follows that $\{-1, -1\} \neq 1$ in $K_2\mathbb{R}$ (by Matsumoto's theorem); indeed, $\{\alpha, \beta\} \neq 1$ in $K_2\mathbb{R}$ whenever $\alpha < 0$ and $\beta < 0$; moreover, if R is any ring with a homomorphic image in $\mathbb{R}$, then $\{-1, -1\} \neq 1$ in K_2R.

Since we have not given a proof of Matsumoto's theorem, we shall now sketch an independent proof that $\{-1, -1\} \neq 1$ in $K_2\mathbb{R}$. We construct a central extension of $E(\mathbb{R}) = SL(\mathbb{R})$, and invoke Proposition 73.

The central extension is constructed topologically, by means of *covering spaces*. (For definitions and proofs, see [10], Chapter 5.) Given a connected topological space X with sufficiently nice local properties, there is a covering projection p : $\tilde{X} \to X$ with $\tilde{X}$ simply connected, that is, $\pi_1(\tilde{X}) = 1$; $\tilde{X}$ is the *universal cover* of X (*loc. cit.*, Theorem 10.2). If $X = G$ is a topological group, it is easy to make $\tilde{G}$ into a topological group also, such that p is a homomorphism and ker p is a (discrete) central subgroup of $\tilde{G}$ (*loc. cit.*, Exercises 5.1, 5.2). Further, if $x \in \ker p$ and u is a path from 1 to x in $\tilde{G}$, then pu is a loop at 1 in G, and we obtain an isomorphism ker $p \to \pi_1(G)$, the fundamental group, by $x \mapsto [pu]$, the homotopy class of pu. (Any loop u at 1 in G can be lifted to a path $\tilde{u}$ starting at 1 in $\tilde{G}$ (*loc. cit.*, Lemma 3.1), and the inverse isomorphism $\pi_1(G) \to$ ker p is given by $[u] \mapsto \tilde{u}(1)$.) So if $\pi_1(G)$ is non-trivial, we obtain a non-trivial central extension of G.

Now $SL_n(\mathbb{R})$ is a connected Lie group, and its local properties are sufficiently nice (it is a manifold) that the above construction produces a central extension p : $\widetilde{SL}_n(\mathbb{R}) \to SL_n(\mathbb{R})$ with kernel $\pi_1(SL_n(R))$. Recall that the injection $SL_n(R) \hookrightarrow SL_{n+1}(R)$ induces an isomorphism of fundamental groups for $n \geq 3$, with $\pi_1(SL_n(R)) \cong \mathbb{Z}/2\mathbb{Z}$, all $n \geq 3$. (See p.115, and the references given there.) The composite map $\widetilde{SL}_n(R) \to SL_n(R) \hookrightarrow SL_{n+1}(R)$ lifts to give a map $\widetilde{SL}_n(R) \to \widetilde{SL}_{n+1}(R)$ (*loc. cit.*, Theorem 5.1), and this is an injection for $n \geq 3$, giving a commutative diagram

$$\begin{array}{ccccccccc} 0 & \to & \mathbb{Z}/2\mathbb{Z} & \longrightarrow & \widetilde{SL}_n(R) & \longrightarrow & SL_n(R) & \longrightarrow & 1 \\ & & \downarrow 1 & & \downarrow & & \downarrow & & \\ 0 & \to & \mathbb{Z}/2\mathbb{Z} & \to & \widetilde{SL}_{n+1}(R) & \to & SL_{n+1}(R) & \to & 1 \end{array}$$

where the rows are central extensions. We can thus form the group

$$\widetilde{SL}(R) = \cup_{n\geq 3} \widetilde{SL}_n(R)$$

and we have a central extension

$$0 \to \mathbb{Z}/2\mathbb{Z} \to \widetilde{SL}(R) \to SL(R) \to 1.$$

It follows immediately from Proposition 73 that $K_2R \neq 1$; further, there is a commutative diagram

$$\begin{array}{ccc} ST(R) & \xrightarrow{\phi} & SL(R) \\ & \psi \searrow \quad \swarrow p & \\ & \widetilde{SL}(R) & \end{array}$$

To see that $\{-1, -1\} \neq 1$ in K_2R, it is thus sufficient to show that $\psi\{-1, -1\} \neq 1$ in $\ker p \subset \widetilde{SL}(R)$.

Given a generator $x_{ij}(a)$ of $ST(R)$, let $\psi(x_{ij}(a)) = z_{ij}(a) \in \widetilde{SL}(R)$. We have

$$\begin{aligned} \{-1, -1\} &= h_{12}(1)h_{12}(-1)^{-1}h_{12}(-1)^{-1} \\ &= h_{12}(-1)^{-2} \\ &= w_{12}(-1)^{-4} \\ &= (x_{12}(1)x_{21}(-1)x_{12}(1))^4 \end{aligned}$$

so

$$\psi\{-1, -1\} = (z_{12}(1)z_{21}(-1)z_{12}(1))^4.$$

Let u_1, u_2 be paths from 0 to 1 in R (to be chosen shortly).

Define

$$u(t) = (z_{12}(u_1(t))z_{21}(-u_2(t))z_{12}(u_1(t)))^4.$$

This is a path from 1 to $\psi\{-1, -1\}$ in $\widetilde{\mathrm{SL}}_3(\mathbb{R})$, and under the isomorphism $\ker p \to \pi_1(\mathrm{SL}_3(\mathbb{R}))$, $\psi\{-1, -1\}$ maps to the homotopy class of the loop pu. Then

$$pu(t) = (B_{12}(u_1(t))B_{21}(-u_2(t))B_{12}(u_1(t)))^4$$

$$= \begin{pmatrix} 1-u_1(t)u_2(t) & 2u_1(t)-u_1(t)^2u_2(t) \\ -u_2(t) & 1-u_1(t)u_2(t) \end{pmatrix}^4.$$

Now choose $u_1(t) = (1 - \cos \frac{1}{2}\pi t)/\sin \frac{1}{2}\pi t$ $(t \neq 0)$, $u_1(0) = 0$, and $u_2(t) = \sin \frac{1}{2}\pi t$, and we have

$$pu(t) = \begin{pmatrix} \cos \frac{1}{2}\pi t & \sin \frac{1}{2}\pi t \\ -\sin \frac{1}{2}\pi t & \cos \frac{1}{2}\pi t \end{pmatrix}^4 = \begin{pmatrix} \cos 2\pi t & \sin 2\pi t \\ -\sin 2\pi t & \cos 2\pi t \end{pmatrix}.$$

But the homotopy class of this loop generates $\pi_1(\mathrm{SL}_2(\mathbb{R})) \simeq \mathbb{Z}$ and maps to the non-trivial element of $\pi_1(\mathrm{SL}_n(\mathbb{R})) \simeq \mathbb{Z}/2\mathbb{Z}$, all $n \geq 3$ (see p.115), so we deduce $\psi\{-1, -1\} \neq 1$, and hence $\{-1, -1\} \neq 1$ in $K_2\mathbb{R}$.

Further, given any $\alpha, \beta \in \mathbb{R}^\bullet$, there are paths from α to $\alpha/|\alpha|$ and from β to $\beta/|\beta|$ in $\mathbb{R}^\bullet$, giving rise to a path from $\psi\{\alpha, \beta\}$ to $\psi\{\alpha/|\alpha|, \beta/|\beta|\}$ in $\ker p \subset \widetilde{\mathrm{SL}}_3(\mathbb{R})$; but $\ker p$ is discrete, so the path is constant, and we deduce that $\psi\{\alpha, \beta\} = 1$ if $\alpha > 0$ or $\beta > 0$, and $\psi\{\alpha, \beta\} \neq 1$ if $\alpha < 0$ and $\beta < 0$. This is precisely the Steinberg symbol with which we started this section.

There is an alternative way of constructing the above central extension $\widetilde{\mathrm{SL}}(\mathbb{R}) \to \mathrm{SL}(\mathbb{R})$. If $G_1 \subset G_2 \subset \ldots$ are topological groups, each a subgroup and a subspace of the next, then $G = \cup_n G_n$ is a group, and can be given the limit topology by defining $X \subset G$ to be open if $X \cap G_n$ is open in G_n for all n. In general, this does *not* make G into a topological group; the inversion map $G \to G$ is the limit of the inversion maps $G_n \to G_n$, and so is continuous, but to make the multiplication map $G \times G \to G$ continuous, we need the product topology on $G \times G$ to coincide with the limit topology from the subspaces $G_n \times G_n$. This does happen, however, when $G_n = \mathrm{SL}_n(\mathbb{R})$ (essentially because we are working inside $\mathbb{R}^\infty = \cup_n \mathbb{R}^n$), and so we can make $\mathrm{SL}(\mathbb{R})$ into a topological group. Further, since $\mathrm{SL}_n(\mathbb{R})$ is closed in $\mathrm{SL}_{n+1}(\mathbb{R})$, every compact subset of $\mathrm{SL}(\mathbb{R})$ is contained in $\mathrm{SL}_n(\mathbb{R})$, some n, and it quickly follows that

$$\pi_1(\mathrm{SL}(\mathbb{R})) \simeq \varinjlim \pi_1(\mathrm{SL}_n(\mathbb{R})) \simeq \mathbb{Z}/2\mathbb{Z}.$$

This enables us to construct the universal cover $\widetilde{\mathrm{SL}}(\mathbb{R})$ as a topological group, with $\ker(\widetilde{\mathrm{SL}}(\mathbb{R}) \to \mathrm{SL}(\mathbb{R}))$ central, of order 2.

8.5 *Generators for K_2 of a semi-local ring*

We now return to the problem of finding generators for K_2R, this time when R is not necessarily commutative. In particular, we shall describe a class of rings, called H-rings, large enough to include all semi-local rings (Proposition 133, below), and for which it is possible to write down a reasonable set of generators for K_2R (Corollary 136, below).

Definition. R is an *H-ring* if $K_2R \subset H(R)$.

Proposition 115. If R is a field or skew field, then R is an H-ring.

Proof. Immediate from Proposition 87 and Corollary 82.

Proposition 116. Let R be any ring, and let $S = M_n(R)$, the ring of all $n \times n$ matrices over R. Then $\mathrm{ST}(R) \simeq \mathrm{ST}(S)$, $K_2R \simeq K_2S$, and if R is an H-ring, so is S.

Proof. We have $\mathrm{GL}(R) \simeq \mathrm{GL}(S)$, by partitioning the matrices, and so $\mathrm{E}(R) \simeq \mathrm{E}(S)$, by the Whitehead lemma (p.111), and the isomorphisms $\mathrm{ST}(R) \simeq \mathrm{ST}(S)$, $K_2R \simeq K_2S$ now follow from Propositions 72 and 73.

We must describe the above isomorphisms more explicitly. For $A \in S$, $A = (a_{ij})$, we have $A = \Sigma_{i,j}\, E_{ij}(a_{ij})$, where $E_{ij}(a) \in S$ is the matrix with a in the i, j position and 0 elsewhere. Thus in $\mathrm{E}(S)$, $B_{k\ell}(A) = \Pi_{i,j}\, B_{k\ell}(E_{ij}(a_{ij}))$. The isomorphism $\mathrm{E}(S) \to \mathrm{E}(R)$ sends $B_{k\ell}(E_{ij}(a))$ to $B_{kn-n+i\ \ell n-n+j}(a)$; for brevity we write in $\mathrm{E}(R)$,

$$B^{ij}_{k\ell}(a) = B_{kn-n+i\ \ell n-n+j}(a).$$

It is clear that $\mathrm{E}(S)$ is generated by the elements $B_{k\ell}(E_{ij}(a))$, and similarly $\mathrm{ST}(S)$ is generated by the elements $x_{k\ell}(E_{ij}(a))$; also $\mathrm{E}(R)$ is generated by the elements $B^{ij}_{k\ell}(a)$ (with $k \neq \ell$ and $1 \leq i \leq n$, $1 \leq j \leq n$), and it is easy to see that $\mathrm{ST}(R)$ is generated by the elements

$$x^{ij}_{k\ell}(a) = x_{kn-n+i\ \ell n-n+j}(a) \quad (i,\ j,\ k,\ \ell \text{ as before}).$$

Examination of the proof of Proposition 73 shows that the isomorphism $\psi : \mathrm{ST}(S) \to \mathrm{ST}(R)$ is given by

$$\psi(x_{k\ell}(E_{ij}(a))) = x^{ij}_{k\ell}(a).$$

Now $\psi(K_2S) = K_2R$, so to show that if R is an H-ring, so is S, we need only show that $H(R) \subset \psi(H(S))$.

Since $h_{ij}(\alpha) = h_{jk}(\alpha)^{-1}h_{ik}(\alpha)$ (see the proof of Lemma 83), $H(R)$ is generated by all

$$h^{ij}_{k\ell}(\alpha) = h_{kn-n+i\ \ell n-n+j}(\alpha)$$

with $\alpha \in R^{\bullet}$, $k \neq \ell$, $1 \leq i \leq n$, $1 \leq j \leq n$. For any permutation π of $\{1, 2, \ldots, n\}$, $\pi : i \mapsto i'$ $(1 \leq i \leq n)$, and any $\alpha_1, \alpha_2, \ldots, \alpha_n \in R^{\bullet}$, we can construct the monomial matrix $A = \Sigma_i\, E_{ii'}(\alpha_i) \in S^{\bullet}$; then $A^{-1} = \Sigma_i\, E_{i'i}(\alpha_i^{-1})$; also, if $B = \Sigma_i\, E_{ii'}(1)$, then $B^{-1} = \Sigma_i\, E_{i'i}(1)$. If $i \neq j$, each of

$$x^{ii'}_{k\ell}(a),\ x^{i'i}_{\ell k}(b) \text{ commutes with each of } x^{jj'}_{k\ell}(c),\ x^{j'j}_{\ell k}(d)$$

$(a, b, c, d \in R)$, so

$$\Pi_i\, h^{ii'}_{k\ell}(\alpha_i)$$

$$= \Pi_i\, (x^{ii'}_{k\ell}(\alpha_i)x^{i'i}_{\ell k}(-\alpha_i^{-1})x^{ii'}_{k\ell}(\alpha_i)x^{ii'}_{k\ell}(-1)x^{i'i}_{\ell k}(1)x^{ii'}_{k\ell}(-1))$$

$$= \psi(w_{k\ell}(A)w_{k\ell}(-B)) = \psi(h_{k\ell}(A)h_{k\ell}(B)^{-1}) \in \psi(H(S)).$$

Given $h_{k\ell}^{ij}(\alpha)$ ($\alpha \in R^{\bullet}$, $k \neq \ell$, $1 \leq i \leq n$, $1 \leq j \leq n$), choose π with $i' = j$, and put $\alpha_i = \alpha$ and $\alpha_m = 1$ for $m \neq i$; then $h_{k\ell}^{mm'}(\alpha_m) = 1$ for $m \neq i$, and it follows that $h_{k\ell}^{ij}(\alpha) \in \psi(H(S))$, whence $H(R) \subset \psi(H(S))$, as required.

Note that the above inclusion is strict if $n > 1$; for instance, $h_{12}(B_{12}(1)) \in H(S)$, but $\psi(h_{12}(B_{12}(1)))$ is not a monomial element of $\mathrm{ST}(R)$, so it cannot belong to $H(R)$.

Exercise: Define R, S, ψ as in Proposition 116. Show that $\mathcal{H}(R) \subset \psi(\mathcal{H}(S))$.

Proposition 117. If R, S are H-rings, so is $R \times S$.
Proof. If we identify $\mathrm{ST}(R) \oplus \mathrm{ST}(S)$ with $\mathrm{ST}(R \times S)$, as in the proof of Proposition 76, we have $K_2R \oplus K_2S = K_2(R \times S)$, and also $H(R) \oplus H(S) = H(R \times S)$, whence the result.

Corollary 118. Let R be a (left) Artinian ring with $\mathrm{rad}(R) = 0$. Then R is an H-ring.
Proof. From the Wedderburn-Artin theorem (p.27, corollary) we can write $R \cong S_1 \times S_2 \times \ldots \times S_m$, where for each i, $S_i = M_{n_i}(R_i)$ for some $n_i \geq 1$ and some field or skew field R_i. The result is now immediate from Propositions 115, 116, and 117.

By the above result, if R is a semi-local ring and $S = R/\mathrm{rad}(R)$, then S is an H-ring. Since $H(S) \subset \mathcal{H}(S)$, it follows by Corollary 101 that $K_2R \subset \mathcal{H}(R)$. We can strengthen this a little:

Proposition 119. Let R be a ring, and let I be a radical ideal of R such that R/I is an H-ring. Then $K_2R \subset \mathcal{H}(R, I)H(R)$.
Proof. The natural map $R \to R/I$ restricts to an epimorphism $R^{\bullet} \to (R/I)^{\bullet}$ (because I is a radical ideal), and hence induces an epimorphism $H(R) \to H(R/I)$. So if $w \in K_2R$, $w \mapsto \bar{w} \in K_2(R/I) \subset$

$H(R/I)$, and we can find $w_1 \in H(R)$ with $\bar{w}_1 = \bar{w}$. But now $ww_1^{-1} \in$ $\mathrm{ST}(R, I)$, so $ww_1^{-1} = t'th$ for some $t' \in T'(R, I)$, $t \in T(R, I)$, $h \in \mathcal{H}(R, I)$, by Proposition 99, and $w = t't(hw_1)$, where $hw_1 \in$ $\mathcal{H}(R, I)H(R)$. Passing to $\mathrm{E}(R)$, an argument similar to the proof of Corollary 100 shows $t' = t = 1$, so $w = hw_1$, as required.

So to show that a semi-local ring is an H-ring, it would suffice to show $\mathcal{H}(R, \mathrm{rad}(R)) \subset H(R)$; but this is false, in general:

Exercise: Let R be the ring of all 2×2 upper triangular matrices over $\mathbb{Z}/2\mathbb{Z}$. Show that R is semi-local, and that if

$$a = \begin{pmatrix} 1 & 0 \\ 0 & 0 \end{pmatrix}, \quad b = \begin{pmatrix} 0 & 1 \\ 0 & 0 \end{pmatrix}$$

then $H_{12}(a, b) \notin H(R)$. (*Hint:* $R^{\bullet}$ is abelian.)

Our approach must therefore be less direct. We need some control on $\ker(K_2R \to K_2(R/\mathrm{rad}(R)))$, and, recalling Proposition 99, we make the following definition:

Definition: Let R be any ring, and $I \lhd R$. Then I is an *H-ideal* of R if $\mathrm{ST}(R, I) = T'(R, I)T(R, I)H(R, I)$.

Proposition 120. Let I be a radical ideal of R. Then I is an H-ideal if and only if $H(R, I) = \mathcal{H}(R, I)$.

Proof. Proposition 99 says $\mathrm{ST}(R, I) = T'(R, I)T(R, I)\mathcal{H}(R, I)$ for I radical, and if $H(R, I) = \mathcal{H}(R, I)$, it is immediate that I is an H-ideal. The proof of the converse is an exercise for the reader.

Proposition 121. Let I be an H-ideal of R. Then

$$\ker(K_2R \to K_2(R/I)) \subset H(R, I).$$

Proof. The proof is similar to the proof of Corollary 100, and details are left to the reader.

When $I = R$, the conclusion of Proposition 121 says $K_2R \subset H(R)$, so one is tempted to say that an H-ring is a ring which is an H-ideal of itself. This is false, in general:

Exercise: Let R be a non-zero ring. Show that $x_{12}(1)x_{21}(-1) \notin T'(R)T(R)H(R)$, and deduce that every H-ideal of R is a proper ideal.

Proposition 122. Let I be a radical H-ideal of R. Then if R/I is an H-ring, so is R.

Proof. Immediate from Propositions 119 and 120.

Let R be a semi-local ring, with $J = \text{rad}(R)$. By Corollary 118, R/J is an H-ring, so by Proposition 122, R will be an H-ring if J is an H-ideal, that is, if $H(R, J) = \mathcal{H}(R, J)$, by Proposition 120. Before seeing whether this is true, we prove some technical results.

Proposition 123. Let R be a ring and let $I \lhd R$. Then $H(R, I)$ is a normal subgroup of $\mathcal{H}(R, I)$.

Proof. Recall that $H(R, I)$ is generated by all $h_{ij}(\alpha)$, $\alpha \in (R, I)^{\bullet} = R^{\bullet} \cap (1 + I)$, and $\mathcal{H}(R, I)$ is generated by all $H_{ij}(a, b)$ with $1 + ab \in R^{\bullet}$ and $a \in I$ or $b \in I$. First note that $h_{ij}(\alpha) = H_{ij}(\alpha - 1, 1)$, by Proposition 96(i), whence $H(R, I) \subset \mathcal{H}(R, I)$. Then, by Propositions 92 and 93,

$$\begin{aligned}
&H_{ij}(a, b)h_{k\ell}(\gamma)H_{ij}(a, b)^{-1} \\
&= H_{ij}(a, b)x_{k\ell}(\gamma)x_{\ell k}(-\gamma^{-1})x_{k\ell}(\gamma)x_{k\ell}(-1)x_{\ell k}(1)x_{k\ell}(-1) \\
&\times H_{ij}(a, b)^{-1} \\
&= x_{k\ell}(\alpha_k\gamma\alpha_\ell^{-1})x_{\ell k}(-\alpha_\ell\gamma\alpha_k^{-1})x_{k\ell}(\alpha_k\gamma\alpha_\ell^{-1})x_{k\ell}(-\alpha_k\alpha_\ell^{-1}) \\
&\times x_{\ell k}(\alpha_\ell\alpha_k^{-1})x_{k\ell}(-\alpha_k\alpha_\ell^{-1}) \\
&= h_{k\ell}(\alpha_k\gamma\alpha_\ell^{-1})h_{k\ell}(\alpha_k\alpha_\ell^{-1})^{-1}
\end{aligned}$$

where $\alpha_i = 1 + ab$, $\alpha_j = 1 + ba$, and $\alpha_s = 1$ if $s \neq i, j$; so if $a \in I$ or $b \in I$, then $\alpha_k, \alpha_\ell \in (R, I)^\bullet$, and the result follows.

We write $\widetilde{\mathcal{H}}(R, I) = \mathcal{H}(R, I)/H(R, I)$; so $\widetilde{\mathcal{H}}(R, I)$ is generated by the images of the $H_{ij}(a, b)$, $1 + ab \in R^\bullet$, $a \in I$ or $b \in I$. We also write $\widetilde{\mathcal{H}}(R) = \widetilde{\mathcal{H}}(R, R) = \mathcal{H}(R)/H(R)$.

Corollary 124. Let I be a radical ideal of R. Then I is an H-ideal of R if and only if $\widetilde{\mathcal{H}}(R, I) = 1$.

Proof. This is just a restatement of Proposition 120.

Proposition 125. Let R be a ring, and let $I \lhd R$. Then $\widetilde{\mathcal{H}}(R, I)$ is abelian.

Proof. From Proposition 94(iii),

$$H_{ij}(a, b)^{-1} H_{jk}(-b, -a)^{-1} H_{ki}(-1, -ab)^{-1} = 1$$

and $H_{ki}(-1, -ab) \in H(R, I)$ by Proposition 96(ii), where $1 + ab \in R^\bullet$ and $a \in I$ or $b \in I$. So

$$\begin{aligned} H_{ij}(a, b) &\equiv H_{jk}(-b, -a)^{1} \pmod{H(R, I)} \\ &\equiv H_{k\ell}(a, b) \pmod{H(R, I)}, \text{ similarly.} \end{aligned}$$

Since $H_{ij}(a, b)$ and $H_{k\ell}(c, d)$ commute when i, j, k, ℓ are distinct, the result follows.

From the above calculation, we can write (unambiguously) $H(a, b)$ for the image of $H_{ij}(a, b)$ in $\widetilde{\mathcal{H}}(R, I)$.

Proposition 126. $\widetilde{\mathcal{H}}(R, I)$ is generated by the (commuting) elements $H(a, b)$ ($1 + ab \in R^\bullet$, $a \in I$ or $b \in I$), and these satisfy the following relations:

(i) $H(a, b) = H(-b, -a)^{-1}$

(ii) $H(a, b)H(a, c) = H(a, b + c + bac)$

(iii) $H(a, bc)H(b, ca)H(c, ab) = 1$

(iv) $H(a, b) = 1$ if $a = 0$ or $b = 0$ or $-a \in (R, I)^\bullet$ or

$b \in (R, I)^{\bullet}$

(v) $H(a, b) = H(a\gamma^{-1}, \gamma b)$, any $\gamma \in (R, I)^{\bullet}$

(vi) $H(a, b) = 1$ if $a, b \in \text{rad}(R) \cap I$

(vii) $\begin{cases} H(a, b)H(a, c) = H(a, b + c) \text{ if } a \in \text{rad}(R) \cap I \\ H(a, c)H(b, c) = H(a + b, c) \text{ if } c \in \text{rad}(R) \cap I. \end{cases}$

Note: in each of the above, the elements must be defined; for instance, in (ii), $a \in I$ or both $b, c \in I$; in (iii), $a \in I$ or $b \in I$ or $c \in I$, or just $bc, ca, ab \in I$ (and of course $1 + abc \in R^{\bullet}$). We are *not* claiming that (i)-(vii) are a set of defining relations for $\tilde{H}(R, I)$.

Proof. We have

$$\begin{aligned} & h_{jk}(\gamma)H_{ij}(a, b)h_{jk}(\gamma)^{-1} \\ &= h_{jk}(\gamma)x_{ji}(-b\alpha^{-1})x_{ij}(a)x_{ji}(b)x_{ij}(-a\beta^{-1})h_{jk}(\gamma)^{-1} \\ &= x_{ji}(-\gamma b\alpha^{-1})x_{ij}(a\gamma^{-1})x_{ji}(\gamma b)x_{ij}(-a\beta^{-1}\gamma^{-1}) \\ &= H_{ij}(a\gamma^{-1}, \gamma b), \text{ where } \alpha = 1 + ab,\ \beta = 1 + ba. \end{aligned}$$

So (v) follows, and now (i)-(iii) are immediate from (v) and Proposition 94, together with Proposition 125. Then (v) reduces (iv) to the case where $a = 0$ or -1 or $b = 0$ or 1, and these cases follow from Proposition 96. So we have proved (i)-(v).

(vi) Let $a, b \in \text{rad}(R) \cap I$. Then

$$H(a, b) = H(a, b)H(a, 1) = H(a, 1 + b + ba) = 1$$

by (ii) and (iv). Note that we need $a \in I$ and $1 + a \in R^{\bullet}$ (and of course $1 + ab \in R^{\bullet}$) so that the terms are defined, that is, we need $1 + a \in (R, I)^{\bullet}$: for this it is sufficient that $a \in \text{rad}(R) \cap I$. We also need $1 + b + ba \in (R, I)^{\bullet}$, and for this it is sufficient that $b \in \text{rad}(R) \cap I$.

(vii) Let $a \in \text{rad}(R) \cap I$. First suppose $b + c \in \text{rad}(R) \cap I$. Then $H(a, b)H(a, c) = H(a, b + c + bac) = 1$, by (ii) and (vi). Now suppose $b - c \in \text{rad}(R) \cap I$; then $H(a, b)H(a, -c) = 1 = H(a, c)H(a, -c)$, so $H(a, b) = H(a, c)$. Finally, for any b, c,

$$H(a, b)H(a, c) = H(a, b + c + bac) = H(a, b + c)$$

since $bac \in \text{rad}(R) \cap I$. The second equation follows, by (i).

Proposition 127. Let R be a ring that is generated additively by its units (that is, a ring such that every element can be written as a sum of units). Let $J = \mathrm{rad}(R)$. Then if R/J is an H-ring, so is R.

Proof. We work in $\tilde{H}(R)$. Given $H(a, b) \in \tilde{H}(R)$, with $a \in J$, write $b = \Sigma_i \beta_i$, where $\beta_i \in R^\bullet$, all i. Then

$$\begin{aligned} H(a, b) &= H(a, \Sigma_i \beta_i) \\ &= \Pi_i H(a, \beta_i) \text{ by Proposition 126(vii)} \\ &= 1 \text{ by Proposition 126(iv)} \end{aligned}$$

where, in the notation of Proposition 126, we are taking $I = R$. Then $H(a, b) = 1$ if $b \in J$, by the above argument together with Proposition 126(i). So we have shown that the image of $H(R, J)$ in $\tilde{H}(R) = H(R)/H(R)$, induced by the inclusion $H(R, J) \subset H(R)$, is trivial, or in other words, $H(R, J) \subset H(R)$; so $H(R, J)H(R) = H(R)$, whence $K_2R \subset H(R)$, by Proposition 119.

Note that the above proof does *not* show that J is an H-ideal of R.

Corollary 128. Let R be a semi-local ring which is generated additively by its units. Then R is an H-ring.

Proof. Immediate, by Proposition 127 and Corollary 118.

Now not every semi-local ring is generated additively by its units: for example, $\mathbb{Z}/2\mathbb{Z} \times \mathbb{Z}/2\mathbb{Z}$, though this is still an H-ring, by Propositions 115 and 117.

Proposition 129. Let $R = M_n(S)$, where $n \geq 1$ and S is a field or skew field. Then R is generated additively by its units, and except in the case $n = 1$, $|S| = 2$, every element of R is the sum of exactly two units.

Proof. The result is clear when $n = 1$, since then $R^\bullet = R - \{0\}$. In fact, if $|S| > 2$, we can find $\alpha, \beta \in S^\bullet$ with $\alpha + \beta \in S^\bullet$, and it is clear that in this case each element of S can be written as the sum of exactly two units (noting that $0 = -1 + 1$). So let $n \geq 1$, $|S| > 2$, and let $A \in R$. By writing each diagonal entry of A as the sum of two units, we can write A as the sum of an upper triangular and a lower triangular unit of R.

It remains to consider the case $S = \mathbb{Z}/2\mathbb{Z}$. If $n = 2$, then each element of R can be written as the sum of exactly two units; for if $A \in R$, we have $A = BDC$, where B, C are units, and

$$D = 0,\ 1,\ \text{or} \begin{pmatrix} 1 & 0 \\ 0 & 0 \end{pmatrix}.$$

But $0 = 1 + 1$,

$$1 = \begin{pmatrix} 1 & 1 \\ 1 & 0 \end{pmatrix} + \begin{pmatrix} 0 & 1 \\ 1 & 1 \end{pmatrix}, \text{ and } \begin{pmatrix} 1 & 0 \\ 0 & 0 \end{pmatrix} = \begin{pmatrix} 1 & 1 \\ 1 & 0 \end{pmatrix} + \begin{pmatrix} 0 & 1 \\ 1 & 0 \end{pmatrix}.$$

Now suppose $n \geq 3$; for $A \in R$ we can write $A = BDC$, where B, C are units, and $D = \text{diag}(\alpha_1, \alpha_2, \ldots, \alpha_n)$, $\alpha_i = 1$, $i \leq j$, and $\alpha_i = 0$, $i > j$, some $j \leq n$. If A is not a unit, then $j < n$, that is $\alpha_n = 0$, and then each diagonal block

$$\begin{pmatrix} \alpha_i & 0 \\ 0 & \alpha_{i+1} \end{pmatrix} \quad (i \text{ odd})$$

can be written as the sum of two units in $M_2(S)$, as above, and if n is odd, we can write $\alpha_n = 1 + 1$; thus D, and hence A, is either a unit or is the sum of two units. The equations

$$\begin{pmatrix} 1 & 0 \\ 0 & 1 \end{pmatrix} = \begin{pmatrix} 1 & 1 \\ 1 & 0 \end{pmatrix} + \begin{pmatrix} 0 & 1 \\ 1 & 1 \end{pmatrix}$$

and

$$\begin{pmatrix} 1 & 0 & 0 \\ 0 & 1 & 0 \\ 0 & 0 & 1 \end{pmatrix} = \begin{pmatrix} 1 & 1 & 0 \\ 1 & 1 & 1 \\ 0 & 1 & 0 \end{pmatrix} + \begin{pmatrix} 0 & 1 & 0 \\ 1 & 0 & 1 \\ 0 & 1 & 1 \end{pmatrix}$$

allow us to write 1, and hence any unit, as the sum of two units in R, and the proof is complete.

Exercise: Show that the semi-local ring R is *not* generated additively by its units if and only if $R/J \simeq \mathbb{Z}/2\mathbb{Z} \times \mathbb{Z}/2\mathbb{Z} \times S$ for some ring S, where $J = \mathrm{rad}(R)$.

Our subsequent calculations will be greatly eased if we concentrate on rings R with $(\mathrm{rad}(R))^2 = 0$; the next proposition shows that this is a harmless restriction.

Proposition 130. Let R be any ring, and let J be a radical ideal of R. Then $H(R, J^2) \subset H(R)$, and if R/J^2 is an H-ring, so is R.

Proof. We work in $\tilde{H}(R)$. From Proposition 126(iii), (vi), with $I = R$, we have $H(a, bc) = 1$ for any $a \in R$, $b, c \in J$. Let $x \in J^2$, $x = b_1c_1 + \dots + b_nc_n$, where $b_i, c_i \in J$, all i. We have $H(a, x) = 1$ if $n = 1$. Then if $b, c \in J$,

$$H(a, x)H(a, (1 + xa)^{-1}bc) = H(a, x + bc)$$

by Proposition 126(ii), since

$$x + (1 + xa)^{-1}bc + xa(1 + xa)^{-1}bc = x + bc.$$

But $(1 + xa)^{-1}b \in J$ and $c \in J$, so $H(a, (1 + xa)^{-1}bc) = 1$ and $H(a, x) = H(a, x + bc)$. It follows by induction on n that $H(a, x) = 1$, all $x \in J^2$, and hence that $H(R, J^2) \subset H(R)$.

If R/J^2 is an H-ring, then $K_2R \subset H(R, J^2)H(R)$ by Proposition 119, and so $K_2R \subset H(R)$, as required.

Now let R be a ring containing orthogonal idempotents e_1, e_2 with $1 = e_1 + e_2$, and put $R_{ij} = e_iRe_j$, each i, j; it is easy to see that $R = R_{11} \oplus R_{12} \oplus R_{21} \oplus R_{22}$ as additive groups. If $r \in R$, then $r = r_{11} + r_{12} + r_{21} + r_{22}$, where $r_{ij} = e_ire_j \in R_{ij}$, each i, j. Put $S = R_{11} \oplus R_{22}$ and $I = R_{12} \oplus R_{21}$. Then S is a subring of R; R_{11}, R_{22} are rings, with identity elements e_1, e_2 respectively, and $S \simeq R_{11} \times R_{22}$ as rings. Further, $I \lhd R$ and $R/I \simeq S$.

Lemma 131. Let R, e_1, e_2, I be as above. If $I^2 = 0$, then every element of $\widetilde{H}(R, I)$ can be written in the form $H(e_1, s)H(e_2, s')$ for some $s, s' \in I$.

Proof. Since $I^2 = 0$, $1 + I \subset R^{\bullet}$, so I is a radical ideal, by Proposition 11. We work in $\widetilde{H}(R, I)$. As above, for any $r \in R$, $r = r_{11} + r_{12} + r_{21} + r_{22}$. For $s \in I$, we have $s = s_{12} + s_{21}$. Then by Proposition 126(vii),

$$H(r, s) = H(r_{11}, s_{12})H(r_{11}, s_{21})H(r_{22}, s_{12})H(r_{22}, s_{21})$$

the other terms vanishing by Proposition 126(vi). Now $s_{12} = e_1 s_{12} = s_{12} e_2$, and $s_{21} = e_2 s_{21} = s_{21} e_1$; by Proposition 126(iii),

$$H(r_{11}, s_{12}e_2)H(s_{12}, e_2 r_{11})H(e_2, r_{11}s_{12}) = 1$$

$$H(r_{11}, e_2 s_{21})H(e_2, s_{21}r_{11})H(s_{21}, r_{11}e_2) = 1$$

$$H(r_{22}, e_1 s_{12})H(e_1, s_{12}r_{22})H(s_{12}, r_{22}e_1) = 1$$

and

$$H(r_{22}, s_{21}e_1)H(s_{21}, e_1 r_{22})H(e_1, r_{22}s_{21}) = 1.$$

But $e_2 r_{11} = r_{11} e_2 = r_{22} e_1 = e_1 r_{22} = 0$, so by Proposition 126 (iv), (i), $\widetilde{H}(R, I)$ is generated by elements of the form $H(e_i, s)$ ($i = 1, 2$, and $s \in I$). By Proposition 126(iv), (vii), for $s \in I$, we have

$$H(-e_1, s)H(-e_2, s) = H(-1, s) = 1$$

and

$$H(e_i, s)H(-e_i, s) = H(0, s) = 1$$

so

$$H(e_1, s)H(e_2, s) = 1.$$

Also, for $s, s' \in I$,

$$H(e_i, s)H(e_i, s') = H(e_i, s + s' + s e_i s')$$

by Proposition 126(ii), so any element of $\widetilde{H}(R, I)$ can be written in the form $H(e_1, s)H(e_2, s')$, for some $s, s' \in I$, as required.

Proposition 132. Let R be a semi-local ring, and let $J = \mathrm{rad}(R)$. If $R/J \simeq \mathbb{Z}/2\mathbb{Z} \times \ldots \times \mathbb{Z}/2\mathbb{Z}$ (n copies), then R is an H-ring.

Proof. Assume that $J^2 = 0$. We use induction on n: if $n = 1$, then R is a local ring, and every element of R is a unit or a sum of two units. The result then follows by Corollary 128.

So now let $n > 1$, and let $R/J \simeq S \times T$, where each of S, T is a product of fewer than n copies of $\mathbb{Z}/2\mathbb{Z}$. Let a be any lifting of $(1, 0) \in S \times T$ to R; note that $1 - a$ lifts $(0, 1)$, and so $a(1 - a) \in J$, and $a^2(1 - a)^2 = 0$, since $J^2 = 0$. Put $e_1 = a^3 + 3a^2(1 - a)$ and $e_2 = 1 - e_1$. Then $3a^2(1 - a) \in J$ and $a^3 \equiv a^2 \equiv a \pmod J$, so e_1 lifts $(1, 0)$ and e_2 lifts $(0, 1)$. Further, $e_2 = 3a(1 - a)^2 + (1 - a)^3$, so $e_1e_2 = 0 = e_2e_1$, and $e_1{}^2 = e_1$, $e_2{}^2 = e_2$. So we are in the situation of Lemma 131; if $R_{ij} = e_iRe_j$ and $I = R_{12} + R_{21}$, as before, then $I \subset J$ (because $(1, 0)$, $(0, 1)$ are *central* orthogonal idempotents in $S \times T$), so $I^2 = 0$. Also the natural map $R \to S \times T$ restricts to epimorphisms $R_{11} \to S$, $R_{22} \to T$, with kernels $\mathrm{rad}(R_{11})$, $\mathrm{rad}(R_{22})$ respectively. By the inductive hypothesis, R_{11} and R_{22} are H-rings, and hence so is $R/I \simeq R_{11} \times R_{22}$, by Proposition 117.

Let $w \in K_2R$. By Proposition 119, $K_2R \subset \tilde{H}(R, I)H(R)$, so by Lemma 131 we can find $s, s' \in I$ and $h \in H(R)$ with

$$w = H_{12}(e_1, s)H_{12}(e_2, s')h.$$

Passing to $\mathrm{E}(R)$, we have

$$\phi(H_{12}(e_1, s)H_{12}(e_2, s')) = \phi(h)^{-1}.$$

But $(R/J)^{\bullet} = 1$, so $R^{\bullet} = 1 + J$, which is abelian, since $J^2 = 0$. So $\phi(h)^{-1}$ is a diagonal matrix whose diagonal entries have product (in any order) equal to 1. So the same is true for $\phi(H_{12}(e_1, s)H_{12}(e_2, s'))$, whence

$$(1 + e_1s)(1 + e_2s') = (1 + se_1)(1 + s'e_2).$$

Since $I^2 = 0$, we obtain $e_1s + e_2s' = se_1 + s'e_2$, or $s_{12} + s'_{21} = s_{21} + s'_{12}$. Thus $s_{12} = s'_{12}$ and $s_{21} = s'_{21}$, whence $s = s'$. But now in $\tilde{H}(R, I)$, $H(e_1, s)H(e_2, s) = 1$, as in the proof of Lemma 131, and so $H_{12}(e_1, s)H_{12}(e_2, s') \in H(R, I) \subset H(R)$, and $w \in H(R)$. Thus R is an H-ring.

In the case where $J^2 \neq 0$, put $\bar{R} = R/J$, $\bar{J} = J/J^2$, so that $R/J \simeq \bar{R}/\bar{J}$, $\bar{J} = \mathrm{rad}(\bar{R})$, and $\bar{J}^2 = 0$. By the above argument, $\bar{R}$ is an H-ring, and so R is an H-ring by Proposition 130.

Proposition 133. Every semi-local ring is an H-ring.

Proof. Let R be semi-local, with $J = \text{rad}(R)$. By the Wedderburn-Artin theorem, $R/J \simeq S_1 \times S_2 \times \ldots \times S_n$, where each S_i is a full matrix ring over a field or skew field.

Case (i). $|S_i| > 2$, all i. Then every element of S_i can be written as a sum of two units (Proposition 129), hence so can every element of R/J, and hence so can every element of R. The result follows by Corollary 128.

Case (ii). $|S_i| = 2$, all i. The result follows by Proposition 132.

Case (iii). $|S_i| > 2$ for $i \le m$ and $|S_i| = 2$ for $m < i \le n$, where $1 \le m < n$. Put $S = S_1 \times \ldots \times S_m$ and $T = S_{m+1} \times \ldots \times S_n$, so that $R/J \simeq S \times T$. Next, assume that $J^2 = 0$. Then exactly as in the proof of Proposition 132, we can lift the elements $(1, 0), (0, 1) \in S \times T$ to orthogonal idempotents $e_1, e_2 \in R$. With notation as for Lemma 131, we have that R_{11} is an H-ring by case (i), and R_{22} is an H-ring by case (ii). So $R/I \simeq R_{11} \times R_{22}$ is an H-ring by Proposition 117.

Let $w \in K_2R$. By Proposition 119, $K_2R \subset H(R, I)H(R)$, so by Lemma 131 we can find $s, s' \in I$ and $h \in H(R)$ with

$$w = H_{12}(e_1, s)H_{12}(e\ , s')h.$$

Now as in case (i), we can write $e_1 = \alpha_1 + \beta_1$ for some $\alpha_1, \beta_1 \in R_{11}^{\bullet}$, so $e_1 = \alpha + \beta$ where $\alpha = \alpha_1 + e_2 \in R^{\bullet}$, $\beta = \beta_1 - e_2 \in R^{\bullet}$. Working in $\tilde{H}(R)$, $H(e_1, s) = H(\alpha, s)H(\beta, s) = 1$, by Proposition 126, and similarly $H(e_1, s') = 1$; then $H(e_1, s')H(e_2, s') = H(1, s') = 1$, by Proposition 126, and so $H(e_2, s') = 1$. Thus $H_{12}(e_1, s)H_{12}(e_2, s') \in H(R)$, and $w \in H(R)$, so R is an H-ring.

Finally, if $J^2 \neq 0$, then the above argument shows that R/J^2 is an H-ring, so R is an H-ring by Proposition 130.

Corollary 134. If R is a commutative semi-local ring (or, more generally, if R is semi-local and $R^{\bullet}$ is abelian), then K_2R is generated by all symbols $\{\alpha, \beta\}$ $(\alpha, \beta \in R^{\bullet})$.

Proof. Immediate from Propositions 133 and 84.

We shall now show one way to generalize Proposition 84: we shall find a set of generators for K_2R, where R is any H-ring, not necessarily commutative, in such a way that in the commutative case we just get the symbols $\{\alpha, \beta\}$ $(\alpha, \beta \in R^{\bullet})$.

Let R be any ring. For $\alpha, \beta \in R^{\bullet}$, write $c_{ijk}(\alpha, \beta) = [h_{ij}(\alpha), h_{ik}(\beta)]$; so $c_{ijk}(\alpha, \beta) = c_{ikj}(\beta, \alpha)^{-1}$. But

$$c_{ijk}(\alpha, \beta) = h_{ik}(\alpha\beta)h_{ik}(\alpha)^{-1}h_{ik}(\beta)^{-1}$$

by Corollary 78, whence $c_{ijk}(\alpha, \beta)$ is independent of the choice of k, and hence is independent of the choice of j also, and we may write unambiguously

$$c_i(\alpha, \beta) = [h_{ij}(\alpha), h_{ik}(\beta)] = h_{ik}(\alpha\beta)h_{ik}(\alpha)^{-1}h_{ik}(\beta)^{-1}.$$

Now $w_{1i}(-1)h_{1k}(\gamma)w_{1i}(-1)^{-1} = h_{ik}(\gamma)$, by Proposition 77(iii), remembering $h_{ik}(1) = 1$, and so

$$w_{1i}(-1)c_1(\alpha, \beta)w_{1i}(-1)^{-1} = c_i(\alpha, \beta).$$

Let $a_i(\gamma) \in \mathrm{GL}(R)$ be the diagonal matrix with γ in the ith diagonal position and all other entries on the diagonal equal to 1. Then $\phi(c_i(\alpha, \beta)) = a_i([\alpha, \beta]) \in \mathrm{E}(R)$, and since $a_i(\gamma)a_i(\delta) = a_i(\gamma\delta)$ $(\gamma, \delta \in R^{\bullet})$, we deduce $a_i(\gamma) \in \mathrm{E}(R)$, all $\gamma \in R^{\bullet\prime}$ (the derived subgroup of $R^{\bullet}$). So for each $\gamma \in R^{\bullet\prime}$, we can choose $b_1(\gamma) \in \mathrm{ST}(R)$ with $\phi(b_1(\gamma)) = a_1(\gamma)$; in particular, we insist $b_1(1) = 1$, and we also insist $b_1(\gamma) \in H(R)$, all $\gamma \in R^{\bullet\prime}$, which is possible since $c_1(\alpha, \beta) \in H(R)$, all α, β. We then define

$$b_i(\gamma) = w_{1i}(-1)b_1(\gamma)w_{1i}(-1)^{-1} \quad (i \neq 1, \gamma \in R^{\bullet\prime})$$

so that $\phi(b_i(\gamma)) = a_i(\gamma)$.

Since $a_i(\gamma)a_i(\delta)a_i(\gamma\delta)^{-1} = 1$, all i, all $\gamma, \delta \in R^{\bullet\prime}$, we may define

$$d(\gamma, \delta) = b_1(\gamma)b_1(\delta)b_1(\gamma\delta)^{-1} \in K_2R$$

and conjugating this by $w_{1i}(-1)$, $i \neq 1$, shows

$$d(\gamma, \delta) = b_i(\gamma)b_i(\delta)b_i(\gamma\delta)^{-1} \quad (\text{any } i, \text{ and } \gamma, \delta \in R^{\bullet\prime}).$$

Note that $d(\gamma, \delta) = 1$ if $\gamma = 1$ or $\delta = 1$, since $b_i(1) = 1$, and so the elements $d(\gamma, \delta)$ are trivial if $R^{\bullet}$ is abelian.

Next, $\phi(c_1(\alpha, \beta)) = a_1([\alpha, \beta]) = \phi(b_1([\alpha, \beta]))$, so we may define

$$e(\alpha, \beta) = c_1(\alpha, \beta)b_1([\alpha, \beta])^{-1} \in K_2R$$

and conjugating this by $w_{1i}(-1)$, $i \neq 1$, shows

$$e(\alpha, \beta) = c_i(\alpha, \beta)b_i([\alpha, \beta])^{-1} \quad (\text{any } i, \text{ and } \alpha, \beta \in R^{\bullet}).$$

Note that if $[\alpha, \beta] = 1$, then $e(\alpha, \beta) = \{\alpha, \beta\}$. Recall that $C(R) = H(R) \cap K_2R$ (Corollary 82); we now generalize Proposition 84 to the non-commutative case:

Proposition 135. Let R be any ring. Then $C(R)$ is generated by all $e(\alpha, \beta)$, $d(\gamma, \delta)$ ($\alpha, \beta \in R^{\bullet}$, $\gamma, \delta \in R^{\bullet\prime}$).

Proof. Let A be the subgroup of K_2R generated by all $e(\alpha, \beta)$, $d(\gamma, \delta)$ ($\alpha, \beta \in R^{\bullet}$, $\gamma, \delta \in R^{\bullet\prime}$); so $A \subset C(R)$. Since $K_2R = \zeta(\mathrm{ST}(R))$, A is a normal subgroup of $\mathrm{ST}(R)$, and we may write

$$b_i(\gamma)b_i(\delta) \equiv b_i(\gamma\delta) \pmod{A}$$

and

$$b_i([\alpha, \beta]) \equiv c_i(\alpha, \beta) \pmod{A}.$$

Now

$$h_{1i}(\lambda)c_1(\alpha, \beta)h_{1i}(\lambda)^{-1}$$
$$= h_{1i}(\lambda)h_{1k}(\alpha\beta)h_{1k}(\alpha)^{-1}h_{1k}(\beta)^{-1}h_{1i}(\lambda)^{-1}$$
$$= h_{1k}(\lambda\alpha\beta)h_{1k}(\lambda)^{-1}h_{1k}(\lambda)h_{1k}(\lambda\alpha)^{-1}h_{1k}(\lambda)h_{1k}(\lambda\beta)^{-1}$$

(by Proposition 77(iii))

$$= h_{1k}(\lambda\alpha\beta)h_{1k}(\lambda\alpha)^{-1}h_{1k}(\beta)^{-1}h_{1k}(\beta)h_{1k}(\lambda)h_{1k}(\lambda\beta)^{-1}$$
$$= c_1(\lambda\alpha, \beta)c_1(\lambda, \beta)^{-1} \quad (\lambda, \alpha, \beta \in R^{\bullet}, 1, i, k \text{ distinct}).$$

It follows that

$$h_{1i}(\lambda)b_1([\alpha, \beta])h_{1i}(\lambda)^{-1} \equiv b_1([\lambda\alpha, \beta])b_1([\lambda, \beta])^{-1} \pmod{A}$$

$$\equiv b_1([\lambda\alpha, \beta][\lambda, \beta]^{-1}) \pmod{A}$$
$$= b_1(\lambda[\alpha, \beta]\lambda^{-1}).$$

Now $H(R)$ is generated by all $h_{1i}(\lambda)$, $\lambda \in R^{\bullet}$ (Lemma 83), so if we write B for the subgroup of $\mathrm{ST}(R)$ generated by A and all $b_1(\gamma)$ $(\gamma \in R^{\bullet\prime})$, then B is normalized by $H(R)$. We then have

$$h_{1j}(\alpha)h_{1k}(\beta) \equiv h_{1k}(\beta)h_{1j}(\alpha) \pmod{B}$$

and

$$h_{1k}(\alpha)h_{1k}(\beta) \equiv h_{1k}(\beta\alpha) \pmod{B}$$

whence, using Lemma 83 again, for any $w \in H(R)$ we can write

$$w \equiv h_{12}(\alpha_2)h_{13}(\alpha_3)\ldots h_{1n}(\alpha_n) \pmod{B}$$

or

$$w \equiv h_{12}(\alpha_2)h_{13}(\alpha_3)\ldots h_{1n}(\alpha_n)b_1(\gamma) \pmod{A}$$

where $\alpha_2, \alpha_3, \ldots, \alpha_n \in R^{\bullet}$ and $\gamma \in R^{\bullet\prime}$. It follows that

$$\phi(w) = \mathrm{diag}(\alpha, \alpha_2^{-1}, \alpha_3^{-1}, \ldots, \alpha_n^{-1}, 1, 1, \ldots)$$

where $\alpha = \alpha_2\alpha_3\ldots\alpha_n\gamma$. If $w \in K_2R$, we deduce that $\alpha_2 = \alpha_3 = \ldots$ $\ldots = \alpha_n = \gamma = 1$, or in other words $C(R) \subset A$, as required.

Since an H-ring is just a ring R for which $C(R) = K_2R$, then from Propositions 133 and 135 we deduce:

Corollary 136. Let R be any semi-local ring (or indeed any H-ring). Then K_2R is generated by all $e(\alpha, \beta)$, $d(\gamma, \delta)$ $(\alpha, \beta \in R^{\bullet}, \gamma, \delta \in R^{\bullet\prime})$.

There is another way to use the above calculations, to obtain an exact sequence involving K_2R, for any H-ring R. Let U_R denote the subgroup of $\mathrm{ST}(R)$ generated by all $c_1(\alpha, \beta)$, $\alpha, \beta \in R^{\bullet}$.

Corollary 137. Let R be any H-ring. Then there is an exact sequence

$$1 \to K_2R \to U_R \to R^{\bullet\prime} \to 1.$$

Proof. By the method of choosing the $b_1(\gamma)$, we have $b_1(\gamma) \in U_R$, all $\gamma \in R^{\bullet\prime}$, and so $e(\alpha, \beta)$, $d(\gamma, \delta) \in U_R$, all $\alpha, \beta \in R^{\bullet}$, $\gamma, \delta \in R^{\bullet\prime}$, so $K_2R \subset U_R$, by Corollary 136. Then the natural map ϕ : $\mathrm{ST}(R) \to \mathrm{E}(R)$ sends $c_1(\alpha, \beta)$ to the diagonal matrix $a_1([\alpha, \beta])$, which we identify with $[\alpha, \beta] \in R^{\bullet\prime}$, and the result follows.

The above corollaries suggest two possible ways to generalize Matsumoto's theorem (Theorem 90). The more direct way is to use the generators for K_2R given in Corollary 136 and try to find enough relations to give a presentation of K_2R. This is attempted in [5], for the case when R is a skew field, but the attempt fails, as not enough relations are given; finding enough relations seems to be a hard problem.

An alternative approach is to use Corollary 137, and to find enough relations to give a presentation of U_R; this is less directly useful in calculating K_2R. We quote without proof the following theorem of Rehmann [12]:

Theorem 138. Let R be a skew field. Then the following is a presentation of U_R:

Generators: $c_1(\alpha, \beta)$ $(\alpha, \beta \in R^{\bullet})$

Relations: $c_1(\alpha, 1 - \alpha) = 1$ $(\alpha \neq 0, 1)$

$$c_1(\alpha\beta, \gamma) = c_1({}^{\alpha}\beta, {}^{\alpha}\gamma)c_1(\alpha, \gamma)$$

$$c_1(\alpha, \beta\gamma)c_1(\beta, \gamma\alpha)c_1(\gamma, \alpha\beta) = 1.$$

Here ${}^{x}y = xyx^{-1}$. If R is commutative, then $R^{\bullet\prime} = 1$, so $K_2R = U_R$, and Theorem 138 reduces to Matsumoto's theorem. Other exact sequences involving K_2R can be obtained from the sequence in Corollary 137, by applying the Hochschild-Serre spectral sequence; for details, see [12].

We have concentrated almost exclusively on the problem of finding generators for K_2R. This is already quite involved; it is equivalent to finding a presentation for $\mathrm{E}(R)$: one takes the elementary matrices and the Steinberg relations, and then any extra relations needed give rise to a generating set for K_2R. Actually calculating K_2R amounts to measuring the independence of these extra relations, and is a much harder problem. What one usually does in practice (for example, to prove Matsumoto's theorem) is to find an explicit way of constructing central extensions of E(R), as we did for E(R) (pp.229-232), and then use the fact that ST(R) is the universal central extension of E(R).

BIBLIOGRAPHY

1. H.Bass, *Topics in Algebraic K-Theory*, Tata Institute of Fundamental Research, Bombay (1966).
2. H.Bass, *Algebraic K-Theory*, Benjamin (1968).
3. H.Bass (ed.), *Algebraic K-Theory* I, II, III (Battelle Institute Conference 1972), *Lecture Notes in Mathematics*, 341, 342, 343, Springer (1973).
4. R.K.Dennis and M.R.Stein, 'The functor K_2: A survey of computations and problems', in [3], vol.II, pp.243-280.
5. S.M.Green, 'Generators and Relations for K_2 of a Division Ring', in [13], pp.74-76.
6. D.Husemoller, *Fibre Bundles*, McGraw-Hill (1966).
7. M.Karoubi, *K-Theory*, *Grundlehren*, 226, Springer (1978).
8. I.G.Macdonald, *Algebraic Geometry*, Benjamin (1968).
9. S.MacLane, *Categories for the Working Mathematician*, *Graduate Texts*, 5, Springer (1971).
10. W.S.Massey, *Algebraic Topology: An Introduction*, *Graduate Texts*, 56, Springer (1971).
11. J.Milnor, *Introduction to Algebraic K-Theory*, *Annals of Mathematics Studies*, 72, Princeton (1971).
12. U.Rehmann, 'Zentrale Erweiterungen der speziellen linearen Gruppe eines Schiefkörpers', *J. Reine Angew. Math.* 301 (1978), pp.77-104.
13. M.R.Stein (ed.), *Algebraic K-Theory* (Evanston 1976), *Lecture Notes in Mathematics*, 551, Springer (1976).
14. R.G.Swan, *Algebraic K-Theory*, *Lecture Notes in Mathematics*, 76, Springer (1968).
15. R.G.Swan, 'Vector bundles and projective modules', *Annals of Mathematics*, 71 (1960), pp.264-277.

INDEX